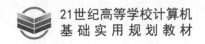

21世纪高等学校计算机
基础实用规划教材

计算机专业英语教程
（第2版）

◎ 江红 余青松 主编

清华大学出版社
北京

内 容 简 介

本书综合计算机信息技术以及英语阅读和应用能力，便于学生日后在科技领域进行国际交流，使学生具有更强的对信息社会快速发展的适应、生存、发展和提升能力。

本书适合于高校大学生以及研究生计算机技术基础（计算机导论）课程的双语或全英语教学。具体内容包括课程知识要点的介绍、参考译文、课程习题和思考等。

本书封面贴有清华大学出版社防伪标签，无标签者不得销售。
版权所有，侵权必究。举报：010-62782989，beiqinquan@tup.tsinghua.edu.cn。

图书在版编目（CIP）数据

计算机专业英语教程 / 江红，余青松主编. —2版. —北京：清华大学出版社，2017（2024.9重印）
（21世纪高等学校计算机基础实用规划教材）
ISBN 978-7-302-47319-0

Ⅰ. ①计… Ⅱ. ①江… ②余… Ⅲ. ①电子计算机 - 英语 - 高等学校 - 教材 Ⅳ. ①TP3

中国版本图书馆 CIP 数据核字（2017）第 124255 号

责任编辑：魏江江　李　晔
封面设计：刘　键
责任校对：李建庄
责任印制：沈　露

出版发行：清华大学出版社
网　　址：https://www.tup.com.cn, https://www.wqxuetang.com
地　　址：北京清华大学学研大厦 A 座　　邮　编：100084
社 总 机：010-83470000　　邮　购：010-83470235
投稿与读者服务：010-62776969，c-service@tup.tsinghua.edu.cn
质 量 反 馈：010-62772015，zhiliang@tup.tsinghua.edu.cn

印 装 者：涿州市般润文化传播有限公司
经　　销：全国新华书店
开　　本：185mm×260mm　　印　张：24.5　　字　数：595千字
版　　次：2012年1月第1版　2017年7月第2版　　印　次：2024年9月第9次印刷
印　　数：14101～14600
定　　价：49.50元

产品编号：075020-01

编者的话

计算机和英语是当代大学生应该掌握的基本技能和工具，而了解、吸收和掌握国外先进的计算机知识，已成为当务之急。学生应该普遍地了解当今世界信息技术基础的知识和发展前沿，掌握一定的理论知识，并具有一定的实践能力。对计算机学科实施双语或全英语教学，便于学生日后进入全球科技领域进行国际交流，使学生具有更强的对信息社会快速发展的适应能力。

计算机专业英语课程的建设目标是使学生了解、吸收和掌握信息技术的理论知识以及实践操作能力，掌握信息技术的英语阅读和使用能力，使学生具有较强的对信息社会快速发展的适应能力，为进一步学习和掌握计算机相关知识点奠定良好的基础。课程主要内容包括：

（1）信息技术基础理论知识中的信息技术基础基本原理、系统软件的基本概念、应用软件的使用、计算机硬件组成、输入输出设备、数据存储和检索、数据通信和网络基础、Internet 和 Web、电子商务、人工智能、移动网络、云计算、大数据、互联网+、物联网、Docker、3D 打印、深度学习以及商业智能/分析等；

（2）信息技术基础实践能力中的 Windows 基本操作、文字处理（Word）、演示文稿的制作（PowerPoint）、电子表格（Excel）的应用、网上信息的浏览和保存、使用 HTML 开发网页、Access 数据库表的创建、查询和报表的应用等。

随着信息技术的发展，并根据本教程第 1 版在各高校使用情况的反馈，第 2 版更新并调整为五大部分：第一部分介绍课程知识要点；第二部分是课程知识要点的部分参考译文；第三部分是课程相关知识的习题与思考；第四部分实践性教学辅导内容，包括实验目的和要求、实验内容和步骤以及课后实践作业等；第五部分是考试复习样题。

本教程适合高校大学生和研究生计算机技术基础（计算机导论）课程的双语或全英语教学。本教程涉及的所有上机实验和题目中的素材、供教师参考的教学电子文稿（PowerPoint 课件）均可以通过清华大学出版社的网站（www.tup.com.cn）、计算中心的服务器（www.cc.ecnu.edu.cn）下载。也可以通过 hjiang@cc.ecnu.edu.cn 直接与作者联系。

本教程的编写者均是长期从事计算机教学和系统研发的教师，在教程编写过程中注意紧扣教学要求、重点突出、简明扼要，注重理论和实践相结合，并力图在教程中介绍信息技术的一些新发展、新概念。本教程由华东师范大学的江红和余青松老师共同编写，上海商学院的黄勇教授和华东师范大学的王行恒副教授主审。

在此，要特别感谢华东师范大学的陈志云副教授、上海戏剧学院的陈永东副教授、井冈山大学的彭蕾老师、华东师范大学的刘艳老师、白玥老师、朱晴婷老师、刘垚老师、蒲鹏老师，在使用本教程第 1 版的过程中提出了宝贵意见和建议。华东师范大学的石明康、李松桓、徐加敏、朱蕾、曾红伟、黄永、龚瑜、钟忠、杨菲菲、杨璀琼等同志验证了实践性教学内容。本教程还得到了郑骏高级工程师、朱敏高级工程师、赵俊逸高级工程师和华东师范大学计算机中心的许多同仁的帮助和支持，以及华东师范大学教务处、研究生院和 La Trobe 大学信息技术基础教研室的各位专家和教师的支持，清华大学出版社的编辑为教程的编写和再版给予了很多的建议和帮助，在此一并致谢。

  本版教程根据第 1 版教程在华东师范大学、同济大学、北京科技大学、西北大学、西安工业大学、陕西师范大学等高校使用情况的反馈，通过较大范围的学生意见征询和任课教师间的交流讨论，对第 1 版教程内容进行了增删、调整和完善。由于时间和编者学识有限，特别是作为一种双语或全英语教程，书中不足之处在所难免，敬请诸位同行、专家和读者指正。

编 者

2017 年 5 月

Assignment 2 (Word)	5 %
Assignment 3 (Excel)	5 %
Assignment 4 (PowerPoint)	5 %
Assignment 5 (HTML)	5 %
Assignment 6 (Access)	5 %
Tests	10 %
Final Exam	60 %
	----------
	100 %
	======

General Informa[tion]

Objectives

This Course is designed to enable students to use Information Technology (IT) systems applications independently to support a range of information processing activities. It is desig[ned] to develop a broad knowledge of the theoretical concepts, principles, boundaries and scope o[f] applications. These activities will be centred on using software applications packages (Windo[ws,] Microsoft Word, Excel, PowerPoint, Access, and HTML) to meet complex informat[ion] requirements while paying attention to security and the needs of other users.

Description

Students should be aware that no assumptions about prior knowledge needed. The topic[s to] be covered are:
- The main hardware units of the computer; how they work, how they are interconnec[ted] and how processing is controlled in the computer.
- Input and output methods.
- How data is stored internally and externally.
- Comparison of storage media.
- Introduction to multiprocessing and multiprogramming computers.
- Introduction to telecommunications.
- Introduction to system software, including operating systems.
- Word Processing.
- PowerPoint Presentation.
- Spreadsheets.
- World Wide Web.
- HTML language.
- DBMS and Microsoft Access.
- The causes of some common software and hardware problems, (e.g. cable connectio[n,] device settings, software option settings) and what actions to take to resolve these.
- Assessing information requirements and designing solutions using IT.
- Computers and society.

Assessment

Assignment 1 (Windows)   5 %

Lecture Program

Week	Lecture Content	Reading / Exercises
Introduction & Theory 1		
1	Introduction to the course	
	Basic computer concepts	Text 1
	CPU: components, system clock, fetch-execute cycle	Text 2
2	Input - needs, types of devices	
	Output - needs, types of devices	Text 3
	Secondary storage: description of devices and media	
	Methods of data organization on disks	Text 4
Theory 2		
3,4	Software: operating systems & application software	Windows Practice Exercises
	Programming, historical development	Text 5
Introduction to Word Processing		
5,6	Getting started with Word	Text 6
	Formatting	
	Graphics, Tables, Textbox & Other Objects	Word Practice Exercises
Introduction to PowerPoint Presentation		
7,8	Getting started with PowerPoint	Text 7
	Formatting a presentation	
	Animate objects	PowerPoint Practice Exercises
9	(Mid-term) Test	
Introduction to Spreadsheets		
10	Definition of a spreadsheet, cell contents	Text 8
	Uses of spreadsheets	
	Range specification, some common functions, formulae,	
	absolute and relative cell addressing	Lab1

11	Functions, spreadsheet design, case study (Orchids shade house)	
	Chart characteristics and design, financial functions	Lab2
	Excel case study	Assignment 1

Introduction to World Wide Web

12	Data communications: needs, fundamentals, equipment, connectivity	Text 10
	Ways of using computers: stand alone, networks	
	Introduction to WWW, computers and society	Text 11,12
13	Constructing web pages—structure of a HTML document	Lab3
	Constructing web pages—tables & images	Lab4
14	HTML case study	Assignment 2

Introduction to Database

15	Databases—concept of a database	Text 9
	Microsoft Access: introduction to Access	Lab5
16	Microsoft Access: queries from Access	Lab6
	Microsoft Access: reports from Access	Lab7
	Microsoft Access case study	Assignment 3
17	Revision	
18	Final examination	

Part I　Knowledge Points（第一部分　知识点）

Text 1　Introduction to Information Technology ································ 2

- 1.1　Computer Literacy ··· 2
- 1.2　Information System ··· 2
- 1.3　Information Technology ··· 2
- 1.4　Information and Communications Technology ····························· 3
- 1.5　What is a Computer ·· 3
- 1.6　Components of a Computer ·· 3
  - 1.6.1　Hardware ··· 4
  - 1.6.2　Software ·· 6
- 1.7　Categories of Computers ·· 7
  - 1.7.1　Supercomputers ··· 7
  - 1.7.2　Mainframe Computers ··· 7
  - 1.7.3　Midrange Computers ··· 8
  - 1.7.4　Minicomputers ··· 8
  - 1.7.5　Personal Computers ·· 8

Text 2　System Unit ··· 11

- 2.1　Central Processing Unit (CPU) ·· 12
  - 2.1.1　Control Unit ·· 12
  - 2.1.2　Arithmetic and Logic Unit ·· 12
  - 2.1.3　Registers ·· 12
  - 2.1.4　Characteristics of CPU ·· 13
- 2.2　Memory ·· 17
  - 2.2.1　Computer Memory Cell ·· 17
  - 2.2.2　Types of Memory ··· 18
  - 2.2.3　RAM ··· 18
  - 2.2.4　Cache ·· 18
  - 2.2.5　ROM ··· 19
  - 2.2.6　Virtual Memory ··· 20

2.3 Ports and Connectors ··· 21
2.3.1 Serial Ports ··· 21
2.3.2 Parallel Ports ·· 22
2.3.3 SCSI Ports ··· 22
2.3.4 USB Ports ·· 22
2.3.5 PC Card Slots ··· 22

Text 3　Input and Output ··· 24
3.1 Input ··· 24
3.1.1 Input Devices ··· 24
3.1.2 Keyboards ·· 24
3.1.3 Pointing Devices ··· 25
3.1.4 Voice Input ·· 28
3.1.5 Digital Cameras ··· 28
3.1.6 Video Input ·· 29
3.1.7 Scanning and Reading Devices ·· 29
3.1.8 Terminals ··· 32
3.1.9 Biometric Input ··· 33
3.2 Output ·· 33
3.2.1 Output Devices ··· 33
3.2.2 Printers ·· 35
3.2.3 Speakers and Headphones ·· 37
3.2.4 Other Output Devices ·· 37
3.2.5 Important Concepts ··· 39

Text 4　Secondary Storage ··· 40
4.1 Floppy Disks ··· 41
4.2 Hard Disks ·· 41
4.2.1 Formatting ·· 42
4.2.2 Capacity ··· 42
4.2.3 Rotational Speed ··· 43
4.2.4 Access Time ··· 43
4.2.5 Characteristics of a Hard Disk ··· 43
4.2.6 Maintaining Data Stored on a Disk ··································· 44
4.2.7 Features of Floppy Disks and Hard Disks ·························· 45
4.3 Flash Memory ··· 45
4.3.1 USB Flash Drives ·· 45
4.3.2 Smart Cards ··· 46
4.4 Optical Storage Technology ·· 46

		4.4.1 CD Disc ·· 47

- 4.4.1 CD Disc ·· 47
- 4.4.2 DVD Disc ·· 47
- 4.4.3 Blu-ray Disc ··· 47
- 4.4.4 MO ·· 48
- 4.5 Tapes ··· 48
- 4.6 RAID Storage Systems ··· 48

Text 5 Software ··· 50

- 5.1 Categories of Software ·· 50
- 5.2 System Software ··· 52
  - 5.2.1 BIOS ·· 52
  - 5.2.2 Operating System ·· 52
  - 5.2.3 Utility Programs ·· 60
- 5.3 Programming Languages ··· 62
  - 5.3.1 Machine language ·· 63
  - 5.3.2 Assembly language ·· 63
  - 5.3.3 Procedural language ··· 63
  - 5.3.4 Non-procedural language ··· 64
  - 5.3.5 Object-oriented programming (OOP) language ··························· 64
  - 5.3.6 Visual programming language ··· 64
  - 5.3.7 Execution of programming languages ······································ 64

Text 6 Introduction to Word Processing ··· 66

- 6.1 Getting Started with Word ··· 66
  - 6.1.1 Select Text ··· 66
  - 6.1.2 Find and Replace ··· 67
  - 6.1.3 Page Setup & Print Preview ·· 68
- 6.2 Formatting ··· 69
  - 6.2.1 Themes, Template, Style and Format Painter ······························ 69
  - 6.2.2 Character Formatting ··· 70
  - 6.2.3 Paragraph Formatting ·· 70
  - 6.2.4 Page Formatting ·· 75
- 6.3 Graphics, Tables, Textbox and Other Objects ······································· 81
  - 6.3.1 Pictures and Graphics ·· 81
  - 6.3.2 Shape ·· 82
  - 6.3.3 SmartArt ·· 82
  - 6.3.4 Symbol ·· 83
  - 6.3.5 Table ·· 83

		6.3.6	Text Box	84
		6.3.7	WordArt	84
		6.3.8	Equations	85

Text 7　Introduction to PowerPoint Presentation ································ 87

- 7.1 Getting Started with PowerPoint ·· 87
  - 7.1.1 PowerPoint Views ·· 87
  - 7.1.2 Start a New Presentation ·· 90
  - 7.1.3 Add Slides ·· 91
  - 7.1.4 Change Slide Order ·· 92
  - 7.1.5 Apply or Change the Slide Layout ·· 92
  - 7.1.6 Present a Slide Show ·· 92
  - 7.1.7 Print Slides or Handouts ·· 93
  - 7.1.8 Tips for Creating an Effective Presentation ·· 95
- 7.2 Formatting a Presentation ·· 95
  - 7.2.1 Apply a Design Template ·· 95
  - 7.2.2 Working with Themes ·· 95
  - 7.2.3 Header and Footer ·· 96
- 7.3 Animating Your Slides ·· 97
  - 7.3.1 Add Slide Transition Effects ·· 97
  - 7.3.2 Animate Objects ·· 97
  - 7.3.3 Hyperlink ·· 98
  - 7.3.4 Action Button ·· 98
  - 7.3.5 Insert a Sound and/or Video Clip on a Slide ·· 100

Text 8　Introduction to Spreadsheets ·· 102

- 8.1 Introduction to Excel ·· 102
  - 8.1.1 Definition of a Spreadsheets ·· 102
  - 8.1.2 Contents of Cells ·· 103
  - 8.1.3 Specifying a Range of Cells ·· 103
- 8.2 Spreadsheet Formulas and Functions ·· 105
  - 8.2.1 Some Common Functions ·· 106
  - 8.2.2 Common Formula Patterns ·· 106
  - 8.2.3 Copying Formulas—Absolute vs. Relative References ·· 107
  - 8.2.4 Standard Spreadsheet Functions ·· 107
  - 8.2.5 Useful Financial Functions ·· 110
- 8.3 Spreadsheet Charts ·· 111
  - 8.3.1 Spreadsheet Chart Elements ·· 111
  - 8.3.2 Creating Spreadsheet Charts ·· 112

Text 9　Introduction to Database ··· 116

- 9.1　Database ·· 116
- 9.2　Database Management Systems ·· 116
  - 9.2.1　Relational Database Hierarchy ··· 117
  - 9.2.2　Examples of a Database File ··· 118
  - 9.2.3　Data Integrity ··· 119
  - 9.2.4　Data Redundancy ··· 119
- 9.3　Introduction to Microsoft Office Access ··· 119
  - 9.3.1　Access Data Files ··· 119
  - 9.3.2　Field Data Types ··· 119
  - 9.3.3　Creating a Database File ··· 120
  - 9.3.4　Create a new table ·· 121
  - 9.3.5　Import an Excel worksheet as a table in a new database ···················· 124
- 9.4　Access Query Design ··· 125
  - 9.4.1　Top Section ··· 125
  - 9.4.2　Bottom Section ·· 126
- 9.5　Access Report Generator ·· 131
  - 9.5.1　Access Reports Overview ··· 131
  - 9.5.2　Format of an Access Report ··· 131

Text 10　Data Communications and Networks ··· 134

- 10.1　Data Communications ·· 134
  - 10.1.1　Data Communications Components ·· 134
  - 10.1.2　MODEM ·· 134
  - 10.1.3　Communication Software ··· 135
  - 10.1.4　Data Transfer ··· 135
  - 10.1.5　Protocol ·· 137
  - 10.1.6　Direction of Data Communications ·· 138
- 10.2　Networks ··· 139
  - 10.2.1　Types of Computer Networks ·· 140
  - 10.2.2　Network Topologies ·· 141
  - 10.2.3　Network Communication Technologies ··· 142
  - 10.2.4　Connecting Networks ·· 145
  - 10.2.5　Network Architecture ·· 146
  - 10.2.6　Communication Channel ·· 146
  - 10.2.7　Data Processing ·· 151

Text 11　Introduction to WWW ·· 153

- 11.1　Introduction to WWW ··· 153

　　　　11.1.1　Web Browsers ·················153
　　　　11.1.2　Uniform Resource Locator ·················153
　　　　11.1.3　Domain Names ·················154
　　　　11.1.4　Web Servers ·················154
　　11.2　Introduction to HTML ·················155
　　　　11.2.1　HTML ·················155
　　　　11.2.2　Requirements ·················155
　　　　11.2.3　Tags ·················156
　　　　11.2.4　Basic HTML Document Structure ·················156
　　　　11.2.5　Some HTML Tags ·················156
　　　　11.2.6　Heading Tags ·················157
　　　　11.2.7　Font Size and Colors ·················157
　　　　11.2.8　Steps to Create a Web Page ·················158
　　11.3　Constructing Web Pages (1)—List, Image, Anchor ·················159
　　　　11.3.1　List ·················159
　　　　11.3.2　Adding Image ·················161
　　　　11.3.3　Anchor Tag ·················162
　　11.4　Constructing Web Pages (2)—Table ·················162
　　　　11.4.1　Table ·················162
　　　　11.4.2　Table Tags ·················162
　　　　11.4.3　Tables Incorporating an Inline Image ·················166

Text 12　Computers and Society ·················170

　　12.1　E-Business ·················170
　　　　12.1.1　E-Business Basics ·················170
　　　　12.1.2　E-Business Models ·················170
　　　　12.1.3　Electronic Shopping Carts ·················171
　　12.2　Electronic Data Interchange ·················171
　　12.3　E-mail ·················172
　　12.4　Instant Messaging ·················172
　　12.5　New Technologies and Patterns ·················172
　　　　12.5.1　Big Data ·················172
　　　　12.5.2　Internet Plus ·················173
　　　　12.5.3　Cloud Computing ·················173
　　　　12.5.4　The Internet of Things ·················174
　　　　12.5.5　Mobile Web ·················174
　　　　12.5.6　Artificial Intelligence ·················175
　　　　12.5.7　Business Intelligence ·················175
　　　　12.5.8　Deep Learning ·················176

	12.5.9	Docker ·· 177
	12.5.10	3D Printing ·· 177
12.6	Social Issues ·· 177
	12.6.1	Computer Crime ··· 177
	12.6.2	Security ··· 180
	12.6.3	Privacy ·· 182
	12.6.4	Computer Ethics ··· 184

Part II Reference Version（第二部分 参考译文）

课文 1 信息技术简介 ··· 186

- 1.1 计算机文化 ··· 186
- 1.2 信息系统 ·· 186
- 1.3 信息技术 ·· 186
- 1.4 信息通信技术 ··· 187
- 1.5 什么是计算机 ··· 187
- 1.6 计算机组成 ··· 187
  - 1.6.1 硬件 ·· 187
  - 1.6.2 软件 ·· 188
- 1.7 计算机的分类 ··· 189
  - 1.7.1 巨型机 ·· 189
  - 1.7.2 大型机 ·· 189
  - 1.7.3 中型机 ·· 189
  - 1.7.4 小型机 ·· 190
  - 1.7.5 个人计算机 ·· 190

课文 2 系统部件 ·· 192

- 2.1 中央处理器 ··· 192
  - 2.1.1 控制单元 ·· 192
  - 2.1.2 算术逻辑单元 ·· 193
  - 2.1.3 寄存器 ·· 193
  - 2.1.4 CPU 的特性 ··· 193
- 2.2 内存 ·· 195
  - 2.2.1 计算机内存单元 ·· 196
  - 2.2.2 内存的分类 ·· 196
  - 2.2.3 随机存储器 ·· 197
  - 2.2.4 高速缓存 ·· 197
  - 2.2.5 只读存储器 ·· 198
  - 2.2.6 虚拟内存 ·· 199

2.3 端口和连接器 ··· 199
    2.3.1 串行口 ··· 200
    2.3.2 并行口 ··· 200
    2.3.3 小型计算机系统接口 ·· 200
    2.3.4 USB 接口 ··· 200
    2.3.5 PC 卡接口 ·· 201

课文 3　输入输出 ·· 202

3.1 输入 ··· 202
    3.1.1 输入设备 ·· 202
    3.1.2 键盘 ··· 202
    3.1.3 定位设备 ·· 202
    3.1.4 语音输入 ·· 204
    3.1.5 数码相机 ·· 204
    3.1.6 视频输入 ·· 205
    3.1.7 扫描和识别设备 ··· 205
    3.1.8 终端 ··· 206
    3.1.9 生物识别输入 ·· 207
3.2 输出 ··· 207
    3.2.1 输出设备 ·· 207
    3.2.2 打印机 ··· 209
    3.2.3 扬声器和耳机 ·· 210
    3.2.4 其他输出设备 ·· 211
    3.2.5 重要概念 ·· 211

课文 4　辅助存储器 ·· 213

4.1 软盘 ··· 213
4.2 硬盘 ··· 214
    4.2.1 格式化 ··· 214
    4.2.2 硬盘容量 ·· 214
    4.2.3 硬盘转速 ·· 214
    4.2.4 访问时间 ·· 215
    4.2.5 硬盘的特性 ··· 215
    4.2.6 维护磁盘上的数据 ·· 216
    4.2.7 软盘和硬盘的特性 ·· 216
4.3 闪存 ··· 216
    4.3.1 USB 闪存盘 ·· 217
    4.3.2 智能卡 ··· 217
4.4 光存储技术 ·· 217

          4.4.1　CD 光盘 ·· 217
          4.4.2　DVD 光盘 ·· 218
          4.4.3　蓝光光盘 ·· 218
          4.4.4　磁光盘 ·· 218
     4.5　磁带 ··· 218
     4.6　RAID 存储系统 ·· 219

课文 5　软件 ·· 220

  5.1　软件的分类 ··· 220
  5.2　系统软件 ·· 221
          5.2.1　BIOS ·· 221
          5.2.2　操作系统 ·· 222
          5.2.3　实用程序 ·· 224
  5.3　编程语言 ·· 225
          5.3.1　机器语言 ·· 225
          5.3.2　汇编语言 ·· 226
          5.3.3　过程程序设计语言 ··· 226
          5.3.4　非过程程序设计语言 ·· 226
          5.3.5　面向对象程序设计语言 ·· 226
          5.3.6　可视化程序设计语言 ·· 226
          5.3.7　程序设计语言的执行 ·· 227

课文 8　电子表格介绍 ··· 228

  8.1　Excel 介绍 ··· 228
          8.1.1　电子表格 ·· 228
          8.1.2　单元格的内容 ··· 228

课文 9　数据库介绍 ·· 230

  9.1　数据库 ··· 230
  9.2　数据库管理系统 ·· 230
          9.2.1　关系数据库 ··· 231
          9.2.2　数据库文件实例 ·· 231
          9.2.3　数据完整性 ··· 232
          9.2.4　数据冗余 ·· 232

课文 10　数据通信和网络 ··· 233

  10.1　数据通信 ·· 233
          10.1.1　数据通信的组成 ·· 233

    10.1.2 调制解调器 ·············· 233
    10.1.3 通信软件 ·············· 234
    10.1.4 数据传输 ·············· 234
    10.1.5 协议 ·············· 234
    10.1.6 数据通信的方向 ·············· 236
  10.2 网络 ·············· 236
    10.2.1 计算机网络的分类 ·············· 236
    10.2.2 网络拓扑结构 ·············· 238
    10.2.3 网络通信技术 ·············· 239
    10.2.4 网络互连 ·············· 240
    10.2.5 网络架构 ·············· 242
    10.2.6 信道 ·············· 242
    10.2.7 数据处理 ·············· 246

课文 11 万维网简介 ·············· 247

  11.1 万维网 ·············· 247
    11.1.1 Web 浏览器 ·············· 247
    11.1.2 统一资源定位器 ·············· 247
    11.1.3 域名 ·············· 248
    11.1.4 Web 服务器 ·············· 248
  11.2 HTML 简介 ·············· 249
    11.2.1 超文本标记语言 ·············· 249
    11.2.2 基本要求 ·············· 249
    11.2.3 标记 ·············· 249
    11.2.4 编制网页的步骤 ·············· 250

课文 12 计算机和社会 ·············· 251

  12.1 电子商务 ·············· 251
    12.1.1 电子商务基本概念 ·············· 251
    12.1.2 电子商务模型 ·············· 251
    12.1.3 电子购物车 ·············· 252
  12.2 电子数据交换 ·············· 252
  12.3 电子邮件 ·············· 252
  12.4 即时消息 ·············· 253
  12.5 新技术和新模式 ·············· 253
    12.5.1 大数据 ·············· 253
    12.5.2 互联网+ ·············· 254
    12.5.3 云计算 ·············· 254
    12.5.4 物联网 ·············· 254

　　　　12.5.5　移动网络 ···255
　　　　12.5.6　人工智能 ···255
　　　　12.5.7　商务智能 ···256
　　　　12.5.8　深度学习 ···256
　　　　12.5.9　Docker ··257
　　　　12.5.10　3D打印 ···257
　　12.6　社会问题 ···258
　　　　12.6.1　网络犯罪 ···258
　　　　12.6.2　信息安全 ···261
　　　　12.6.3　隐私权 ··262
　　　　12.6.4　计算机伦理 ··264

Part III　Exercises（第三部分　习题与思考）

Tutorial 1　Introduction to Information Technology ························268

Tutorial 2　System Unit ···269

Tutorial 3　Input and Output Devices ·······································270

Tutorial 4　Secondary Storage ··271

Tutorial 5　Software ···272

Tutorial 6　Data Communications ··273

Tutorial 7　Networks ··274

Part IV　Labs（第四部分　实践指导）

Warm-up Exercise 1　Windows Practice Exercises ··························276

Warm-up Exercise 2　Word Practice Exercises ·······························278

Warm-up Exercise 3　PowerPoint Practice Exercises ························282

Lab 1　Introduction to Spreadsheets—Excel ································285

Lab 2　Orchids Shade House Case Study ·····································290

Assignment 1 ··· 293

Lab 3　Creating Your Webpage (1) ·· 295

Lab 4　Creating Your Webpage (2) ·· 298

Assignment 2 ··· 304

Lab 5　Creating a New Datafile ·· 305

Lab 6　Access Queries ·· 309

Lab 7　Access Report Generation ··· 319

Assignment 3 ··· 323

Part V　Revisions　（第五部分　复习题）

Revision 1 ·· 330
- SECTION　A ·· 330
- SECTION　B ·· 333
- SECTION　C ·· 334

Revision 2 ·· 340
- SECTION　A ·· 340
- SECTION　B ·· 343
- SECTION　C ·· 344

Revision 3 ·· 349
- SECTION　A ·· 349
- SECTION　B ·· 352
- SECTION　C ·· 353

Revision 4 ·· 358
- SECTION　A ·· 358
- SECTION　B ·· 361
- SECTION　C ·· 362

参考文献 ··· 368

Part I Knowledge Points

（第一部分 知识点）

Text 1
Introduction to Information Technology

1.1 Computer Literacy

Computer literacy is the knowledge and ability to use computers and technology efficiently. Computer literacy can also refer to the comfort level someone has with using computer programs and other applications that are associated with computers. Another valuable component of computer literacy is to know how computers work and operate.

1.2 Information System

An information system (IS) is any combination of information technology and people's activities using that technology to support operations, management, and decision-making. An information system has five parts: people, procedures, software, hardware, and data.
- People: end users who use computers to make themselves more productive;
- Procedures: specify rules or guidelines for computer operations;
- Software: provides step-by-step instructions for computer hardware;
- Hardware: equipments that process the data to create information, including keyboard, mouse, monitor, system unit, and other devices;
- Data: raw, unprocessed facts, including text, numbers, images, audio, and video, etc.

1.3 Information Technology

Information is the result of processing, manipulating and organizing data in a way that adds to the knowledge of the receiver. In other words, it is the context in which data is taken.

Information Technology (IT) is the study, design, development, implementation, support or management of computer-based information systems, particularly software applications and computer hardware. In short, IT deals with the use of electronic computers and computer-based tools to convert, store, protect, process, transmit and retrieve information, securely.

1.4 Information and Communications Technology

Information and communications technology (ICT) is an extended term for information technology (IT) which stresses the role of unified communications and the integration of telecommunications (telephone lines and wireless signals), computers as well as necessary enterprise software, middleware, storage, and audio-visual systems, which enable users to access, store, transmit, and manipulate information.

In the past few decades, information and communication technologies have provided society with a vast array of new communication capabilities. For example, people can communicate in real-time with others in different countries using technologies such as instant messaging, voice over IP (VoIP), and video-conferencing.

Modern information and communication technologies have created a "global village," in which people can communicate with others across the world as if they were living next door. For this reason, ICT is often studied in the context of how modern communication technologies affect society.

1.5 What is a Computer

A computer is an electronic device, operating under the control of instructions stored in its own memory, which can accept data, manipulate the data according to specified rules, produce results, and store the results for future use.

1.6 Components of a Computer

The electric, electronic, and mechanical components of a computer, or hardware, include a system unit, input devices, output devices, storage devices, and communications devices. The system unit is a case that contains the electronic components of a computer that are used to process data. An input device allows users to enter data or instructions into a computer. An output device conveys information to one or more people. A storage device records and/or retrieves items to and from storage media. A communications device enables a computer to send and receive data, instructions, and information to and from one or more computers.

1.6.1 Hardware

Hardware is the electronic circuits and mechanical parts that physically make up the computer. These components include a system unit, input devices, output devices, storage devices, and communications devices. Figure I-1-1 shows some common computer hardware components.

1.6.1.1 System Unit

System unit is also known as base unit. It is the main unit of a personal computer, typically consisting of a metal or (rarely) plastic enclosure containing the motherboard, power supply, cooling fans, internal disk drives, and the memory modules and expansion cards that are plugged into the motherboard, such as video and network cards, but does not include the keyboard or monitor, or any peripheral devices.

1.6.1.2 Input Devices

An input device is any hardware component that allows users to enter data or instructions into a computer. Six widely used input devices are the keyboard, mouse, microphone, scanner, digital camera, and PC video camera (ref. Figure I-1-1).

1.6.1.3 Output Devices

An output device is any hardware component that conveys information to one or more people. Three commonly used output devices are monitors, printers, and speakers (ref. Figure I-1-1).

1.6.1.4 Storage Devices

A storage device is capable of storing data. The term usually refers to mass storage devices, such as disks and tape drives.

1.6.1.5 Communications Devices

A communications device is a hardware component that enables a computer to send (transmit) and receive data, instructions, and information to and from one or more computers. A widely used communications device is modem (ref. Figure I-1-1).

1.6.1.6 Buses

The above five units (system unit, input devices, output devices, storage devices, and communications devices) are interconnected through various buses. A bus is a set of wires (a physical path) through which data and instructions are transmitted from one part of a computer to another (ref. Figure I-1-2).

Part I　Knowledge Points（第一部分　知识点）

Figure I-1-1　Hardware Components of a Computer

Figure I-1-2　Bus (Data Path)

1) Data Bus

Data bus transfers actual data whereas the address bus transfers information about where the data should go.

2) Address Bus

Address bus is the collection of wires connecting the CPU with main memory that is used to identify particular locations (addresses) in main memory.

3) Control Bus

Control bus is the physical connections that carry control information between the CPU and other devices within the computer. Whereas the data bus carries actual data that is being processed, the control bus carries signals that report the status of various devices. For example, one line of the bus is used to indicate whether the CPU is currently reading from or writing to main memory.

1.6.2 Software

Computer software, or just software, is the collection of computer programs and related data that provide the instructions to control the computer. We can also say software refers to one or more computer programs and data held in the storage of the computer for some purposes. Program software performs the function of the program it implements, either by directly providing instructions to the computer hardware or by serving as input to another piece of software. There are two general categories of software: system software and application software.

1.6.2.1 System Software

System software consists of the programs that manage and control the hardware so that application software can perform a task. It is an essential part of the computer system. Two types of system software are the operating system and utility programs.

1) Operating System

Operating system (OS) is the most important program that runs on a computer. Every general-purpose computer must have an operating system to run other programs. Operating systems perform basic tasks, such as recognizing input from the keyboard, sending output to the display screen, keeping track of files and directories on the disk, and controlling peripheral devices such as disk drives and printers.

E.g., Windows 7/8/10, Windows XP, Windows 2000/2003, Linux, etc.

2) Utility Programs

Utility program (also known as service program, service routine, tool, or utility routine) is a type of computer software. It is specifically designed to help manage and tune the computer hardware, operating system or application software, and perform a single task or a small range of tasks. Utility software has long been integrated into most major operating systems.

E.g., disk defragmenters, virus scanners, compression utilities, etc.

1.6.2.2 Application Software

Application software (also called end-user programs) is a subclass of computer software that employs the capabilities of a computer directly and thoroughly to a task that the user wishes to perform. This should be contrasted with system software which is involved in integrating a computer's various capabilities, but typically does not directly apply them in the performance of tasks that benefit the user. In this context the term application refers to both the application software and its implementation.

Typical examples of software applications are database programs, word processors, spreadsheets, and media players, etc.

Figuratively speaking, application software sits on top of system software because it is unable to run without the operating system and system utilities.

1.7 Categories of Computers

- Supercomputers;
- Mainframe Computers;
- Midrange Computers;
- Minicomputers;
- Personal Computers.

1.7.1 Supercomputers

A supercomputer (ref. Figure I-1-3, the Columbia Supercomputer, located at the NASA Ames Research Center) is a computer that led the world (or was close to doing so) in terms of processing capacity, particularly speed of calculation, at the time of its introduction. It is the fastest, most powerful and most expensive computer.

1.7.2 Mainframe Computers

A mainframe computer (ref. Figure I-1-4, Honeywell-Bull DPS 7 mainframe, circa 1990) is a very large and expensive computer capable of supporting hundreds, or even thousands, of users simultaneously.

Figure I-1-3 Supercomputer   Figure I-1-4   Mainframe Computer

1.7.3 Midrange Computers

A midrange computer (ref. Figure I-1-5) refers to computers that are more powerful and capable than personal computers but less powerful and capable than mainframe computers. In the past, midrange computers were known as minicomputers.

1.7.4 Minicomputers

A minicomputer (ref. Figure I-1-6) is a midsized computer. In size and power, minicomputers lie between workstations and mainframes. In the past decade, the distinction between large minicomputers and small mainframes has blurred, however, as has the distinction between small minicomputers and workstations.

Figure I-1-5　Midrange Computer

Figure I-1-6　Minicomputer

1.7.5 Personal Computers

A personal computer (PC) is a small, single-user computer based on a microprocessor. It can perform all of its input, processing, output, and storage activities by itself and contains a processor, memory, one or more input and output devices, and storage devices.

1.7.5.1 Workstations

A workstation is a high-end personal computer designed for technical or scientific applications. Intended primarily to be used by one person at a time, they are commonly connected to a local area network and run multi-user operating systems. Workstations are used for tasks such as computer-aided design (CAD), drafting and modelling, computation-intensive scientific and engineering calculations, image processing, architectural modelling, and computer graphics for animation and motion picture visual effects.

1.7.5.2　Desktop Computers

A desktop computer (ref. Figure I-1-7) is a computer made for use on a desk in an office or at home and is distinguished from portable computers such as laptops or PDAs. Desktop computers are also known as microcomputers.

1.7.5.3　Notebook Computers

A notebook computer (also called notebook or laptop, ref. Figure I-1-8) is a small mobile computer, small enough that it can sit on your lap.

Figure I-1-7　Desktop Computer　　　　　Figure I-1-8 Notebook Computer

1.7.5.4　Tablet PCs

A tablet PC (ref. Figure I-1-9) is a notebook or slate-shaped mobile computer. Its touchscreen or graphics tablet/screen hybrid technology allows the user to operate the computer with a stylus or digital pen, or a fingertip, instead of a keyboard or mouse. The tablet computer market was invigorated by Apple through the introduction of the iPad device in 2010.

1.7.5.5　Handheld Computers

A handheld computer (also known as a handheld device, mobile device or simply handheld) is a small computing device, typically, small enough to hold and operate in the hand and having an operating system capable of running mobile apps. These may provide a diverse range of functions. Typically, the device will have a display screen with a small numeric or alphanumeric keyboard or a touchscreen providing a virtual keyboard and buttons (icons) on-screen. Many such devices can connect to the Internet and interconnect with other devices such as car entertainment systems or headsets via Wi-Fi, Bluetooth or near field communication (NFC). Integrated cameras, digital media players, mobile phone and GPS capabilities are common.

Power is typically provided by a lithium battery. Smartphones, PDAs, and e-books are popular handheld devices.

A smartphone is a mobile phone that offers more advanced computing ability and connectivity than a contemporary basic feature phone. The Nokia 9 210 was the first color screen Communicator model which was the first true smartphone with an open operating system. Another example is the iPhone introduced by Apple Inc.

A personal digital assistant (PDA) is a most widely used handheld device that combines computing, telephone/fax, Internet and networking features. A typical PDA can function as a cellular phone, fax sender, Web browser and personal organizer. Figure I-1-10 shows a PDA from Palm, Inc.

An electronic book (or e-book) (ref. Figure I-1-11) is a book publication made available in digital form, consisting of text, images, or both, readable on the flat-panel display of computers or other electronic devices. Although sometimes defined as "an electronic version of a printed book", some e-books exist without a printed equivalent. Commercially produced and sold e-books are usually intended to be read on dedicated e-reader devices. However, almost any sophisticated computer device that features a controllable viewing screen can also be used to read e-books, including desktop computers, laptops, tablets and smartphones.

Figure I-1-9　Apple's iPad

Figure I-1-10　PDA

Figure I-1-11　Amazon's Kindle

Text 2

System Unit

A system unit, also known as a base unit, is the main body of a personal computer, typically consisting of a metal or (rarely) plastic enclosure containing the chassis, microprocessor, main memory, bus, and ports, but does not include the keyboard, monitor, or any peripheral devices.

1) Motherboard

A motherboard (sometimes called a system board) (ref. Figure I-2-1) is the main circuit board of a microcomputer. The motherboard contains the connectors for attaching additional boards. Typically, the motherboard contains the CPU, BIOS, memory, mass storage interfaces, serial and parallel ports, expansion slots, and all the controllers required to control standard peripheral devices, such as the display screen, keyboard, and disk drive. Collectively, all these chips that reside on the motherboard are known as the motherboard's chipset.

Figure I-2-1　ASUS CUSL2-C Motherboard

2) Computer Chip

A computer chip is a small piece of semiconducting material (usually silicon) on which an integrated circuit is embedded. A typical chip can contain millions of electronic components (transistors). Computers consist of many chips placed on electronic boards called printed circuit boards. There are different types of chips. For example, CPU chips (also called microprocessors) contain an entire processing unit, whereas memory chips contain blank memory.

Two main components of the system unit are the CPU and memory.

2.1 Central Processing Unit (CPU)

The Central Processing Unit (CPU) interprets and carries out the basic instructions that operate the computer. Sometimes referred to simply as the processor or central processor, the CPU is where most calculations take place. In terms of computing power, the CPU is the most important element of a computer system.

On large machines, CPUs require one or more printed circuit boards. On personal computers and small workstations, the CPU is housed in a single chip called a microprocessor.

CPU contains three different parts: the control unit, the arithmetic and logic unit, and the registers.

2.1.1 Control Unit

The control unit (CU), which extracts instructions from memory and decodes and executes them, calls on the ALU when necessary. Control unit:
- Directs and co-ordinates most of the operations in the computer;
- Interprets each instruction and then initiates the action.

2.1.2 Arithmetic and Logic Unit

The arithmetic logic unit (ALU) is a digital circuit that calculates an arithmetic operation (addition, subtraction, multiplication, etc.) and logic operations (AND, OR, NOT, XOR, etc.) between two numbers. The ALU is a fundamental building block of the central processing unit of a computer.

2.1.3 Registers

A register is a special, high-speed, temporary storage area within the CPU. All data must be represented in a register before it can be processed. For example, if two numbers are to be multiplied, both numbers must be in registers, and the result is also placed in a register. (The register can contain the address of a memory location where data is stored rather than the actual data itself.)

There are three main registers.
- Accumulator (Acc): The register that stores the results supplied by the ALU;

- Program Counter (PC): The register that contains the address of the next machine code instruction to be expected;
- Instruction Register (IR): The register that holds the instruction to be executed.

2.1.4 Characteristics of CPU

2.1.4.1 Instruction Cycle

An instruction cycle (ref. Figure I-2-2) (sometimes called machine cycle, fetch-and-execute cycle, fetch-decode-execute cycle, or FDX) is the basic operation cycle of a computer instruction. It is the process by which a computer retrieves a program instruction from its memory, determines what actions the instruction requires, and carries out those actions.

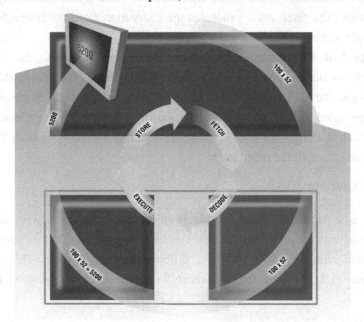

Figure I-2-2 An Instruction Cycle

Each computer's CPU can have different cycles based on different instruction sets, but will be similar to the following cycle.

1) Fetch the instruction from memory

The next instruction is fetched from the memory address that is currently stored in the Program Counter (PC), and stored in the Instruction Register (IR).

2) Decode the instruction

The instruction decoder interprets the instruction. If the instruction has an indirect address, the effective address is read from main memory, and any required data is fetched from main memory is to be processed and then placed into data registers. During this phase the instruction inside the IR gets decoded.

3) Execute the instruction

The CU passes the decoded information as a sequence of control signals to the relevant function units of the CPU to perform the actions required by the instruction such as reading values from registers, passing them to the ALU to perform mathematical or logic functions on them, and writing the result back to a register. If the ALU is involved, it sends a condition signal back to the CU.

4) Store the result

The result generated by the operation is stored in the main memory, or sent to an output device. Based on the condition of any feedback from the ALU, Program Counter may be updated to a different address from which the next instruction will be fetched.

The cycle is then repeated.

Steps 1 and 2 of the Instruction Cycle are called the Fetch Cycle. These steps are the same for each instruction. The fetch cycle processes the instruction from the instruction word which contains an opcode and an operand.

Steps 3 and 4 of the Instruction Cycle are part of the Execute Cycle. These steps will change with each instruction. The first step of the execute cycle is the Process-Memory. Data is transferred between the CPU and the I/O module. Next is the Data-Processing which uses mathematical operations as well as logical operations in reference to data. Central alterations is the next step, which is a sequence of operations, for example a jump operation. The last step is a combined operation from all the other steps.

2.1.4.2 Pipelining

Pipelining is a technique used in advanced microprocessors where the CPU begins executing a second instruction before the first has been completed. That is, several instructions are in the pipeline simultaneously, each at a different processing stage. This results in faster processing. Figure I-2-3 shows the difference between machine cycle with and without pipelining.

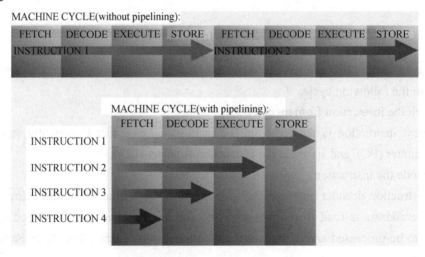

Figure I-2-3 Machine Cycle with and without Pipelining

2.1.4.3 RISC and CISC

Instruction set is the set of instructions that CPU can execute. CPU's instruction set is either RISC or CISC.

1) RISC (Reduced Instruction Set Computing)

The reduced instruction set computer, or RISC (pronounced /'risk/), is a CPU design strategy that favors an instruction set reduced both in size and complexity of addressing modes, in order to enable easier implementation, greater instruction level parallelism, and more efficient compilers. RISC is used by many workstations and also some personal computers. The features of RISC are:
- A reduced number of instructions to only those used more frequently;
- Longer programs;
- Less instructions;
- Quicker than CISC.

2) CISC (Complex Instruction Set Computing)

CISC (pronounced /'sisk/) stands for complex instruction set computer. Most personal computers use a CISC architecture, in which the CPU supports as many as two hundred instructions. The features of CISC are:
- Supports a large number of instructions;
- Shorter programs;
- More complex instructions.

2.1.4.4 System Clock

System clock is small quartz crystal circuit that is used by CPU to control the timing of all computer operations. Clock speed, also called clock rate, is the speed at which a microprocessor executes instructions. Every computer contains an internal clock that regulates the rate at which instructions are executed and synchronizes all the various computer components. The CPU requires a fixed number of clock ticks (or clock cycles) to execute each instruction. The faster the clock, the more instructions the CPU can execute per second. Clock speeds are expressed in megahertz (MHz) or gigahertz (GHz). In short, system clock:
- Synchronizes the timing of all computer operations;
- Each tick → execute one instruction (older computers);
- Many CPU's today execute more than one instruction per clock cycle;
- Clock speed is measured in hertz (1 hertz – 1 clock tick per second).

2.1.4.5 Word Size

Word is the basic storage unit around which a computer is designed. Word size or word length is the number of bits a computer can process at one time. The size of a word varies from one computer to another, depending on the CPU. For computers with a 16-bit CPU, a word is 16 bits (2 bytes). On large mainframes, a word can be as long as 64 bits (8 bytes). Some computers and programming languages distinguish between shortwords and longwords. A shortword is usually 2 bytes long, while a longword is 4 bytes.

2.1.4.6　Speed of a Computer

The speed of a computer is measured by:
- Word size;
- Bus size;
- Clock speed.

2.1.4.7　Heat Sink and Heat Pipe

A heat sink (ref. Figure I-2-4) is a component designed to lower the temperature of an electronic device by dissipating heat into the surrounding air. All modern CPUs require a heat sink. Some also require a fan. A heat sink without a fan is called a passive heat sink; a heat sink with a fan is called an active heat sink. Heat sinks are generally made of an aluminum alloy and often have fins.

A heat pipe (ref. Figure I-2-5) is a heat transfer mechanism that can transport large quantities of heat with a very small difference in temperature between the hot and cold interfaces.

Figure I-2-4　Heat Sink with Fan Attached　　　　Figure I-2-5　Heat Sink with Heat Pipe

2.1.4.8　Parallel Processing

Parallel processing is the simultaneous use of more than one CPU to execute a program. Ideally, parallel processing makes a program run faster because there are more engines (CPUs) running it. In practice, it is often difficult to divide a program in such a way that separate CPUs can execute different portions without interfering with each other.

2.2 Memory

Memory is the internal storage areas in the computer. Memory consists of electronic components that store instructions waiting to be executed by the processor, data needed by those instructions, and the results of processed data (information). Memory usually consists of one or more chips on the motherboard or some other circuit board in the computer. Most memory keeps data and instructions temporarily, although some forms of memory are permanent.

(Alias: main memory, internal memory, primary memory)

2.2.1 Computer Memory Cell

Computers are electronic machines that all data are stored by means of electrical charges. All numbers, characters, control signals, addresses etc. are coded into patterns of 0's and 1's.

A bit is a binary digit (0 or 1), the smallest unit of data the computer can process.

(1) A byte (pronounced /'baɪt/) is the basic storage unit in memory. One byte consists of 8 bits (e.g.: 01010110 represents the capital letter T).

- Each byte resides temporarily in a location in memory called an address.
- An address is a unique number that identifies the location of the byte in memory.

(2)
- 1KB = 1024 B = 2^{10} Bytes
- 1MB = 1024KB = 2^{20} Bytes
- 1GB = 1024MB = 2^{30} Bytes
- 1TB = 1024GB = 2^{40} Bytes

(3) Binary coding schemes: There are three most commonly used types of coding schemes.
- ASCII (American Standard Code for Information Interchange; pronounced /'æski/) is an 8-bit character-encoding scheme. ASCII includes definitions for 128 characters (Ref. Appendix A): 33 of them are non-printing control characters (now mostly obsolete) that affect how text and space is processed; 94 of them are printable characters, and the space is considered an invisible graphic. Most modern character-encoding schemes are based on ASCII.
- EBCDIC (Extended Binary Coded Decimal Interchange Code) is an 8-bit character-encoding scheme developed by IBM and is used primarily for large computers.
- Unicode is a character encoding standard developed by the Unicode Consortium that represents almost all of the written languages of the world. The Unicode character has multiple representation forms, including UTF-8 (8-bit), UTF-16 (16-bit), and UTF-32

(32-bit). Most Windows interfaces use the UTF-16 form.

2.2.2 Types of Memory

The system unit contains two different types of memory: volatile and nonvolatile.

2.2.2.1 Volatile Memory

Volatile memory is computer memory that loses its contents when the power is turned off. All RAM except the CMOS RAM used for the BIOS is volatile.

2.2.2.2 Nonvolatile Memory

Nonvolatile memory, or non-volatile memory, is computer memory that can retain the stored information even when not powered. Nonvolatile memory is typically used for long-term persistent storage.

Secondary storage devices (e.g., hard disks, floppy disk drives, magnetic tape, and optical discs) are all nonvolatile storage.

2.2.3 RAM

RAM (Random Access Memory) is a type of computer memory that can be accessed randomly; that is, any byte of memory can be read from and written to by the processor and other devices. RAM is volatile, which means as soon as the power is turned off; whatever data was in RAM is lost. RAM is the most common type of memory found in computers and other devices, such as printers.

There are two basic types of RAM: dynamic RAM (DRAM) and static RAM (SRAM). The two types differ in the technology they use to hold data, dynamic RAM being the more common type. Dynamic RAM needs to be refreshed thousands of times per second. Static RAM does not need to be refreshed, which makes it faster; but it is also more expensive than dynamic RAM.

2.2.4 Cache

Cache (pronounced /kæʃ/) is a special high-speed storage mechanism. It can be either a reserved section of main memory or an independent high-speed storage device. Two types of caching are commonly used in personal computers: memory caching and disk caching.

2.2.4.1 Memory Cache

A memory cache, sometimes called a cache store or RAM cache, is a portion of memory made of high-speed static RAM (SRAM) instead of the slower and cheaper dynamic RAM (DRAM) used for main memory. Memory caching is effective because most programs access the same data or instructions over and over. By keeping as much of this information as possible in SRAM, the computer avoids accessing the slower DRAM. Some memory caches are built into the architecture of microprocessors. The Intel 80486 microprocessor, for example, contains an 8K memory cache, and the Pentium has a 16K cache. Such internal caches are often called Level 1 (L1) caches. Most modern PCs also come with external cache memory, called Level 2 (L2) caches. These caches sit between the CPU and the DRAM. Like L1 caches, L2 caches are composed of SRAM but they are much larger.

(1) Memory cache helps speed the processes of the computer because it stores frequently used instructions and data.

(2) Three types of memory cache: L1, L2, L3 (servers).

(3) When the processor needs an instruction or data, it searches memory in this order:
- L1;
- L2;
- L3;
- RAM;
- Hard disk/CD/DVD.

2.2.4.2 Disk Cache

Disk cache is a portion of RAM used to speed up access to data on a disk. The RAM can be part of the disk drive itself (sometimes called a hard disk cache or buffer) or it can be general-purpose RAM in the computer that is reserved for use by the disk drive (sometimes called a soft disk cache). Hard disk caches are more effective, but they are also much more expensive, and therefore smaller. Nearly all modern disk drives include a small amount of internal cache.

2.2.5 ROM

Computers almost always contain a small amount of read-only memory (ROM) that holds permanent data and instructions (e.g., BIOS and start-up programs). ROM memory cannot be written to and altered and can only be read, and is not cleared when the power is turned off. ROM is referred to as being nonvolatile.

2.2.5.1 Firmware

Firmware is a kind of software (programs or data) that has been written onto ROM. Firmware is a combination of software and hardware. ROMs, PROMs and EPROMs that have data or programs recorded on them are firmware. Typical examples of devices containing firmware range from end-user products such as remote controls or calculators, through computer parts and devices like hard disks, keyboards, or memory cards, all the way to scientific instrumentation and industrial robotics. Also more complex consumer devices, such as mobile phones, digital cameras, television remote control (ref. Figure I-2-6), etc., contain firmware to enable the device's basic operation as well as implementing higher-level functions.

2.2.5.2 PROM

A PROM (Programmable Read-Only Memory) is a memory chip on which you can store a program. PROMs are manufactured as blank chips on which data can be written with a special device called a PROM programmer. But once the PROM has been used, you cannot wipe it clean and use it to store something else. Like ROMs, PROMs are nonvolatile.

Figure I-2-6 Firmware-controlled Device

2.2.5.3 EEPROM

An EEPROM (Electrically Erasable Programmable Read-Only Memory, also called an E²PROM, and pronounced "e-e-prom", "double-e prom" or simply "e-squared") is a special type of PROM that can be erased by exposing it to an electrical charge. Like other types of PROM, EEPROM retains its contents even when the power is turned off. Also like other types of ROM, EEPROM is not as fast as RAM. EEPROM is similar to flash memory (sometimes called flash EEPROM). The principal difference is that EEPROM requires data to be written or erased one byte at a time whereas flash memory allows data to be written or erased in blocks. This makes flash memory faster.

2.2.5.4 CMOS

CMOS (pronounced /'siːmɔs/) is short for Complementary Metal Oxide Semiconductor. CMOS is a widely used type of semiconductor. CMOS semiconductors use both NMOS (negative polarity) and PMOS (positive polarity) circuits. Since only one of the circuit types is on at any given time, CMOS chips require less power than chips using just one type of transistor. This makes them particularly attractive for use in battery-powered devices, such as portable computers. Personal computers also contain a small amount of battery-powered CMOS memory to hold the date, time, and system setup parameters. In short, CMOS usually
- Stores a computer's startup information;
- Stores configuration information about the computer (e.g., type of disk drives, keyboard, monitor etc.).

2.2.6 Virtual Memory

Virtual memory is a memory management technique developed for multitasking kernels (ref. Figure I-2-7).
- RAM can seemingly be expanded by using some of a hard disk to act as extra "main memory".

- The purpose of virtual memory is to enlarge the address space, the set of addresses a program can utilize.
- To facilitate copying virtual memory into real memory, the operating system divides virtual memory into pages, each of which contains a fixed number of addresses. Each page is stored on a disk until it is needed. When the page is needed, the operating system copies it from disk to main memory, translating the virtual addresses into real addresses.
- The process of translating virtual addresses into real addresses is called mapping. The copying of virtual pages from disk to main memory is known as paging or swapping.

2.3 Ports and Connectors

External devices often attach to the computer by a port. Ports have different types of connectors. A connector joins a cable to a peripheral.

A port is an interface on a computer to which you can connect a device. Personal computers have various types of ports (ref. Figure I-2-8). Internally, there are several ports for connecting disk drives, display screens, and keyboards. Externally, personal computers have ports for connecting modems, printers, mouse, and other peripheral devices.

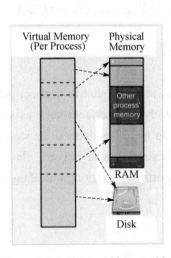

Figure I-2-7   Virtual Memory Maps to Physical Memory

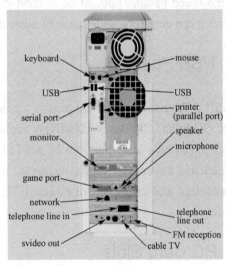

Figure I-2-8   Ports on a Personal Computer

A connector is the part of a cable that plugs into a port or interface to connect one device to another. Most connectors are either male (containing one or more exposed pins) or female (containing holes in which the male connector can be inserted).

2.3.1 Serial Ports

A serial port is a port, or interface, that can be used for serial communication, in which only

1 bit is transmitted at a time. Most serial ports on personal computers conform to the RS-232C or RS-422 standards.
- One type of serial port—COM port;
- Usually connect mouse, keyboard, and modem.

2.3.2 Parallel Ports

A parallel port (also known as a printer port or Centronics port) is a parallel interface for parallel communication, in which more than 1 bit is transmitted at a time. The IEEE 1284 standard defines the bi-directional version of the port, which allows the transmission and reception of data bits at the same time.
- One type of parallel port—LPT port;
- Usually connect printer.

2.3.3 SCSI Ports

SCSI (pronounced /ˈskʌzi/) is short for Small Computer System Interface. A SCSI port is a special high-speed parallel port. SCSI is most commonly used for hard disks, tape drives, and printers, but it can connect a wide range of other devices, including scanners and CD drives.

2.3.4 USB Ports

A USB (Universal Serial Bus) port is an external bus standard that to connect peripheral devices to computers. USB has effectively replaced a variety of interfaces such as serial and parallel ports. A single USB port can be used to connect up to 127 peripheral devices, such as mouse, modems, keyboards, digital cameras, printers, personal media players, flash drives, network adapters, and external hard drives. USB also supports hot plugging and Plug and Play installation.

2.3.5 PC Card Slots

PC Card (originally PCMCIA Card) is the form of a peripheral interface designed for laptop computers. The PC Card standard (as well as its successor ExpressCard) was defined and developed by a group of companies called the Personal Computer Memory Card International Association (PCMCIA). The United States computer industry created the Personal Computer Memory Card International Association to challenge the Japanese JEIDA memory card devices by offering a competing standard for memory-expansion cards. In 1991 the two standards merged as JEIDA 4.1 or PCMCIA 2.0 (PC Card).

PC Card was originally designed for computer storage expansion, but the existence of a usable general standard for notebook peripherals led to many kinds of devices being made available in this form. Typical devices included network cards, modems, and hard disks. The cards were also used in early digital SLR cameras, such as the Kodak DCS 300 series. The original use, as storage expansion, is no longer common.

Many notebooks in the 1990s came with two type-II slots with no barrier in between (allowing installation of two type-II cards or one, double-sized, type-III card). The PC card port has been superseded by the faster ExpressCard interface, but some modern portable computers still use them.

Text 3

Input and Output

Input/output, or I/O, refers to the communication between an information processing system (such as a computer), and the outside world. Inputs are the signals or data sent to the system, and outputs are the signals or data sent by the system to the outside.

3.1 Input

3.1.1 Input Devices

An input device is a hardware mechanism that transforms information in the external world for consumption by a computer. Often, input devices are under direct control by a human user, who uses them to communicate commands or other information to be processed by the computer, which may then transmit feedback to the user through an output device. Input and output devices together make up the hardware interface between a computer and the user or external world.

(1) Input device provides:
- Entry of data;
- Conversion into computer format.

(2) Popular input devices include the keyboard, mouse, stylus, digital pen, microphone, digital camera and scanner.

3.1.2 Keyboards

Computer keyboard (ref. Figure I-3-1) is the set of typewriter-like keys that enables users to enter data into a computer. The keys on computer keyboards are often classified as follows.
- Alphanumeric keys—letters and numbers.
- Punctuation keys—comma, period, semicolon, and so on.
- Special keys—function keys, control keys, arrow keys, Caps Lock key, and so on.

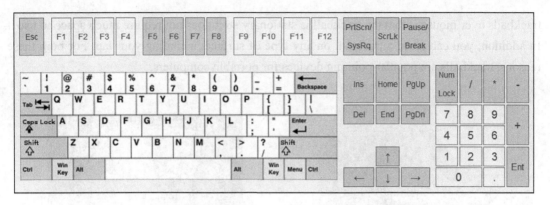

Figure I-3-1 IBM/Windows Keyboard (US layout)

3.1.3 Pointing Devices

A pointing device is an input device with which users can control the movement of the pointer to select items on a display screen. Examples of pointing devices include mouse, trackballs, touchpad, joysticks, touch screen, and light pens, etc.

3.1.3.1 Mouse

A mouse (ref. Figure I-3-2, a computer mouse with the most common standard features: two buttons and a scroll wheel, which can also act as a third button) is a device that controls the movement of the cursor or pointer on a display screen. Physically, a mouse consists of an object held under one of the user's hands, with one or more buttons. As users move the mouse, the pointer on the display screen moves in the same direction.

There are three basic types of mouse.

- Mechanical: This is a type of computer mouse that has a rubber or metal ball on its underside that can roll in all directions. Mechanical sensors within the mouse detect the direction the ball is rolling and move the screen pointer accordingly.
- Optomechanical: This type is the same as the mechanical mouse except that it uses optical sensors to detect the motion of the ball. A mouse pad should be used under the mouse to run on.
- Optical: This type uses a laser to detect the mouse's movement. Optical mice have no mechanical moving parts. This type responds more quickly and precisely than mechanical and optomechanical mouse.

3.1.3.2 Trackballs

Essentially, a trackball is a mouse lying on its back (ref. Figure I-3-3). The user rotates the ball with the thumb, fingers, or the palm of the hand to move a cursor. There are usually one to three buttons next to the ball, which users use just like mouse buttons. The advantage of trackballs over mouse is that the trackball is stationary so it does not require much space to use it. In addition, you can place a trackball on any type of surface, including your lap. For both these reasons, trackballs are popular pointing devices for portable computers.

Figure I-3-2 Computer Mouse   Figure I-3-3 The Kensington Expert Mouse Trackball

3.1.3.3 Touchpads

A touchpad (also trackpad) is a small, touch-sensitive pad (ref. Figure I-3-4) used as a pointing device on some portable computers. By moving a finger or other objects along the pad, the user can move the pointer on the display screen. Touchpads are a common feature of laptop computers and also used as a substitute for a computer mouse where desk space is scarce. They can also be found on personal digital assistants (PDAs) and some portable media players, such as the iPod using the click wheel (ref. Figure I-3-5).

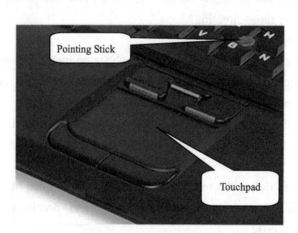

Figure I-3-4 Touchpad and a Pointing Stick on an IBM Laptop   Figure I-3-5 iPod Classic

3.1.3.4 Joysticks and Wheels

A joystick (ref. Figure I-3-6) is an input device consisting of a handheld stick that pivots on a base and reports its angle or direction to the device it is controlling. Joysticks are often used to control video games, and usually have one or more push-buttons whose state can also be read by the computer.

Figure I-3-6 Joystick and Wheel

A wheel (also steering wheel, ref. Figure I-3-6) is a circular device capable of rotating on its axis, facilitating movement or transportation or performing labor in machines.

Joysticks and wheels are used mostly for computer games, but they are also used occasionally for CAD/CAM systems and other applications.

3.1.3.5 Touch Screens

A touch screen is a type of display screen that has a touch-sensitive transparent panel covering the screen. Instead of using a pointing device such as a mouse or light pen, users can use the finger or hand to point directly to objects on the screen.

3.1.3.6 Light Pens

A light pen is an input device that utilizes a light-sensitive detector to select objects on a display screen. A light pen is similar to a mouse, except that with a light pen the user can move the pointer and select objects on the display screen by directly pointing to the objects with the pen.

3.1.3.7 Stylus or Digital Pens

A stylus or digital pen (ref. Figure I-3-7) is a pointing device that looks like a ballpoint pen, but uses pressure, instead of ink, to write text and draw lines.

Figure I-3-7 Several Styluses

3.1.3.8 Pointing Stick

The pointing stick is a pointing device first developed by IBM for its notebook computers that consists of a miniature joystick, usually with a rubber eraser-head tip, positioned somewhere between the keys on the keyboard (ref. Figure I-3-4). Most pointing sticks are pressure-sensitive, so the pointer moves faster when more pressure is applied.

3.1.4 Voice Input

Voice input is the processing of entering data by speaking into a microphone.

3.1.4.1 Voice Recognition

Voice recognition, also called speech recognition, is the computer's capability of distinguishing spoken words. Voice recognition is the field of computer science that deals with designing computer systems that can recognize spoken words. Note that voice recognition implies only that the computer can take dictation, not that it understands what is being said. Comprehending human languages falls under a different field of computer science called natural language processing.

3.1.4.2 Audio input

Audio input is the process of entering any sound, such as speech, music, and sound effects, into the computer.

3.1.5 Digital Cameras

A digital camera (or digicam, ref. Figure I-3-8) is an electronic device used to capture and store photographs digitally, instead of using photographic film like conventional cameras, or recording images in an analog format to magnetic tape like many video cameras. Modern compact digital cameras are typically multifunctional, with some devices capable of recording sound and/or video as well as photographs.

Figure I-3-8 Canon PowerShot A95

3.1.6 Video Input

Video input is to capture full-motion images into a computer and store them on a storage medium such as a hard disk or DVD.

3.1.6.1 Digital Video Cameras

A digital video (DV) camera is a new generation of video camera (or camcorder) that records video as digital signals instead of as analog signals.

3.1.6.2 Web Cams

A web camera (or webcam) (ref. Figure I-3-9) is a real-time camera (usually, though not always, a video camera) whose images can be accessed using the World Wide Web, instant messaging, or a PC video calling application. The term webcam is also used to describe the low-resolution digital video cameras designed for such purposes, but which can also be used to record in a non-real-time fashion.

Figure I-3-9 Webcam

3.1.6.3 Video Conferencing

A videoconference or video conference (also known as a videoteleconference) is a set of interactive telecommunication technologies which allow two or more locations to interact via two-way video and audio transmissions simultaneously. It has also been called visual collaboration and is a type of groupware. Videoconferencing differs from videophone in that it is designed to serve a conference rather than individuals.

3.1.7 Scanning and Reading Devices

3.1.7.1 Optical Scanners

An optical scanner (ref. Figure I-3-10), usually called a scanner, is a light-sensing input device that scans images, printed text, handwriting, or an object, and converts them to digital images. Optical scanners do not distinguish text from illustrations; they represent all images as bit maps. Therefore, users cannot directly edit text that has been scanned. To edit text read by an optical scanner, users need an optical character recognition (OCR) system to translate the image into ASCII characters. Most optical scanners sold today come with OCR packages.

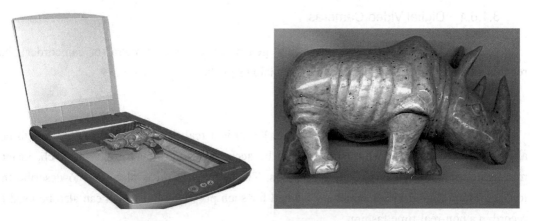

Figure I-3-10 Desktop Scanner and Scanned Object

3.1.7.2 Optical Readers

An optical reader is a device that uses a light source to read characters, marks, and codes and then converts them into digital data that a computer can process.

Two technologies used by optical readers are optical character recognition and optical mark recognition.

1) Optical Character Recognition

Optical Character Recognition (OCR) is a type of computer software designed to translate images of handwritten or typewritten or printed text (usually captured by a scanner) into machine-editable text, or to translate pictures of characters into a standard encoding scheme representing them (e.g., ASCII or Unicode). OCR is a field of research in pattern recognition, artificial intelligence and computer vision.

2) Optical Mark Recognition

Optical Mark Recognition (also called Optical Mark Reading and OMR) is the technology of electronically extracting intended data from marked fields, such as checkboxes and fill-infields, on printed forms. OMR technology scans a printed form and reads predefined positions and records where marks are made on the form. This technology is useful for applications in which large numbers of hand-filled forms need to be processed quickly and with great accuracy, such as surveys, reply cards, questionnaires and ballots. One of the most familiar applications of optical mark recognition is the use of #2 (HB in Europe) pencil bubble optical answer sheets in multiple choice question examinations (ref. Figure I-3-11). Students mark their answers, or other personal information, by darkening circles marked on a pre-printed sheet. Afterwards the sheet is automatically graded by a scanning machine.

3.1.7.3 Barcode Scanners

A barcode scanner (or barcode reader) (ref. Figure I-3-12) is an optical reader that uses laser beams to read bar codes. It is an identification code that consists of a set of vertical lines and spaces of different widths. The bar code represents data that identifies the manufacturer and the item.

Figure I-3-11   OMR Test Form

Figure I-3-12   Barcode Scanner

3.1.7.4 MICR Readers

A MICR (Magnetic Ink Character Recognition) reader is a device that converts MICR characters into a form that a computer can process.

MICR is a character recognition technology used primarily by the banking industry to facilitate the processing of cheques (ref. Figure I-3-13). The technology allows computers to read information (such as account numbers) off of printed documents. Unlike barcodes or similar technologies, however, MICR codes can be easily read by humans. MICR provides a secure, high-speed method of scanning and processing information.

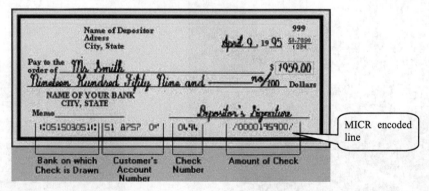
Figure I-3-13   Cheque Sample

3.1.8 Terminals

A terminal is a device that enables the user to communicate with a computer. Generally, a terminal is a combination of keyboard and display screen.

The function of a terminal is confined to display and input of data; a device with significant local programmable data processing capability may be called a "smart terminal" or fat client. A terminal that depends on the host computer for its processing power is called a thin client. A personal computer can run software that emulates the function of a terminal, sometimes allowing concurrent use of local programs and access to a distant terminal host system.

3.1.8.1 Point-of-Sale Terminals

Point-of-Sale (POS, ref. Figure I-3-14) or checkout refers to the capturing of data and customer payment information at a physical location when goods or services are bought and sold. The POS transaction is captured using a variety of devices which include computers, cash registers, optical and bar code scanners, magnetic card readers, or any combination of these devices.

3.1.8.2 Automated Teller Machine

An Automated Teller Machine (ATM, ref. Figure I-3-15), also known as a automated banking machine (ABM) or automated transaction machine or Cash Machine, is a computerized telecommunications device that provides the customers of a financial institution with access to financial transactions in a public space without the need for a human clerk or bank teller. On most modern ATMs, the customer is identified by inserting a plastic ATM card with a magnetic stripe or a plastic smartcard with a chip, which contains a unique card number and some security information, such as an expiration date. Security is provided by the customer entering a personal identification number (PIN).

Figure I-3-14 Point-of-Sale at a Wal-Mart store

Figure I-3-15 ATM

3.1.8.3 Smart Display

Smart display was a Microsoft initiative to use a portable touchscreen LCD monitor as a thin client for PCs, connecting via Wi-Fi.

3.1.9 Biometric Input

3.1.9.1 Biometrics

Biometrics is the study of measurable biological characteristics. In computer security, biometrics refers to authentication techniques that rely on measurable physical characteristics that can be automatically checked. There are several types of biometric identification schemes.
- Face: the analysis of facial characteristics.
- Fingerprint: the analysis of an individual's unique fingerprints.
- Hand geometry: the analysis of the shape of the hand and the length of the fingers.
- Retina: the analysis of the capillary vessels located at the back of the eye.
- Iris: the analysis of the colored ring that surrounds the eye's pupil.
- Signature: the analysis of the way a person signs his name.
- Vein: the analysis of pattern of veins in the back of the hand and the wrist.
- Voice: the analysis of the tone, pitch, cadence and frequency of a person's voice.

Classification of some biometric traits is showed in Figure I-3-16.

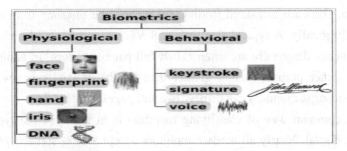

Figure I-3-16 Classification of Some Biometric Traits

3.1.9.2 Biometric Devices

A biometric device is a device that translates a personal characteristic into a digital code that is compared with a digital code stored in a computer.

3.2 Output

3.2.1 Output Devices

An output device is any hardware component that conveys information to one or more people. Output devices include display screens, printers, speakers, and headphones/headsets, etc.

3.2.1.1　Screens and Monitors

Screen is also called VDU (Visual Display Unit), terminal, monitor, which is an electronic visual display for computers. The term monitor, however, usually refers to the entire box, which comprises the display device, circuitry, and an enclosure, whereas display screen can mean just the screen.

3.2.1.2　Classification of Monitors

There are many ways to classify monitors.

(1) In terms of color capabilities, monitors can be classified into three classes.

- Monochrome: Monochrome monitors actually display two colors, one for the background and one for the foreground. The colors can be black and white, green and black, or amber and black.
- Gray-scale: A gray-scale monitor is a special type of monochrome monitor capable of displaying different shades of gray.
- Color: Color monitors can display over 1 million different colors. Color monitors are sometimes called RGB monitors because they accept three separate signals—red, green, and blue.

(2) After above classification, the most important aspect of a monitor is its screen size. Like televisions, screen sizes are measured in diagonal inches, the distance from one corner to the opposite corner diagonally. A typical size for small VGA monitors is 14 inches. Monitors that are 16 or more inches diagonally are often called full-page monitors. In addition to their size, monitors can be either portrait (height greater than width) or landscape (width greater than height). Larger landscape monitors can display two full pages, side by side.

(3) Another common way of classifying monitors is in terms of the type of signal they accept: analog or digital. Nearly all modern monitors accept analog signals, which are required by the VGA, SVGA, and other high-resolution color standards.

3.2.1.3　Types of Screens

1) Cathode Ray Tube (CRT)

CRT (ref. Figure I-3-17(a)) is a desktop monitor that is similar to a standard television because it contains a cathode-ray tube. A CRT works by moving an electron beam back and forth across the back of the screen. Each time the beam makes a pass across the screen, it lights up phosphor dots on the inside of the glass tube, thereby illuminating the active portions of the screen. By drawing many such lines from the top to the bottom of the screen, it creates an entire screenful of images.

2) Liquid Crystal Display (LCD)

LCD (ref. Figure I-3-17(b)) is a type of display used in digital watches and many portable computers. LCD displays utilize two sheets of polarizing material with a liquid crystal solution between them. An electric current passed through the liquid causes the crystals to align so that light cannot pass through them. Each crystal, therefore, is like a shutter, either allowing light to pass through or blocking the light. LCD provides sharp and flicker-free displays.

3) Gas Plasma

Gas plasma display (ref. Figure I-3-17(c)) is a type of flat-panel display that uses gas plasma technology, which substitutes a layer of gas for the liquid crystal material used in a flat-panel monitor. When voltage is applied, the gas releases ultraviolet (UV) light, which causes the pixels on the screen to glow and form an image.

(a) CRT (b) LCD (c) Gas plasma display

Figure I-3-17　Types of Screens

3.2.2　Printers

A printer is an output device that produces text and graphics on a physical medium such as paper or transparency film. There are many different types of printers.

(1) In terms of the technology utilized, printers fall into the following categories.
- Daisy-wheel: Similar to a ball-head typewriter, this type of printer has a plastic or metal wheel on which the shape of each character stands out in relief. A hammer presses the wheel against a ribbon, which in turn makes an ink stain in the shape of the character on the paper. Daisy-wheel printers produce letter-quality print but cannot print graphics.
- Dot-matrix: Creates characters by striking pins against an ink ribbon. Each pin makes a dot, and combinations of dots form characters and illustrations.
- Ink-jet: An inkjet printer (ref. Figure I-3-18) is a type of computer printer that sprays ink at a sheet of paper. Ink-jet printers are capable of producing high quality print approaching that produced by laser printers. In general, the price of ink-jet printers is lower than that of laser printers. However, they are also considerably slower. Another drawback of ink-jet printers is that they require a special type of ink that is apt to smudge on inexpensive copier paper. Because ink-jet printers require smaller mechanical parts

than laser printers, they are especially popular as portable printers. In addition, color ink-jet printers provide an inexpensive way to print full-color documents.
- Laser: A laser printer (ref. Figure I-3-19) is a type of printer that utilizes a laser beam to produce an image on a drum. It is a common type of computer printer that rapidly produces high quality text and graphics on plain paper.
- LCD & LED: A type of printer similar to a laser printer, but uses liquid crystals or light-emitting diodes rather than a laser to produce an image on the drum.
- Line printer: A high-speed printer capable of printing an entire line at one time. A fast line printer can print as many as 3 000 lines per minute. The disadvantages of line printers are that they cannot print graphics, the print quality is low, and they are very noisy.
- Thermal printer (ref. Figure I-3-20): An inexpensive printer that works by pushing heated pins against heat-sensitive paper. Thermal printers are widely used in calculators and fax machines.

Figure I-3-18　Epson Inkjet Printer

Figure I-3-19　HP LaserJet1 012 Laser Printer

Figure I-3-20　Thermal Printer

- Photo printer: A photo printer is a color printer that is specifically designed to print high quality digital photos on photo paper. These printers usually have a very high number of nozzles and are capable of printing droplets as small as 1 picoliter.

(2) Printers are also classified by the following characteristics.
- Quality of type: The output produced by printers is said to be either letter quality (as good as a typewriter), near letter quality, or draft quality. Only daisy-wheel, ink-jet, and laser printers produce letter-quality type.

- Speed: Measured in characters per second (cps) or pages per minute (ppm), the speed of printers varies widely. Daisy-wheel printers tend to be the slowest, printing about 30 cps. Line printers are fastest (up to 3 000 lines per minute). Dot-matrix printers can print up to 500 cps, and laser printers range from about 4 to 20 pages per minute.
- Impact or non-impact: Impact printers include all printers that work by striking an ink ribbon. Daisy-wheel, dot-matrix, and line printers are impact printers. Non-impact printers include laser printers and ink-jet printers. Non-impact printer has no physical contact between paper and printing device. The important difference between impact and non-impact printers is that impact printers are much noisier.
- Graphics: Some printers (daisy-wheel and line printers) can print only text. Other printers can print both text and graphics.
- Fonts: Some printers, notably dot-matrix printers, are limited to one or a few fonts. In contrast, laser and ink-jet printers are capable of printing an almost unlimited variety of fonts.

3.2.3 Speakers and Headphones

Speaker (or "loudspeaker") is an electro-acoustic transducer that converts electrical signals into sounds loud enough to be heard at a distance.

Headphones (ref. Figure I-3-21) are a pair of small loudspeakers, or less commonly a single speaker, with a way of holding them close to a user's ears and a means of connecting them to a signal source such as an audio amplifier, radio, CD player or portable media player. They are also known as stereophones or headsets. The in-ear versions are known as earphones or earbuds. In the context of telecommunication, the term headset is used to describe a combination of headphone and microphone used for two-way communication.

    Handset   Handset   Headset   Headphone   Earphone   Earset

Figure I-3-21　Different Kinds of Headphones

3.2.4 Other Output Devices

3.2.4.1 Fax Machines

A fax (short for facsimile) is a document (pictures and text) sent over a telephone line. A fax machine (ref. Figure I-3-22) is a device that can send or receive faxes over a telephone line.

3.2.4.2 Plotters

A plotter (ref. Figure I-3-23) is a device that draws pictures on paper based on commands from a computer. Plotters differ from printers in that they draw lines using a pen. As a result, they can produce continuous lines, whereas printers can only simulate lines by printing a closely spaced series of dots. Multicolor plotters use different-colored pens to draw different colors.

Figure I-3-22   Samsung Fax Machine

Figure I-3-23   Inkjet Plotter

3.2.4.3 Data Projectors

A data projector (ref. Figure I-3-24) is an output device that takes the text and images displaying on a computer screen and projects them on a larger screen so an audience can see the image clearly. Projectors are widely used for conference room presentations, classroom training, home theatre and live events applications.

3.2.4.4 High-definition television (HDTV)

High-Definition Television (HDTV) is a new type of television that provides much better resolution than current televisions based on the NTSC standard. HDTV

Figure I-3-24   Data Projector

is a digital TV broadcasting format where the broadcast transmits widescreen pictures with more detail and quality than found in a standard analog television, or other digital television formats. HDTV is a type of Digital Television (DTV) broadcast, and is considered to be the best quality DTV format available. Types of HDTV displays include direct-view, plasma, rear screen, and front screen projection.

3.2.4.5 Multifunction Peripherals

A multifunction peripheral (MFP, sometimes called as all-in-one device) is a single device that looks like a copy machine but provides the functionality of a printer, scanner, copy machine, and perhaps a fax machine. These devices are becoming a popular option for SOHO (Small Office/Home Office) users because they're less expensive than buying three or four separate devices.

3.2.5 Important Concepts

3.2.5.1 Pixel

In digital imaging, each spot is called a pixel (short for picture element). The pixel is the smallest addressable screen element, and it is the smallest unit of picture that can be controlled. Graphics monitors display pictures by dividing the display screen into thousands (or millions) of pixels, arranged in rows and columns.

The number of distinct colors that can be represented by a pixel depends on the number of bits per pixel (bpp).

- 1 bpp, $2^1 = 2$ colors (monochrome)
- 2 bpp, $2^2 = 4$ colors
- 3 bpp, $2^3 = 8$ colors
- $\vdots$
- 8 bpp, $2^8 = 256$ colors
- 16 bpp, $2^{16} = 65\ 536$ colors ("Highcolor")
- 24 bpp, $2^{24} \approx 16.8$ million colors ("Truecolor")
- 32 bpp, $2^{32} \approx 4.3$ billion colors

3.2.5.2 Resolution

Resolution refers to the sharpness and clarity of an image, usually defined in pixels. Resolution is most often used to describe monitors, printers, and bit-mapped graphic images.

(1) In the case of dot-matrix and laser printers, the resolution indicates the number of dots per inch. For example, a 300-dpi (dots per inch) printer is one that is capable of printing 300 distinct dots in a line 1 inch long. This means it can print 90 000 dots per square inch.

(2) For graphics monitors, the screen resolution signifies the number of dots (pixels) on the entire screen. For example,

- CGA: 640×200 pixels;
- EGA: 640×350 pixels;
- VGA: 640×480 pixels;
- Super VGA: 800×600 resolution with 256 colors needs 512KB graphics board, 1024×768 resolution with 256 colors needs a 1MB graphics board.

Text 4
Secondary Storage

Secondary storage (ref. Figure I-4-1), also called storage medium or auxiliary storage or external memory, differs from primary storage in that it is not directly accessible by the CPU. The computer usually uses its input/output channels to access secondary storage and transfers the desired data using intermediate area in primary storage. Secondary storage does not lose the data when the device is powered down—it is non-volatile. Per unit, it is typically less expensive than primary storage.

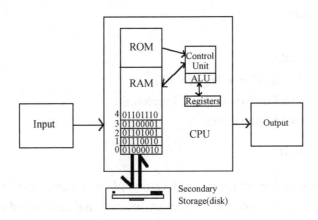

Figure I-4-1   Computer Components

1) Characteristics of secondary storage
- Non-volatile (compared to RAM);
- Cheap per MB (compared to RAM/ROM);
- Convenient (compared to filing cabinet);
- Relatively fast (compared to filing cabinet);
- Slow compared to RAM.

2) Examples of secondary storage media and devices
- Floppy disks;
- Hard disks;
- Optical disks (e.g., CD-ROM, CD-R, CD-RW; DVD-ROM, DVD-R, DVD-RW; Blu-ray Disc);

- Flash memory (e.g., USB flash drives);
- Tape;
- Cartridge tape;
- Magnetic-Optical drive;
- PC cards;
- Smart cards.

4.1 Floppy Disks

A floppy disk (often called floppy or diskette, ref. Figure I-4-2) is a kind of soft magnetic disk. It is called floppy because it flops if you wave it (at least, the 8-inch and 5¼-inch floppy disks do). Unlike most hard disks, floppy disks are portable, because users can remove them from a disk drive. Disk drives for floppy disks are called floppy drives. Floppy disks are slower to access than hard disks and have less storage capacity, but they are much less expensive. And most importantly, they are portable:

Figure I-4-2 8-inch, 5¼-inch, and 3½-inch floppy disks

- Circular piece of plastic;
- Made up of tracks & sectors;
- 512 bytes in each sector.

The capacity of a floppy disk:

E.g., for a 3½" Macintosh HD 1.44MB floppy disk (double density)

= 2 sides × 80 tracks × 18 sectors × 512 bytes/sector

= 1 474 560 bytes ≈ 1.44MB capacity

4.2 Hard Disks

A hard disk is a magnetic disk on which users can store computer data, instructions, and information. The term hard is used to distinguish it from a soft or floppy disk. Hard disks hold more data and are faster than floppy disks.

Figure I-4-3 shows a typical hard disk drive (HDD) design. A single hard disk usually consists of a spindle that holds flat circular disks called platters, onto which the data are recorded. Each platter requires two read/write heads, one for each side. All the read/write heads are attached to a single actuator arm (or access arm). An actuator arm moves the heads on an arc (roughly radially) across the platters as they spin, allowing each head to access almost the entire surface of the platter as it spins. Each platter has the same number of tracks, and a track location that cuts across all platters is called a cylinder.

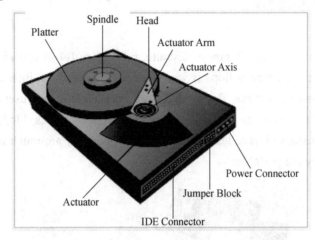

Figure I-4-3　Diagram of a computer hard disk drive

4.2.1　Formatting

Disk formatting is the process of preparing a hard disk or other storage medium for use, including setting up an empty file system.
- Defines the tracks and sectors on the surface of the disk;
- Generally erases all data ("cleans" the disk);
- Creates FAT (FAT contains information on sectors for each file plus free sectors) and root directory structure.

4.2.2　Capacity

The capacity of an HDD can be calculated by multiplying the number of cylinders by the number of heads by the number of sectors by the number of bytes/sector (most commonly 512). For example, the capacity of a hard disk with eight platters (sixteen heads), 16 383 cylinders and 63 sectors is $16 \times 16,383 \times 63 \times 512 \approx 8G$.

Current disk capacity can be 36GB, 40GB, 45GB, 60GB, 75GB, 80GB, 120GB, 150GB, 160GB, 200GB, 250GB, 300GB, 320GB, 400GB, 500GB, 640GB, 750GB, 808GB, 1TB, 1.5TB,

2TB, 2.5TB, 3TB, or 1PB.

4.2.3 Rotational Speed

Rotational speed (sometimes called speed of revolution) indicates how fast an HDD is running. It tells how many full rotations (i.e. revolutions or cycles) completed in one minute around a fixed axis. It is therefore a cyclic frequency, measured in revolutions per minute (rpm).

Computers' hard drives typically rotate at 5 400 (4 200 for notebook computer) or 7 200 rpm and some high-performance drives rotate at 10 000 or 15 000 rpm.

4.2.4 Access Time

Access time defines the speed of storage devices. It is the time a program or device takes to locate a single piece of information and make it available to the computer for processing.

(1) Access time measures
- The amount of time it takes a storage device to locate an item on a storage medium;
- The time required to deliver an item from disk to RAM;
- Usually between 5 to 80 milliseconds.

(2) Access Time = Seek Time + Rotational Delay + Data Transfer Time
- Seek Time: Time for the read/write head to find the right track;
- Rotational Delay (average 1/2 revolution): Time for the right sector to rotate under the read/write head. Average rotational delay is the half time it takes to do a full rotation;
- Data Transfer Time: Time for the data to transfer from disk to RAM.

E.g., Calculate the average access time of a hard drive which has a seek time of 25 milliseconds, data transfer time of 2 milliseconds and rotates at 4500 rpm.

4500 rev takes 1 min

1 rev takes $\frac{1}{4500}$ min = $\frac{1}{4500} \times 60 \times 1000$ ms ≈ 13.33ms

0.5 rev takes 6.67 ms

Access time = seek time + rotational delay + data transfer time

≈ 25ms + 6.67ms + 2ms

≈ 33.67ms

4.2.5 Characteristics of a Hard Disk

- Made up of platters, cylinders and sectors;
- Rotation speed 4200rpm or above;
- Head "floats" on surface;

- Bad sectors & head crash;
- SCANDISK (or CHKDSK) command detects and deallocates bad sectors;
- Capacities range from 36GB to 1PB or above.

4.2.6　Maintaining Data Stored on a Disk

4.2.6.1　Backup

Backup is to copy a file, program, or disk to a second medium as a precaution in case the first medium fails.

4.2.6.2　Fragmentation

Fragmentation is the status of the contents of a file are scattered across two or more noncontiguous sectors. This is entirely invisible to users, but it can slow down the speed at which data is accessed because the disk drive must search through different parts of the disk to put together a single file.

4.2.6.3　Defragmentation

Defragmentation (ref. Figure I-4-4) is the process of physically reorganizing the files so they are stored in contiguous sectors and the operating system can access data more quickly.

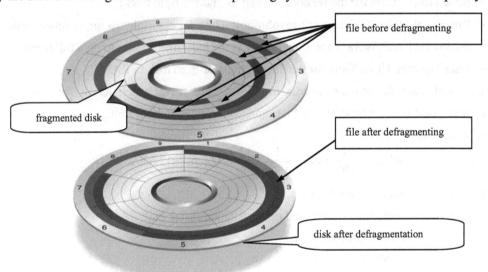

Figure I-4-4　Defragmentation Process

4.2.6.4　Data compression

Data compression is to store data in a format that requires less space than usual.

4.2.7 Features of Floppy Disks and Hard Disks

(1) A floppy disk is a type of magnetic media that allows users to read from and write on a disk any number of times. Magnetic media use magnetic particles to store items such as data, instructions, and information on a disk's surface.

(2) Characteristics of a hard disk include its platters, read/write heads, cylinders, sectors, capacity, revolutions per minute, access time, and disk cache (ref. §2.2.4.2).

(3) Key differences between floppy disks and hard disks.
- Floppy disk: Head is in close contact with surface of the disk.
- Hard disk: Heads "fly" on a film of air a fraction of an mm above the disk's surface. If head touches the disk, the surface of disk and disk head will be destroyed. This is called "head crash".

Hard Disk	Floppy Disk
rotates constantly	stops/starts
high speed (3600 rpm~15 000 rpm)	low speed (300 rpm)
High storage capacity (36GB~1PB)	Low storage capacity (1.44MB)
Access time approx. 5 to 80 ms or less	Access time approx. 175 to 300 ms

4.3 Flash Memory

Flash memory is a non-volatile computer storage chip that can be electrically erased and reprogrammed. A flash memory card (also called flash card, ref. Figure I-4-5) is an electronic flash memory data storage device used for storing digital information. Flash memory is primarily used in memory cards, USB flash drives, digital cameras, mobile phones, laptop computers, MP3 players, and video game consoles. They are small, re-recordable, and they can retain data without power.

(a)Compact Flash (CF)　　(b)Secure Digital (SD) Card　　(c)xD-Picture Card　　(d)Sony Memory Stick PRO Duo

Figure I-4-5　Different Types of Flash Memory Cards

4.3.1 USB Flash Drives

A USB flash drive (also called thumb drive, jump drive, pen drive, key drive, token, or simply USB drive) consists of a flash memory data storage device integrated with a USB interface. USB flash drives are typically removable and rewritable, and physically much smaller than a floppy disk. Storage capacities in 2010 can be as large as 256 GB with steady improvements in size and price per capacity expected. Figure I-4-6 shows an example of a USB flash drive.

4.3.2 Smart Cards

A smart card, chip card, or integrated circuit card (ICC), is a small electronic device about the size of a credit card that contains electronic memory, and possibly an embedded integrated circuit (IC). Figure I-4-7 shows a smart card used for health insurance in France.

Smart cards are used for a variety of purposes, including:
- Prepaid telephone card;
- Employee time and attendance tracking;
- Patient's medical records;
- Digital cash.

Figure I-4-6 USB Flash Drive   Figure I-4-7 Smart Card

4.4 Optical Storage Technology

Optical storage refers to the storage of data on an optically readable medium. Data is recorded by making marks in a pattern that can be read back with the aid of light, usually a beam of laser light precisely focused on a spinning disc.

An optical disc is designed to support one of three recording types: read-only (e.g.: CD, CD-ROM and DVD-ROM), recordable (write-once, e.g.: CD-R and DVD-R), or re-recordable (rewritable, e.g.: CD-RW and DVD-RW). A standard Blu-ray disc can hold about 25 GB of data, a DVD about 4.7 GB, and a CD about 700 MB. Table I-4-1 shows the types and capacity of different optical discs.

Table I-4-1 Types and Capacity of Different Optical Discs

	Recording Types			Capacity
	ROM	R	RW	
CD	CD-ROM	CD-R	CD-RW	700 MB
DVD	DVD-ROM	DVD+R, DVD-R	DVD+RW, DVD-RW, DVD RAM	4.7 GB
BD	BD-ROM	BD-R	BD-RE	25 GB

4.4.1 CD Disc

A Compact Disc (also known as a CD) is an optical disc used to store digital data. It was originally developed to store sound recordings exclusively, but later it also allowed the preservation of other types of data.

Standard CDs have a diameter of 120 mm and can hold up to 80 minutes of uncompressed audio (700 MB of data). The Mini CD has various diameters ranging from 60 to 80 mm; they are sometimes used for CD singles or device drivers, storing up to 24 minutes of audio.

4.4.2 DVD Disc

DVD, also known as Digital Video Disc or Digital Versatile Disc, is a type of optical disk technology similar to the CD-ROM. Its main uses are video and data storage. DVDs are of the same dimensions as compact discs (CDs), but are capable of storing almost seven times as much data. The DVD specification supports disks with capacities of from 4.7GB to 17GB.

Variations of the term DVD often indicate the way data is stored on the discs: DVD-ROM (read only memory) has data that can only be read and not written; DVD-R and DVD+R (recordable) can record data only once, and then function as a DVD-ROM; DVD-RW (re-writable), DVD+RW, and DVD-RAM (random access memory) can all record and erase data multiple times. The wavelength used by standard DVD lasers is 650 nm; thus, the light has a red color.

4.4.3 Blu-ray Disc

A Blu-ray Disc (also called BD) is a high-density optical disc format that uses a 405nm-wavelength blue-violet laser technology, in contrast to the 650nm-wavelength red laser technology used in traditional DVD formats. The rewritable Blu-ray disc, with a data transfer rate of 36Mbps, can hold up to 27GB of data on a single-sided single layer disc (compared to the traditional DVD's 4.7GB capacity), which amounts to about 12 hours of standard video or more than 2 hours of high-definition video.

4.4.4 MO

MO (Magneto-Optical, ref. Figure I-4-8) is a type of data storage technology that combines magnetic disk technologies with optical technologies, such as those used in CD-ROMs. Like magnetic disks, MO disks can be read and written to. And like floppy disks, they are removable. However, their storage capacity can be more than 200 megabytes, much greater than magnetic floppies. In terms of data access speed, they are faster than floppies but not as fast as hard disk drives.

(a) A 130 mm 2.6GB MO   (b) A 90 mm 640MB MO   (c) A 90 mm 230 MB MO

Figure I-4-8   MO Disks

4.5 Tapes

A tape is a magnetically coated strip of plastic on which data can be encoded. Tapes for computers are similar to tapes used to store music.

(1) Tapes are useful for:
- Backups of hard disks;
- Distributing large programs.

(2) Data on tapes can only be accessed sequentially (compared to disk).

(3) QIC (Quarter Inch Cartridges,) and DAT (Digital Audio Tape) forms are most common now.

Because tapes are so slow, they are generally used only for long-term storage and backup. Data to be used regularly is almost always kept on a disk.

4.6 RAID Storage Systems

RAID is short for Redundant Array of Independent (or Inexpensive) Disks, a category of disk drives that employ two or more drives in combination for fault tolerance and performance. RAID disk drives are used frequently on servers. The various designs of RAID systems involve two key goals: increase data reliability and increase input/output performance.

- Level 0—Striped Disk Array without Fault Tolerance: Provides data striping (spreading out blocks of each file across multiple disk drives) but no redundancy. This improves performance but does not deliver fault tolerance. If one drive fails then all data in the array is lost.
- Level 1—Mirroring and Duplexing: Provides disk mirroring. Level 1 provides twice the read transaction rate of single disks and the same write transaction rate as single disks.
- Level 5—Block Interleaved Distributed Parity: Provides data striping at the byte level and also stripe error correction information. This results in excellent performance and good fault tolerance. Level 5 is one of the most popular implementations of RAID.

Text 5

Software

Software, consisting of programs, enables a computer to perform specific tasks, as opposed to its physical components (hardware) which can only do the tasks they are mechanically designed for.

5.1 Categories of Software

Two general categories of software are showed in Figure I-5-1.

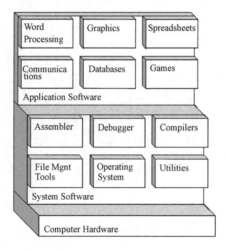

Figure I-5-1 Categories of Software

Other examples of software include:
- Programming languages define the syntax and semantics of computer programs. For example, Pascal, C, C++, VB/VB.NET, C#, Java, etc.
- Middleware controls and co-ordinates distributed systems. Middleware is computer software that connects software components or some people and their applications. The software consists of a set of services that allows multiple processes running on one or more machines to interact. This technology evolved to provide for interoperability in support of the move to coherent distributed architectures, which are most often used to

support and simplify complex distributed applications. It includes web servers, application servers, and similar tools that support application development and delivery. Middleware is especially integral to modern information technology based on XML, SOAP, Web services, and service-oriented architecture.

- Testware is software for testing hardware or a software package. Generally speaking, Testware is a sub-set of software with a special purpose, that is, for software testing, especially for software testing automation. Testware is produced by both verification and validation testing methods.
- Firmware is low-level software often stored on electrically programmable memory devices. Firmware is given its name because it is treated like hardware and run ("executed") by other software programs. Typical examples of devices containing firmware range from end-user products such as remote controls or calculators, through computer parts and devices like hard disks, keyboards, or memory cards, all the way to scientific instrumentation and industrial robotics. Also more complex consumer devices, such as mobile phones, digital cameras, etc., contain firmware to enable the device's basic operation as well as implementing higher-level functions.
- Device drivers control parts of computers such as disk drives, printers, CD drives, or computer monitors. A driver typically communicates with the device through the computer bus or communications subsystem to which the hardware connects. When a calling program invokes a routine in the driver, the driver issues commands to the device. Once the device sends data back to the driver, the driver may invoke routines in the original calling program. Drivers are hardware-dependent and operating-system-specific. They usually provide the interrupt handling required for any necessary asynchronous time-dependent hardware interface.

System software is a set of programs that interacts directly with hardware. System software is a generic term referring to any computer software which manages and controls the hardware so that application software can perform a task. System software serves as the interface between the user, the application software, and the computer's hardware.

The most basic types of system software are:
- The computer BIOS and device firmware, which provide basic functionality to operate and control the hardware connected to or built into the computer;
- The operating system (prominent examples being Microsoft Windows, Mac OS X and Linux), which allows the parts of a computer to work together by performing tasks like transferring data between memory and disks or rendering output onto a display device. It also provides a platform to run high-level system software and application software;
- Utility programs, which help to analyze, configure, optimize and maintain the computer.

Application software, also called end-user programs, is a subclass of computer software that employs the capabilities of a computer directly to a task that the user wishes to perform. This should be contrasted with system software which is involved in integrating a computer's various capabilities, but typically does not directly apply them in the performance of tasks that benefit the user.

E.g., database programs (Access), word processors (Word), and spreadsheets (Excel).

5.2 System Software

5.2.1 BIOS

BIOS (Basic Input/Output System) is the built-in software that determines what a computer can do without accessing programs from a disk. On PCs, the BIOS contains all the code required to control the keyboard, display screen, disk drives, serial communications, and a number of miscellaneous functions. The BIOS is typically placed in a ROM chip that comes with the computer (it is often called a ROM BIOS). This ensures that the BIOS will always be available and will not be damaged by disk failures. It also makes it possible for a computer to boot itself. In short, BIOS:

- Stored in the ROM chip;
- Contains the computer's startup instructions (ref. Figure I-5-2).

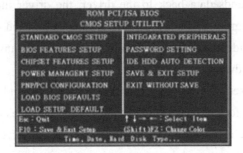

Figure I-5-2 CMOS Setup Utility on a PC

5.2.2 Operating System

Operating system (OS) is the most important program that runs on a computer. Every general-purpose computer must have an operating system to run other programs. Operating systems (ref. Figure I-5-3) perform basic tasks, such as recognizing input from the keyboard, sending output to the display screen, keeping track of files and directories on the disk, and controlling peripheral devices such as disk drives and printers.

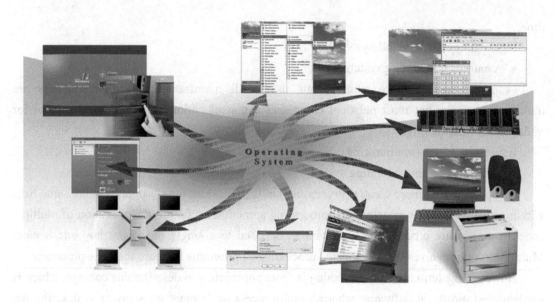

Figure I-5-3 Functions of OS

5.2.2.1 Booting-Up

Booting up (also known as "booting") is a bootstrapping process that starts operating systems when the user turns on a computer system. A boot sequence is the initial set of operations that the computer performs when power is switched on. The bootloader typically loads the main operating system for the computer. The main steps in booting up are:

(1) Turn on the computer. Power supply sends electrical signals to all parts of the computer;

(2) The processor chip looks for the BIOS in the ROM chip;

(3) BIOS executes POST (power on self test), checks components, e.g., keyboard, drives etc.;

(4) POST results are compared to data in CMOS chip;

(5) BIOS searches for the first available boot disk;

(6) Load MBR (Master Boot Record) from the first available boot disk;

(7) Load bootstrap program from the active partition;

(8) Finally, the OS is loaded into memory (RAM).

5.2.2.2 Categories of OS

(1) users/tasks.

- A single user/single tasking operating system:

~allows only one user to run one program at a time.

E.g., DOS.

- A single user/multitasking operating system:

~allows a single user to work on two or more programs that reside in memory at the same time.

E.g., Windows 3.X, Windows 95/98.

- A multiuser operating system:

~supports two or more simultaneous users. All mainframes and minicomputers are multi-user systems, but most personal computers and workstations are not. Another term for multi-user is time sharing.

E.g., Unix, Linux (multitasking).

- A multiprocessing operating system:

~supports two or more Central Processing Units (CPUs) running programs at the same time within a single computer system. Multiprocessing sometimes refers to the execution of multiple concurrent software processes in a system as opposed to a single process at any one instant. Multiprocessing involves the coordinated processing of programs by more than one processor.

However, the term multiprogramming is more appropriate to describe this concept, which is implemented mostly in software, whereas multiprocessing is more appropriate to describe the use of multiple hardware CPUs. A system can be both multiprocessing and multiprogramming, only one of the two, or neither of the two.

E.g., Windows 2000/2003, Windows XP/Vista, Windows 7/8/8.1/10, Microsoft server operating systems (Windows NT 3.1/3.5/3.51/4.0, Windows 2000/2003/2008/2012/2016 Server and Windows Home Server), UNIX, Linux.

(2) stand-alone/network/embedded.

- stand-alone:

A stand-alone operating system is a complete operating system that works on a desktop computer, notebook computer, or mobile computing device and that also works in conjunction with a network operating system. Some stand-alone operating systems are called client operating systems because they also work in conjunction with a network operating system. Client operating systems can operate with or without a network. Other stand-alone operating systems include networking capabilities, allowing the home and small business user to set up a small network.

E.g., DOS, Windows XP, Windows 10, Mac OS X, OS/2 Warp Client, UNIX, and Linux.

- Network:

A network operating system (NOS) is an operating system that is designed specifically to control a network and its message (e.g. packet) traffic and queues, controls access by multiple users to network resources such as files, and provides for certain administrative functions, including security.

E.g., NetWare, Windows 2003/XP, Vista, Windows 7/8/10, OS/2 Warp Server for e-business, UNIX, Linux, and Solaris.

- Embedded:

An embedded operating system is an operating system for embedded computer systems.

These operating systems are designed to be very compact and efficient, forsaking many functions that non-embedded computer operating systems provide, and which may not be used by the specialized applications they run. They are frequently also real-time operating systems. The operating systems on most PDAs and small devices are embedded operating system, reside on a ROM chip. Popular embedded operating systems include Pocket PC (P/PC, PPC), Windows Mobile 2003/2003 SE, Windows Mobile 5/6, Windows Phone CE/7/8, Windows 10 Mobile, Palm OS, Symbian OS, Windows IoT (formerly Windows Embedded), VxWorks, Android, and iOS (originally iPhone OS).

5.2.2.3 Command Based vs. Graphical OS

1) Command Based OS
- Command based has limited vocabulary of commands;
- Unusable if do not know commands.

E.g., copy A:\afile.txt C:\dir\newname.txt.

2) Graphical OS
- Graphical presents commands via pull-down menus;
- Generally commands are faster for expert user.

5.2.2.4 Graphical OS Manipulation

1) Start Menu / Start Screen

The Start menu is a user interface element used in Microsoft Windows since Windows 95 and in some other operating systems. It provides a central launching point for computer programs and performing other tasks.

Traditionally, the Start menu provided a customizable nested list of programs for the user to launch, as well as a list of most recently opened documents, a way to find files and get help, and access to the system settings. Later enhancements via Windows Desktop Update included access to special folders like "My Documents" and "Favorites" (browser bookmarks). Windows XP's Start menu was expanded to encompass various My Documents folders (including My Music and My Pictures), and transplanted other items like My Computer and My Network Places from the Windows desktop. Until Windows Vista, the Start menu was constantly expanded across the screen as the user navigated through its cascading sub-menus.

Windows 10 (ref. Figure I-5-4) re-introduced the Start menu in a revised form. It uses a two column design similar to Windows 7's version, except that the right side is populated by tiles, similarly to Windows 8's Start screen. Applications can be pinned to the right half, and their respective tiles can be resized and grouped into user-specified categories. The left column displays a vertical list, containing frequently-used applications, and links to the "All apps" menu, File Explorer, Settings, and Power options. Some of these links, and additional links to folders such as Downloads, Pictures, and Music, can be added through Settings' "Choose which folders appear on Start" page. The Start menu can be resized, or be placed in a full-screen display resembling the Windows 8/8.1 Start screen (although scrolling vertically instead of horizontally). The Start menu also enters this state when "Tablet mode" is enabled.

Figure I-5-4 The Start menu in Windows 10

2) File Manager/ File Browser /File Explorer

A file manager or file browser is a computer program that provides a graphical user interface to manage files and folders. It is also the component of the operating system that presents many user interface items on the monitor such as the taskbar and desktop.

The most common operations performed on files or groups of files include creating, opening (e.g. viewing, playing, editing or printing), renaming, moving or copying, deleting and searching for files, as well as modifying file attributes, properties and file permissions. Folders and files may be displayed in a hierarchical tree based on their directory structure. Some file managers contain features inspired by web browsers, including forward and back navigational buttons.

Some file managers provide network connectivity via protocols, such as FTP, NFS, SMB or WebDAV. This is achieved by allowing the user to browse for a file server (connecting and accessing the server's file system like a local file system) or by providing its own full client implementations for file server protocols.

The file manager is renamed File Explorer from Windows 8 and Windows Server 2012 onwards, and introduces new features such as a redesigned interface incorporating a ribbon toolbar (ref. Figure I-5-5), and a redesigned file operation dialog that displays more detailed progress and allows for file operations to be paused and resumed.

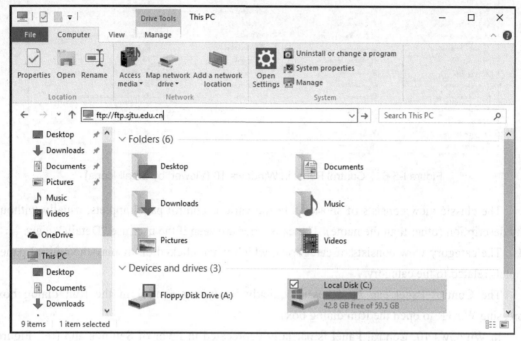

Figure I-5-5　File Explorer in Windows 10

3) Control Panel/Settings app

The Control Panel (ref. Figure I-5-6) is a part of the Microsoft Windows graphical user interface which allows users to view and manipulate basic system settings and controls via applets, such as adding, removing and setting hardware (Printer, Fax, Monitor, Microphone, Speaker, Mouse, Keyboard, Disk Drive, Hub, Sensor, etc.), adding and removing software, controlling user accounts, and changing accessibility options. Additional applets can be provided by third party software.

The Control Panel has been an inherent part of the Microsoft Windows operating system since Windows 2.0, with many of the current applets being added in later versions. Beginning with Windows 95, the Control Panel is implemented as a special folder, i.e. the folder does not physically exist, but only contains shortcuts to various applets such as Add or Remove Programs and Internet Options.

In recent versions of Windows, the Control Panel has two views, Classic View and Category View, and it is possible to switch between these through an option that appears on either the left side or top of the window.

Many of the individual Control Panel applets can be accessed in other ways. For instance, Display Properties can be accessed by right-clicking on an empty area of the desktop and choosing Properties.

Figure I-5-6   Control Panel in Windows 10 (Viewed by Small Icons)

The classic view consists of shortcuts to the various control panel applets, usually without any description (other than the name). The categories are seen if the user use "Details" view.

The category view consists of categories, which when clicked on display the control panel applets related to the category.

The Control Panel can be accessed quickly by typing control in the Run dialog box (pressing Win+R to open the Run dialog box).

In Windows 10, Control Panel is partially deprecated in favor of Settings app (ref. Figure I-5-7), which was originally introduced on Windows 8 as "PC Settings" to provide a touchscreen-optimized settings area using its Metro-style app platform. The Settings app has been significantly enhanced and now offers most of the traditional Control Panel features. It includes every feature a casual or nontechnical PC user will need.

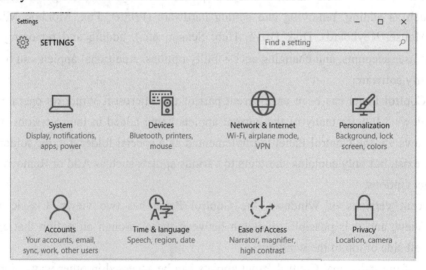

Figure I-5-7   Settings app in Windows 10

5.2.2.5 Some New Features in Windows 10

1) Cortana Personal Assistant

Cortana (ref. Figure I-5-8) is an intelligent personal assistant created by Microsoft for Windows 10, Windows 10 Mobile, Windows Phone 8.1 (where it now supersedes Bing Mobile), Microsoft Band, Xbox One, iOS and Android. It has been launched as a key ingredient of Microsoft's planned "makeover" of the future operating systems for Windows Phone and Windows.

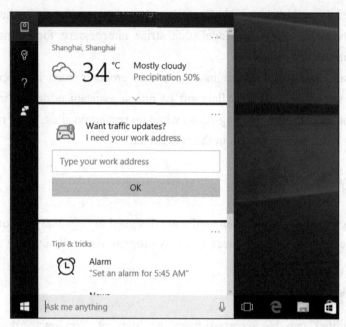

Figure I-5-8 Cortana in Windows 10

Cortana is currently available in English, French, German, Italian, Spanish, Chinese, and Japanese language editions, depending on the software platform and region in which it is used. Cortana mainly competes against assistants such as Apple's Siri and Google's Google Now.

Cortana can set reminders, recognize natural voice without the requirement for keyboard input, and answer questions using information from the Bing search engine (e.g., current weather and traffic conditions, sports scores, biographies). If Firefox is the default browser, Cortana uses the Firefox default search engine instead of Bing. Searches will only be made with Microsoft Bing search engine and all links will open with Microsoft Edge. Windows 8.1's universal Bing SmartSearch features are incorporated into Cortana, which replaces the previous Bing Search app which was activated when a user presses the "Search" button on their device. Cortana includes a music recognition service. Cortana can simulate rolling dice and flipping a coin. It integrates with the Microsoft Band watch band for Windows Phone devices if connected via Bluetooth, it can make reminders and phone notifications. Because Skype is a Microsoft product, Cortana provides smooth activation of Skype video calls from voice commands.

2) Windows Edge

Microsoft Edge is the default web browser developed by Microsoft and included in Windows 10, Windows 10 Mobile, and Xbox One consoles, replacing Internet Explorer as the default web browser on all device classes. Microsoft Edge adds new features such as integration with Cortana digital assistant, annotation tools, and a reading mode.

Microsoft Edge integrates with Microsoft's online platforms: it integrates with the Cortana digital assistant to provide voice control, search functionality, and dynamic, personalized information related to searches within the address bar. Users can make annotations to web pages that can be stored to and shared with OneDrive. It also integrates with the "Reading List" function and provides a "Reading Mode" that strips unnecessary formatting from pages to improve their legibility.

Microsoft has integrated Cortana into numerous products such as Microsoft Edge, the browser bundled with Windows 10. Microsoft's Cortana assistant is deeply integrated into its Edge browser. Cortana can find opening-hours when on restaurant sites, show retail coupons for websites, or show weather information in the address bar.

5.2.3 Utility Programs

A utility program is a kind of system software designed to help analyze, configure, optimize and maintain the computer. A single piece of utility program is usually called a utility (abbr. util) or tool.

5.2.3.1 Windows Utilities

Windows utilities can be accessed from the Start → Windows Administrative Tools command, ref. Figure I-5-9.

1) Disk Cleanup

Disk cleanup (ref. Figure I-5-10) is a computer program tool included in Microsoft Windows designed to help free up space on the computer user's hard drive. It searches and analyzes the hard drive, and then it determines which files on a hard drive may no longer be needed and deletes those files.

There are a number of different types of file categories that disk cleanup targets when it performs the initial disk analysis:

- Downloaded program files;
- Temporary Internet files;
- System queued Windows error reporting files;
- Delivery optimization files;
- Recycle bin;
- Temporary files;
- Thumbnails.

Figure I-5-9　System Tools　　　　Figure I-5-10　Disk Cleanup

2) Defragment and Optimize Drives

One of the best ways you can improve your PC's performance is by optimizing the drive. Windows includes features to help optimize the different types of drives that PCs use today. No matter which type of drive your PC uses, Windows automatically chooses the optimization that's right for your drive.

By default, Optimize Drives, previously called Disk Defragmenter, runs automatically on a weekly schedule at the time set in automatic maintenance. But you can also optimize drives on your PC manually, ref. Figure I-5-11.

5.2.3.2　Utility Suites

1) Personal Computer Maintenance

Personal computer maintenance utility is a utility program that identifies and fixes operating system problems, detects and repairs disk problems, and includes the capability of improving a computer's performance.

E.g., Norton Utilities.

2) Antivirus programs

Antivirus (or anti-virus) program is a utility program that searches a hard disk for viruses and removes any that are found. Most antivirus programs include an auto-update feature that enables the program to download profiles of new viruses so that it can check for the new viruses as soon as they are discovered.

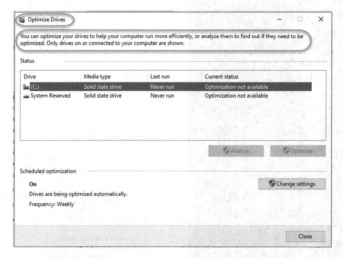

Figure I-5-11　Defragment and Optimize Drives

E.g., Norton Antivirus, McAfee VirusScan.

3) File Compression

File compression utility is a utility program that shrinks the size of a file(s), so the file takes up less storage space than the original file.

E.g., WinRar, WinZip, PKZIP.

5.3　Programming Languages

A programming language (Figure I-5-12) is a vocabulary and set of grammatical rules for instructing a computer to perform specific tasks. Each language has a unique set of keywords (words that it understands) and a special syntax for organizing program instructions. Languages improve in parallel with hardware developments.

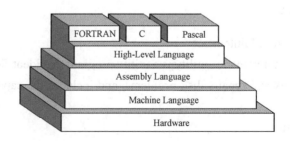

Figure I-5-12　Hierarchy of Programming Languages

Types of programming languages are low-level and high-level.
- Low-level programming languages are machine dependent. A machine-dependent language runs on only one particular type of computer. These programs are not easily portable to other types of computers. Each instruction in a low-level language usually equates to a single machine instruction.
- High-level programming languages are machine independent. A machine-independent language can run on many different types of computers and operating systems.

Machine languages and assembly languages are low-level languages. Procedural languages, non-procedural languages, object-oriented programming languages, and visual programming languages are high-level languages.

5.3.1 Machine language

- 1st. Generation Language.
- Written in the language of the computer—0's and 1's.
- The language to which all other generations of languages must be converted.
- Each different type of CPU has its own unique machine language.

5.3.2 Assembly language

- 2nd. Generation Language.
- Uses simple abbreviations or symbols to represent a number of machine language instructions.
- Specific to a certain physical (or virtual) computer architecture. This is in contrast to most high-level programming languages, which are ideally portable.
- Codes are converted to machine language by a special program called an assembler.
E.g., SUB   STA   LDA.

5.3.3 Procedural language

- High level language.
- 3rd. Generation Language (3GL).
- More English like.
- Allows concentration on logic of solving problem rather than controlling computer.
- Compiler program converts instructions into machine language, stored as an EXEcutable file.
E.g., BASIC, COBOL, FORTRAN, Ada, Pascal, C.
PRINT("Hello, My Friends!")

5.3.4 Non-procedural language

- Very high-level language.
- 4th. Generation Language (4GL).
- Programmer tells computer what to do but not how to do it.
- Programmer can be 10 times more productive than in 3rd. generation language.

E.g., database (FoxBASE, FoxPro) query language SQL (Structured Query Language), Prolog, LISP.

SQL statements: select ID, name, score
from StudentFile
where score>=90   and   gender="female"
will output the following results:

ID	name	score
1011011	Mary	96
1011005	Jenny	91
1011020	Melissa	98

5.3.5 Object-oriented programming (OOP) language

- Uses "objects"—data structures consisting of data fields and methods together with their interactions—to design applications and computer programs.
- Uses event-driven program to check for and respond to events.
- Examples of events: press a key, click a button, etc.
- Programming techniques: data abstraction, encapsulation, modularity, polymorphism, and inheritance.

E.g., C++ (pronounced SEE-plus-plus), Java, C# (pronounced SEE-sharp).

5.3.6 Visual programming language

- 5th. Generation Language.
- Provide a visual or graphical interface for creating source codes.
- Developers drag and drop objects to build programs.

E.g., VB/VB.NET, Delphi, PowerBuilder.

5.3.7 Execution of programming languages

Regardless of what language is used, the program need eventually be converted into machine language so that the computer can understand and execute it. There are two ways to do

this:
- Compile the program: to transform a program written in a high-level programming language from source code into object code. Programmers write programs in a form called source code. Source code must go through several steps before it becomes an executable program. The first step is to pass the source code through a compiler, which translates the high-level language instructions into object code. The final step in producing an executable program—after object code is produced by the compiler, it is passed through a linker. The linker combines modules and gives real values to all symbolic addresses, thereby producing machine code. A compiler is likely to perform many or all of the following operations: lexical analysis, preprocessing, parsing, semantic analysis, code generation, and code optimization.
- Interpret the program: to translate high-level instructions into an intermediate form, which it then executes. In contrast, a compiler translates high-level instructions directly into machine language. Compiled programs generally run faster than interpreted programs. The advantage of an interpreter, however, is that it does not need to go through the compilation stage during which machine instructions are generated. This process can be time-consuming if the program is long. The interpreter, on the other hand, can immediately execute high-level programs. For this reason, interpreters are sometimes used during the development of a program, when a programmer wants to add small sections at a time and test them quickly. In addition, interpreters are often used in education because they allow students to program interactively.

Many languages have been implemented using both compilers and interpreters, such as Java and C#.

Text 6
Introduction to Word Processing

6.1 Getting Started with Word

Word is designed to help you create professional-quality documents. It can also help you organize and write documents more efficiently.

When you create a document in Word, you can choose to start from a blank document or let a template do much of the work for you. Word's powerful editing and reviewing tools can help you work with others to make your document great.

6.1.1 Select Text

In Microsoft Office Word, you can select text or items by using the mouse or the keyboard. Table I-6-1 summarized how to select text in the body of a document.

Table I-6-1　Select text in the body of a document

To select	Do this
Any amount of text	Click where you want to begin the selection, hold down the left mouse button, and then drag the pointer over the text that you want to select; or hold down the Shift key while pressing the [Right], [Left], [Down], [Up] direction keys
A word	Double-click anywhere in the word
A line of text	Move the pointer to the left of the line until it changes to a right-pointing arrow, and then click
A sentence	Hold down Ctrl, and then click anywhere in the sentence
A paragraph	Triple-click anywhere in the paragraph
Multiple paragraphs	Move the pointer to the left of the first paragraph until it changes to a right-pointing arrow, and then press and hold down the left mouse button while you drag the pointer up or down
A large block of text	Click at the start of the selection, scroll to the end of the selection, and then hold down Shift while you click where you want the selection to end
An entire document	Move the pointer to the left of any text until it changes to a right-pointing arrow, and then triple-click; or press Ctrl + A

Continued

To select	Do this
Headers and footers	In Print Layout view (Print Layout view: A view of a document or other object as it will appear when you print it. For example, items such as headers, footnotes, columns, and text boxes appear in their actual positions.), double-click the dimmed header or footer text. Move the pointer to the left of the header or footer until it changes to a right-pointing arrow, and then click
Footnotes and endnotes	Click the footnote or endnote text, move the pointer to the left of the text until it changes to a right-pointing arrow, and then click
A vertical block of text	Hold down ALT while you drag the pointer over the text
A text box or frame	Move the pointer over the border of the frame or text box until the pointer becomes a four-headed arrow, and then click

6.1.2 Find and Replace

You can use Microsoft Office Word to find and replace text, formatting, paragraph marks, page breaks, and other items. You can also find and replace noun or adjective forms or verb tenses.

You can extend your search by using wildcards and codes to find words or phrases that contain specific letters or combinations of letters.

6.1.2.1 Find Text

You can quickly search for every occurrence of a specific word or phrase by the following steps.

(1) On the Home tab, in the Editing group, click Find.

(2) In the Find what box, type the text that you want to search for.

(3) Do one of the following:

- To find each instance of a word or phrase, click Find Next.
- To find all instances of a specific word or phrase at one time, click Find All, and then click Main Document.

6.1.2.2 Find and Replace Text

You can replace a word or phrase with another — for example, you can replace College with University.

(1) On the Home tab, in the Editing group, click Replace.

(2) Click the Replace tab.

(3) In the Find what box, type the text that you want to search for.

(4) In the Replace with box, type the replacement text.

(5) Do one of the following:

- To find the next occurrence of the text, click Find Next.

- To replace an occurrence of the text, click Replace. After you click Replace, Office Word moves to the next occurrence of the text.
- To replace all occurrences of the text, click Replace All.

6.1.3 Page Setup & Print Preview

Word's Page Layout dialog box lets you define margins, layout, and so on.

6.1.3.1 Change or Set Page Margins

Page margins are the blank space around the edges of the page. In general, you insert text and graphics in the printable area between the margins. However, you can position some items in the margins—for example, headers, footers, and page numbers.

Word automatically sets a one-inch page margin around each page. With a few clicks you can choose a different margin that's been pre-defined, or create your own.

To change or set page margins, on the Page Layout tab, in the Page Setup group, click Margins (ref. Figure I-6-1). Click the margin type that you want. You can also specify your own margin settings by choosing Custom Margins command, and then in the Top, Bottom, Left, and Right boxes, enter new values for the margins.

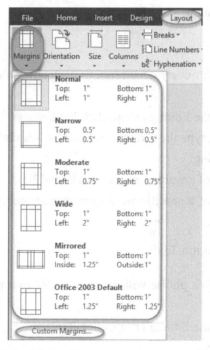

Figure I-6-1 Change or set page margins

6.1.3.2 Select Page Orientation

You can choose either portrait (vertical) or landscape (horizontal) orientation for all or part of your document. When you change the orientation, the galleries of predesigned page and cover page options also change to offer pages that have the orientation that you choose.

To set page orientation, on the Page Layout tab, in the Page Setup group, click Orientation (ref. Figure I-6-2), and then click Portrait or Landscape.

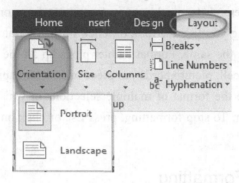

Figure I-6-2　Select page orientation

6.2　Formatting

6.2.1　Themes, Template, Style and Format Painter

6.2.1.1　Template

Templates are files that help you design interesting, compelling, and professional-looking documents. They contain content and design elements that you can use as a starting point when creating a document. All the formatting is complete; you add what you want to them. Examples are resumes, invitations, and newsletters.

6.2.1.2　Theme

To give your document a designer-quality look—a look with coordinating theme colors and theme fonts—you'll want to apply a theme. You can use and share themes among the Office for Mac applications that support themes, such as Word, Excel, and PowerPoint. For example, you can create or customize a theme in PowerPoint, and then apply it to a Word document or Excel sheet. That way, all of your related business documents have a similar look and feel.

6.2.1.3　Word styles

Themes provide a quick way to change the overall color and fonts. If you want to change text formatting quickly, Word styles are the most effective tools. After you apply a style to different sections of text in your document, you can change the formatting of this text simply by changing the style. Word includes many types of styles, some of which can be used to create reference tables in Word. For example, the Heading style, which is used to create a Table of Contents.

6.2.1.4 Format Painter

Use the Format Painter 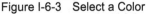 to quickly copy formatting from one thing in a document to another. Just select the thing you like the look of, click Format Painter, and then click the thing you want to change to look the same. Format Painter picks up all the formatting from your first thing, whether it's a shape, cell, picture border, or piece of text, and applies it to the second.

If you want to change the format of multiple selections in your document, you must first double-click Format Painter. To stop formatting, press Esc or click on the Format Painter button once to turn it off.

6.2.2 Character Formatting

6.2.2.1 Change the Font Color (Text Color)

To change the font color, do the following:

(1) Select the text that you want to change.

(2) On the Home tab, in the Font group, click the arrow next to Font Color (ref. Figure I-6-3), and then select a color.

You can also use the formatting options on the Mini toolbar (ref. Figure I-6-4) to quickly format text. The Mini toolbar appears automatically when you select text.

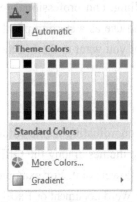

Figure I-6-3 Select a Color

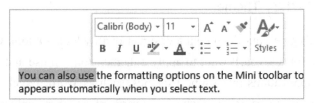

Figure I-6-4 Mini Toolbar

6.2.3 Paragraph Formatting

6.2.3.1 Spacing and Indent

Line spacing determines the amount of vertical space between the lines of text in a paragraph. Paragraph spacing determines the amount of space above or below a paragraph.

To change the line spacing for only part of a document:

(1) Select the paragraphs you want to change.

(2) On the Home tab, click the Line and Paragraph Spacing button (ref. Figure I-6-5).

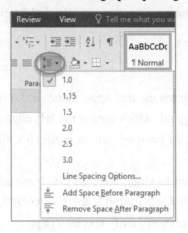

Figure I-6-5　Line and Paragraph Spacing

(3) Choose the number of line spaces you want or click Line Spacing Options at the bottom of the menu, and then select the options you want in the Paragraph dialog box (ref. Figure I-6-6) under Spacing.

Figure I-6-6　Indents and Spacing

On the Home tab, in the Paragraph group, click the Paragraph Launcher button , you can also open the Paragraph dialog box.

If you want to change the spacing before or after the selected paragraphs, click the arrow next to Before or After and enter the amount of space that you want.

Indentation determines the distance of the paragraph from either the left or the right margin. Within the margins, you can increase or decrease the indentation of a paragraph or group of paragraphs.

6.2.3.2 Alignment

Horizontal alignment determines the appearance and orientation of the edges of the paragraphs. Text can be left-aligned, which means the left edge of the text is flush with the left margin, right-aligned, centered, or justified, which means it's aligned evenly along the left and right margins.

Select the text that you want to align, then, on the Home tab, in the Paragraph group, click Align Left Button (▤) or Align Right Button (▤), Center Button (▤), or Justify Button (▤) to align text left or right, center text, or justify text on a page.

6.2.3.3 Borders and Shading

Microsoft Word allows you to place a border on any or all of the four sides of selected text, paragraphs, and pages. You can also add many type of shading to the space occupied by selected text, paragraphs, and pages (ref. Figure I-6-7).

If you want to customize borders by changing the border settings, line style, color, and width preferences, and shading by changing the shading settings: fill, style and color. You can make these changes in the Borders and Shading dialog box (ref. Figure I-6-8 and Figure I-6-9).

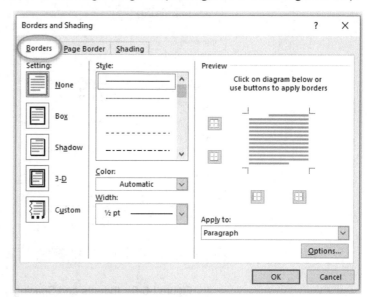

Figure I-6-7 Borders and Shading   Figure I-6-8 Borders

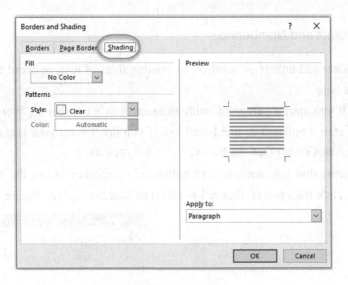

Figure I-6-9   Shading

6.2.3.4   Drop Caps

A Drop Cap allows you to offset the text (usually the first letter) of the sentence or paragraph. The size of a drop cap is usually two or more lines. It's a nice way to bring attention to the article or paragraph.

Microsoft Word allows you to choose two different positions—Dropped or in the Margin.

- Dropped—formats the first character in the paragraph as a dropped capital letter; aligns it to the left; paragraph text wraps around the dropped cap.
- In the Margin—formats the first character in the paragraph as a dropped capital letter; places it in the left margin, beginning at the first line of the paragraph.

On the Insert tab, in the Text group, click on the small triangle next to the Drop Cap button, you can choose Dropped or in Margin style, or click Drop Cap Options to open the Drop Cap dialog box (ref. Figure I-6-10) to customize Dropped or in Margin settings.

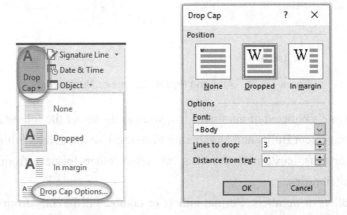

Figure I-6-10   The Drop Cap options and dialog box

6.2.3.5 Bullets and Numbering

You can quickly add bullets or numbers to existing lines of text, or Word can automatically create lists as you type.

By default, if you start a paragraph with an asterisk or a number 1., Word recognizes that you are trying to start a bulleted or numbered list. If you don't want your text turned into a list, you can click the AutoCorrect Options button that appears.

Select the items that you want to add bullets or numbering to, on the Home tab, in the Paragraph group, click Bullets (ref. Figure I-6-11(a)) or Numbering (ref. Figure I-6-11(b)).

(a) Bullets   (b) Numbering

Figure I-6-11 Add and Format Bullets or Numbering

When you create a bulleted or numbered list, you can do any of the following:
- Use the convenient Bullet and Numbering libraries: Use the default bullet and numbering formats for lists, customize the lists, or select other formats from the Bullet and Numbering libraries.
- Format bullets or numbers: Format bullets or numbers differently from the text in a list. For example, click a number and change the number color for the entire list, without

making changes to the text in the list.
- Use pictures or symbols: Create a picture bulleted list to add visual interest to a document or a Web page.

6.2.4 Page Formatting

6.2.4.1 Columns of Text

Select what you want your columns to start. On the **Layout** tab, in the **Page Setup** group, click **Columns** button. You may choose from the list of presets (ref. Figure I-6-12(a)) or you may click "More Columns" to manually select the settings for your columns (ref. Figure I-6-12 (b)).

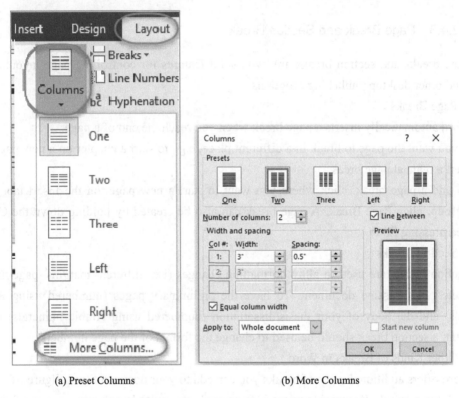

(a) Preset Columns (b) More Columns

Figure I-6-12 Add and Format Bullets or Numbering

6.2.4.2 Page Background

A background or a page color is primarily used to create a more interesting background for your Word document. Backgrounds are visible in Web Layout view and Full Screen Reading view.

You can use gradients, patterns, pictures, solid colors, or textures for backgrounds.

Gradients, patterns, pictures, and textures are tiled or repeated to fill the page. If you save a document as a Web page, the pictures, textures, and gradients are saved as JPEG files and the patterns are saved as GIF files.

On the Design tab, in the Page Background group, select either the Watermark Background or Page Color options or Page Borders options (ref. Figure I-6-13). You can also choose a combination of the three for a more interesting and striking page background settings.

Figure I-6-13　Page Background

6.2.4.3　Page Break and Section Break

Page breaks and section breaks are two useful features for controlling page layout in MS Word and other desktop publishing programs.

1. Page Breaks

Word automatically inserts a page break when you reach the end of a page.

If you want the page to break in a different place, e.g., to start a chapter on a new page, you can insert a manual page break.

To add a page break, click where you want to start a new page, on the Insert tab, in the Pages group, click Page Break. A page break can also be created by holding down the Control key, then pressing Enter.

2. Section Breaks

Section breaks are used to allow formatting changes (i.e., different margins, page number styles, etc.) in the same document. To have the preliminary pages (numbered using Roman numerals) and the body of your thesis/dissertation (numbered using Arabic numerals) in one document, a section break should be used to change the format of the page numbers.

3. Other Kinds of Breaks in Word

Word offers additional types of breaks you can add to your document (ref. Figure I-6-14).

- Column Break: If your document is formatted into multiple columns, you can force text from the first column to move to the next one. This is a better option than pressing Enter a few times to move the text to the next column, since doing that could mess up your formatting if the font size changes or you edit the text.
- Text Wrapping: If you have a picture or other object in your document and want to add a caption, the text wrapping break will keep the caption with the object while making the rest of the document flow around both the object and your caption.
- Next Page: This works just like the page break under the Insert menu, except it also

creates a new section with the option to use entirely different formatting from the previous sections. For example, you can use a different section to rotate a page to landscape or portrait mode, add a different header or footer, or format the section into columns without affecting the rest of the document.

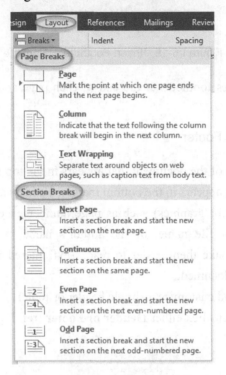

Figure I-6-14 Word Break Types

- Continuous Break: Creates a new section, like the next page break, but doesn't start you on a new page.
- Even and Odd Page Breaks: Insert a section break and also take you to the next even or odd page (depending on which break type you select) so you can format your alternating pages in a document differently (e.g., right or left pages in a book).

6.2.4.4 Footnotes and Endnotes

Footnotes and endnotes are used in printed documents to explain, comment on, or provide references for text in a document. You might use footnotes for detailed comments and endnotes for citation of sources.

Microsoft Office Word automatically numbers footnotes and endnotes for you, after you specify a numbering scheme. When you add, delete, or move notes that are automatically numbered, Word renumbers the footnote and endnote reference marks.

On the References tab, in the Footnotes group, click Insert Footnote or Insert Endnote (ref. Figure I-6-15). Then the cursor moves there where you enter the reference while the footnote or endnote number is located where you told Word to put the endnote or footnote.

Figure I-6-15　Insert Footnote or Endnote

By default, Word places footnotes at the end of each page and endnotes at the end of the document.

6.2.4.5　Header and Footer

The header is a section of the document that appears in the top margin, while the footer is a section of the document that appears in the bottom margin. Headers and footers generally contain additional information (text or graphics) such as page numbers, the time and date, a company logo, the document title or file name, or the author's name, which can help keep longer documents organized and make them easier to read. Text entered in the header or footer will appear on each page of the document.

Here's how to create and customize a simple header or footer.

(1) Click **Insert** and then click either **Header or Footer** (ref. Figure I-6-16).

Figure I-6-16　Insert Header or Footer or Page Number

(2) Dozens of built-in layouts appear. Scroll through them and click the one you want.

(3) The header and footer space will open in your document, along with the Header & Footer Tools. You won't be able to edit the body of your document again until you close the Header & Footer Tools.

(4) Type the text you want in the header or footer. Most headers and footers have placeholder text that you can type right over.

(5) When you're done, click Close Header and Footer (ref. Figure I-6-17).

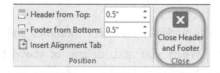

Figure I-6-17　Close Header and Footer

Tip: Whenever you want to open the Header & Footer Tools, double-click inside the header or footer area.

6.2.4.6 Page Number

Page numbers can be used to automatically number each page in your document. They come in a wide range of number formats and can be customized to suit your needs. Page numbers are usually placed in the header, footer, or side margin. Microsoft Office Word has many preformatted page number designs so that you can quickly insert headers or footers that display the page number.

Word can automatically label each page with a page number and place it in a header, footer, or side margin. If you have an existing header or footer, it will be removed and replaced with the page number.

To add page numbers, on the Insert tab, in the Header & Footer group, click Page Number (ref. Figure I-6-18). Choose a location (such as Top of Page), and then pick a style. Word automatically numbers every page.

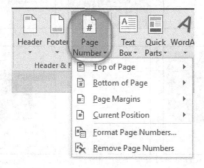

Figure I-6-18   Insert Page Number

To change the header or footer or the information in the page margins, do the following: Double-click the header or footer, and then click the Headers & Footers tab under Header & Footer Tools.

6.2.4.7 Front Cover

A Word cover page introduces the document with a title, an image or both, providing relevant information about the document. For example, a periodic report on business development may open with a cover page that includes your company logo.

Cover pages are always inserted at the beginning of a document, regardless of where the cursor appears in the document.

Microsoft Word offers a gallery of convenient predesigned cover pages. Choose a cover page and replace the sample text with your own.

To add a cover page, on the Insert tab, in the Pages group, click Cover Page (ref. Figure I-6-19), and choose a cover page layout from the gallery of options.

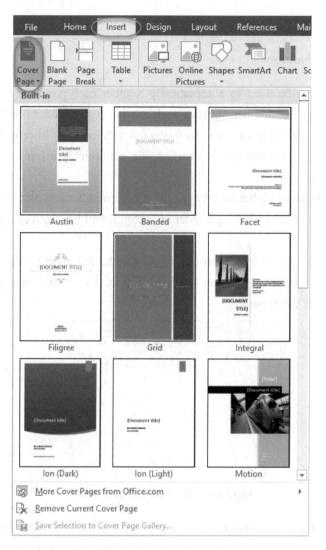

Figure I-6-19　Add a Cover Page

After you insert a cover page, you can replace the sample text with your own text by clicking to select an area of the cover page, such as the title, and typing your text.

6.2.4.8　Table of Contents

A Table of Contents (ToC) is just like the list of chapters at the beginning of a book. It lists each section in the document and the page number where that section begins.

To create a table of contents, first apply heading styles—Heading 1 and Heading 2, for example—to the text that you want to include in the table of contents. Click where you want to insert the table of contents—usually at the beginning of a document. On the References tab, in the Pages group, click Table of Contents (ref. Figure I-6-20), and then click the table of contents style that you want.

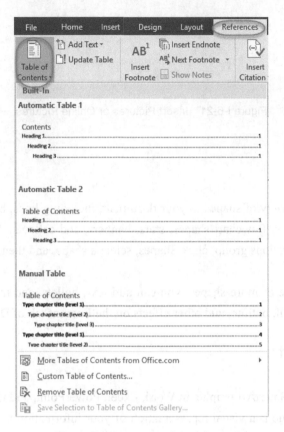

Figure I-6-20   Add a Cover Page

You can customize the way a table of contents appears. For example, you can change the font, how many heading levels to show, and whether to show dotted lines between the entries and the page numbers.

6.3   Graphics, Tables, Textbox and Other Objects

6.3.1   Pictures and Graphics

Pictures can be inserted (or copied) from many different places, including a computer, an online source like Bing.com, or a Web page. You can also change how a picture is positioned with text within a document.

To insert a picture, on the Insert tab, in the Illustrations group, click Picture or Online Pictures (ref. Figure I-6-21). Locate the picture that you want to insert. Double-click the picture that you want to insert.

You can also change how a picture is positioned with text within a document by using the Position and Wrap Text commands. You can also format the picture by the command options in the Picture Tools bar.

Figure I-6-21　Insert Pictures or Online Pictures

6.3.2　Shape

You can add a variety of shapes to your document, including lines, basic geometric shapes, arrows, equation shapes, flowchart shapes, stars, banners, and callouts. To add a shape, on the Insert tab, in the Illustrations group, click Shapes, select a shape, and then click and drag to draw the shape.

After you add one or more shapes, you can add text, bullets, and numbering to them, and you can change their fill, outline, and other effects on the **Format** tab of Drawing Tools bar.

6.3.3　SmartArt

You can create a SmartArt graphic in Word, Excel, PowerPoint, and Outlook.

A SmartArt graphic is a visual representation of your information that you can quickly and easily create, choosing from among many different layouts, to effectively communicate your message or ideas.

To create a SmartArt graphic, on the Insert tab, in the Illustrations group, click SmartArt. In the Choose a SmartArt Graphic dialog box (ref. Figure I-6-22), click the type and layout that you want. Enter your text by doing one of the following:

- Click in a shape in your SmartArt graphic, and then type your text.
- Click [Text] in the Text pane, and then type or paste your text.
- Copy text from another program, click [Text], and then paste into the Text pane.

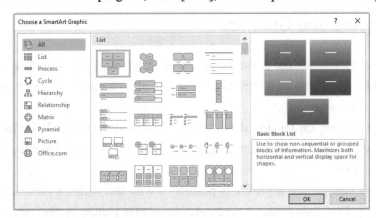

Figure I-6-22　Choose a SmartArt Graphic

You can change the look of your SmartArt graphic by changing the fill of its shape or text; by adding effects, such as shadows, reflections, glows, or soft edges; or by adding three-dimensional (3-D) effects, such as bevels or rotations.

6.3.4 Symbol

Sometimes you need a character that's not on your keyboard, like a foreign currency symbol (£), a trademark (™), or mathematical constant (π).

To insert a symbol, on the Insert tab, in the Symbols group, click Symbol (ref. Figure I-6-23). Do one of the following:

- Click the symbol that you want in the drop-down list.

Figure I-6-23　Insert Symbol

- If the symbol that you want to insert is not in the list, click More Symbols. In the Font box, click the font that you want, click the symbol that you want to insert, and then click Insert.

6.3.5 Table

You can insert a table by choosing from a selection of preformatted tables or by selecting the number of rows and columns that you want. You can also design your own table if you want more control over the shape of your table's columns and rows.

To insert a table, on the Insert tab, in the Tables group, click Table and move the cursor over the grid until you highlight the number of columns and rows you want, or choose Insert Table command (ref. Figure I-6-24).

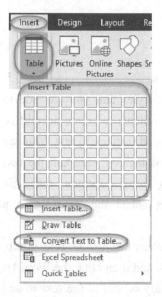

Figure I-6-24　Insert Table

If you need to make adjustments, you can add columns and rows in a table, delete rows or columns or merge cells.

When you click in the table, the Table Tools appear (ref. Figure I-6-25).

Figure I-6-25　Table Tools

Use Table Tools to choose different colors, table styles, add a border to a page or remove borders from a table. You can even insert a formula to provide the sum for a column or row of numbers in a table.

If you already have text in your document that would look better as a table, Word can convert text to a table (ref. Figure I-6-24).

If you want a table with rows and columns in different sizes, you can use the cursor to draw it by using Draw Table command in Figure I-6-24.

6.3.6　Text Box

Text boxes can be useful for drawing attention to specific text. They can also be helpful when you need to move text around in your document. Word allows you to format text boxes and the text within them with a variety of styles and effects.

To add a text box, on the Insert tab, in the Text group, click the Text Box (ref. Figure I-6-26). You can choose a built-in text box, or using Draw Text Box command to create the text box.

6.3.7　WordArt

WordArt is a gallery of text styles that you can add to your Microsoft Office system documents to create decorative effects, such as shadowed or mirrored (reflected) text. You can make changes to WordArt, such as the font size and the text color, by using the drawing tools options available automatically after you insert or select the WordArt in a document.

Figure I-6-26　Insert Text Box

To insert WordArt, on the Insert tab, in the Text group, click WordArt (ref. Figure I-6-27), and then click the WordArt style that you want. You can modify or add to the text in an existing WordArt object whenever you want.

6.3.8 Equations

Figure I-6-27   Insert WordArt

Word includes equations that you can drop into your documents—no formatting required. If the built-in equations don't meet your needs, you can modify them or you can build your own sophisticated equation from scratch.

With the ink-to-math feature in Word 2016, you can also write out equations with your stylus, finger, or mouse, and have Word convert them to text.

To insert an equation, on the Insert tab, in the Symbols group, click Equation (ref. Figure I-6-28), and choose the equation you want from the gallery, or choose Insert New Equation command.

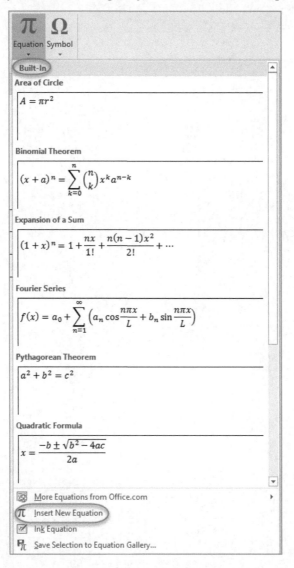

Figure I-6-28   Insert Equation

After you insert the equation or the placeholder for your new equation, the Equation Tools Design tab (ref. Figure I-6-29) opens with many symbols and options for customizing and adding to your equation. Select the equation whenever you want the tab to reappear.

Figure I-6-29 Equation Tools

Text 7
Introduction to PowerPoint Presentation

7.1 Getting Started with PowerPoint

7.1.1 PowerPoint Views

The views in Microsoft PowerPoint 2016 that you can use to edit, print, and deliver your presentation are as follows:
- Normal view.
- Outline view.
- Slide Sorter view.
- Notes Page view.
- Reading view.
- Slide Show view (which includes Presenter view).
- Master views: Slide, Handout, and Notes.

You can find PowerPoint views in two places:
- On the View tab, in the Presentations Views and Master Views groups, as shown in Figure I-7-1.

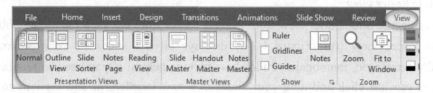

Figure I-7-1　View menu

- You can also find the most frequently used views (Normal, Slide Sorter, Reading View, and Slide Show) on the task bar at the bottom right of the slide window, as shown in Figure I-7-2.

Figure I-7-2　View buttons at the bottom of the screen

1) Views for creating your presentation
- Normal view.

You can get to Normal view from the task bar, or from the View tab on the ribbon.

Normal view is the main editing view, where you write and design your presentations. As shown in Figure I-7-3, Normal view displays slide thumbnails on the left, a large window showing the current slide, and a section below the current slide where you can type your speaker notes for that slide.

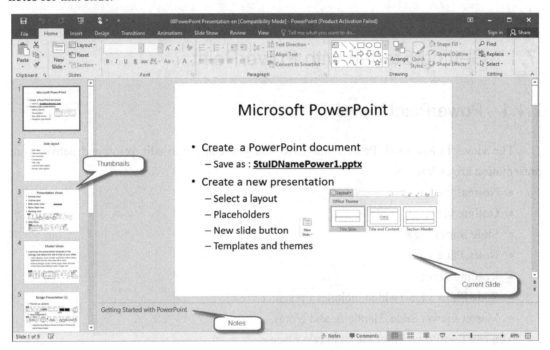

Figure I-7-3　Normal View

- Slide Sorter view.

You can get to Slide Sorter view from the task bar, or from the View tab on the ribbon.

Slide Sorter view (ref. Figure I-7-4) displays all the slides in your presentation in horizontally sequenced, thumbnails. Slide show view is helpful if you need to reorganize your slides—you can just click and drag your slides to a new location, or add sections to organize your slides into meaningful groups.

- Notes Page view.

You can show or hide your speakers notes with the Notes button, or you can get to Notes Page view from the View tab on the ribbon.

The Notes pane is located beneath the slide window. You can print your notes or include the notes in a presentation that you send to the audience, or just use them as cues for yourself while you're presenting.

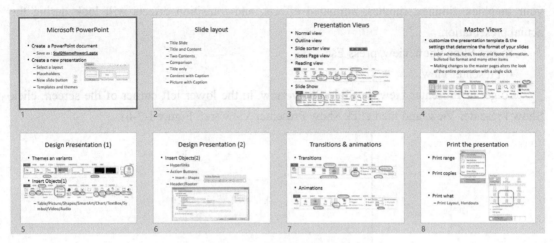

Figure I-7-4 Shows Slide Sorter View

• Outline view.

You can get to Outline view from the View tab on the ribbon.

NOTE: In PowerPoint 2013 and later, you can no longer get to Outline view from Normal view. You'll need to get to it from the View tab.

Use Outline view (ref. Figure I-7-5) to create an outline or story board for your presentation. It displays only the text on your slides.

• Master views.

To get to a master view, on the View tab, in the Master Views group, choose the master view that you want.

Master views include, Slide, Handout, and Notes. They are the main slides that store information about the presentation, including background, color, fonts, effects, placeholder sizes and positions. The key benefit to working in a master view is that you can make universal style changes to every slide, notes page, or handout associated with your presentation.

2) Views for delivering and viewing a presentation

• Slide Show view.

You can get to Slide Show view from the task bar.

Use Slide Show view to deliver your

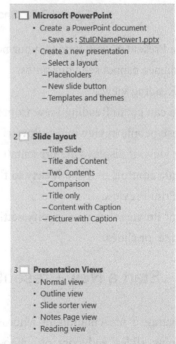

Figure I-7-5 Shows Outline View in PowerPoint

presentation to your audience. Slide Show view occupies the full computer screen, exactly the way your presentation will look on a big screen when your audience sees it. You can see how your graphics, timings, movies, animated effects, and transition effects will look during the actual presentation.

To exit Slide Show view, press Esc.

- Presenter view.

To get to Presenter view, in Slide Show view, in the lower left corner of the screen, click Show Presenter View, and then click Show Presenter View (ref. Figure I-7-6).

Figure I-7-6 Show PowerPoint presenter view menu

Use Presenter view to view your notes while delivering your presentation. In Presenter view, your audience cannot see your notes.

- Reading view.

You can get to Reading view from the task bar.

Most people reviewing a PowerPoint presentation without a presenter will want to use Reading view. It displays the presentation in a full screen like Slide Show view, and it includes a few simple controls to make it easy to flip through the slides.

- Print Preview.

Print Preview lets you specify settings for what you want to print — handouts, notes pages, and outline, or slides.

7.1.2 Start a New Presentation

Creating a presentation in Microsoft PowerPoint involves starting with a basic design; adding new slides and content; choosing layouts; modifying slide design, if you want, by changing the color scheme or applying different design templates; and creating effects such as animated slide transitions. The information below focuses on the options available to you when you start the process.

The following steps show us how to create a blank presentation.

(1) Start PowerPoint. PowerPoint opens in Backstage view, where you can choose from one of the built-in themes and templates.

(2) To select the theme you want, press Enter. A preview pane opens, and the theme name is announced. The focus is on the Create button.

(3) To create the presentation, press Enter or Spacebar. The presentation opens in Normal view with one title slide.

(4) Press the Tab key to move the focus to the title placeholder on the first slide, and type the text for the title.

(5) To move to the next placeholder, which is for the subtitle, press the Tab key and then start typing.

7.1.3 Add Slides

The following steps show us how to add a new slide.

(1) On the Home tab, click New Slide (ref. Figure I-7-7).

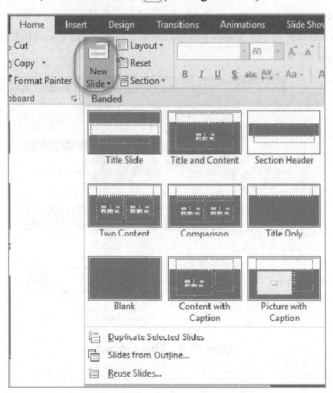

Figure I-7-7 New Slide layouts

(2) In the gallery of layouts, click the layout that you want for your new slide. Each option

in the gallery is a different slide layout that may contain placeholders for text, videos, pictures, charts, shapes, clip art, a background, and theme formatting, such as theme colors, fonts, and effects. Learn more about slide layouts.

(3) Your new slide is inserted, and you can click inside a placeholder to begin adding content.

7.1.4 Change Slide Order

To rearrange the order of slides, in the pane on the left, click the thumbnail of the slide that you want to move, and then drag it to the new location.

TIP: To select multiple slides, press and hold Ctrl while you click each slide that you want to move, and then drag them as a group to the new location.

7.1.5 Apply or Change the Slide Layout

To change the layout of an existing slide, do the following:

(1) In Normal view, on the pane that contains the Outline and Slides tab, click the Slides tab, and then click the slide that you want to apply a new layout to.

(2) On the Home tab, in the Slides group, click Layout, and then click the new layout that you want (ref. Figure I-7-8).

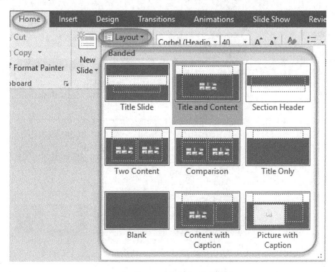

Figure I-7-8　Slide Layouts

7.1.6 Present a Slide Show

Once your slide show is complete, you'll need to learn how to present it to an audience.

PowerPoint offers several tools and features to help make your presentation smooth, engaging, and professional.

The following steps show us how to display a slide show onscreen.

- Click the Start From Beginning command ![icon] on the Quick Access toolbar (ref. Figure I-7-9), or press the F5 key at the top of your keyboard. The presentation will appear in full-screen mode.

Figure I-7-9 From Beginning command on the Quick Access toolbar

- You can also click the Play Slide Show command at the bottom of the PowerPoint window (ref. Figure I-7-10) to begin a presentation from the current slide.

Figure I-7-10 Play Slide Show command

- Click the Slide Show tab (ref. Figure I-7-11) on the Ribbon to access even more options. From here, you can start the presentation from the current slide and access advanced presentation options.

Figure I-7-11 Slide Show tab

- To advance and reverse slides:

You can advance to the next slide by clicking your mouse or pressing the spacebar on your keyboard. Alternatively, you can use or arrow keys on your keyboard to move forward or backward through the presentation.

To get out of Slide Show view at any time, on the keyboard, press Esc.

7.1.7 Print Slides or Handouts

You can use PowerPoint to print your slides (one slide per page), print slides with presenter notes, or print an outline. You can also print handouts of your presentation — with one, two, three, four, six, or nine slides on a page. Your audience can then use these handouts to follow along as you give your presentation or they can keep them for future reference.

Follow these steps to set the printing options (including number of copies, printer, slides to print, number of slides per page, color options, and more), and then print your slides.

(1) Click the File tab.

(2) Click Print, and then under Print, in the Copies box, enter the number of copies that you want to print.

(3) Under Printer, select the printer that you want to use. (If you want to print in color, be sure to select a color printer.)

(4) Under Settings, select the slides you want to print:

- To print all slides, click Print All Slides.
- To print one or more slides that you select, click the back arrow and on the Home tab, and in Normal view, select your slides from the Slides tab in the left pane. Press and hold Ctrl while you click the slides that you want to print. When you're done with your selection, click File > Print > Print All Slides > Print Selection.
- To print only the slide that is currently displayed, click Print All Slides > Print Current Slide.
- To print specific slides by number, click Print All Slides > Custom Range, and then enter a list of individual slides, a range, or both. Use commas to separate the numbers and no spaces; for example, 1,3,5-12.

(5) When you expand the Full Page Slides list, you can do the following:

- Under Print Layout, click Full Page Slides to print one slide per page. To print slides with presenter notes, click Notes Pages. To print an outline of your presentation, click Outline.
- To print one or multiple slides per page in handout format, under Handouts, click the number of slides you want per page, and whether you want them to appear in order vertically or horizontally.
- To print a thin border around your slides, select Frame Slides. (Click it again to prevent a border from printing.)
- To print your slides on the paper that you selected for your printer, click Scale to Fit Paper.
- To increase resolution, blend transparent graphics, and print soft shadows in your print job, click High quality.

(6) Click the Collated list, and then choose whether you want your slides to print collated or uncollated.

(7) Click the Color list, and then click one of the following:

- Color—This option prints in color on a color printer.
- Grayscale—This option prints images that contain variations of gray tones between black and white. Background fills are printed as white, so that the text will be more legible. (Sometimes grayscale appears the same as Pure Black and White.)
- Pure Black and White—This option prints the handout without gray fills.

(8) To include or change headers and footers, click the Edit Header and Footer link, and then make your selections in the Header and Footer dialog box that appears.

(9) When you are finished with your selections, click Print.

You can select a layout for your handouts in print preview or in the Print dialog box.

7.1.8 Tips for Creating an Effective Presentation

The followings are the tips for creating an effective presentation:
- Minimize the number of slides.
- Choose a font style that your audience can read from a distance.
- Choose a font size that your audience can read from a distance.
- Keep your text simple by using bullet points or short sentences.
- Use graphics to help convey your message.
- Make labels for charts and graphs understandable.
- Make labels for charts and graphs understandable.
- Use high contrast between background color and text color. You can use suitable Themes, as they can automatically set the contrast between a light background with dark colored text or dark background with light colored text.
- Use high contrast between background color and text color.

7.2 Formatting a Presentation

7.2.1 Apply a Design Template

A PowerPoint template is a pattern or blueprint of a slide or group of slides that you save as a .potx file. Templates can contain layouts, theme colors, theme fonts, theme effects, background styles, and even content.

To apply a template, on the File tab, click New, select an existing template. You can also search for online template and themes.

7.2.2 Working with Themes

A theme is a predefined combination of colors, fonts, and effects (like shadows, reflections, 3-D effects, and more). Different themes also use different slide layouts. You've already been using a theme, even if you didn't know it: the default Office theme. You can choose from a variety of new themes at any time, giving your entire presentation a consistent, professional look.

Every theme you use in your presentation includes a slide master and a related set of layouts. If you use more than one theme in your presentation, you'll have more than one slide master and multiple sets of layouts.

The following steps show us how to apply a theme (ref. Figure I-7-12).

(1) Select the Design tab on the Ribbon, then locate the Themes group. Each image represents a theme.

(2) Click the More drop-down arrow to see all available themes.

Figure I-7-12　Apply a theme

(3) Select the desired theme.

(4) Once you've applied a theme, you can also select a variant for that theme from the Variants group. Variants use different theme colors while preserving a theme's overall look.

In PowerPoint, you can apply a document theme to all slides, to selected slides only, or to the master slide. Right-click the document theme, and then click the option that you want.

7.2.3　Header and Footer

To add information such as slide numbers, the time and date, a company logo, the presentation title or file name, the presenter's name, and more to the top of each handout or notes page in your presentation, or to bottom of each slide, handout or notes page, we can use headers and footers.

The following steps show us how to add a header or footer to a presentation.

(1) Select the slide with the footer you want to change.

(2) On the Insert tab, click Header & Footer (ref. Figure I-7-13).

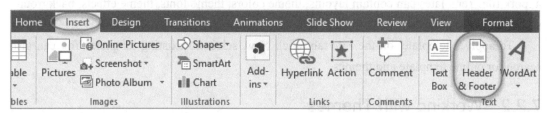

Figure I-7-13　Insert Header & Footer

(3) In the Header & Footer box (ref. Figure I-7-14), on the Slide tab, check the Footer box, and set the options such as Date and time, Slide number, Footer as well as Don't show on title slide.

(4) Click either Apply (only this slide) or Apply to All (all slides) button.

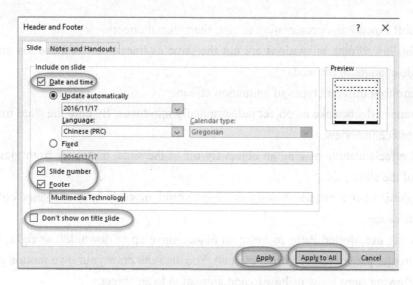

Figure I-7-14 Header and Footer Dialog

7.3 Animating Your Slides

7.3.1 Add Slide Transition Effects

Slide transitions are animation-like effects that occur in Slide Show view when you move from one slide to the next during a presentation. You can control the speed, add sound, and even customize the properties of transition effects.

The following steps show us how to add a transition to a slide.

(1) On the Transitions tab, in the Transition To This Slide group, click the slide transition effect (ref. Figure I-7-15) that you want for that slide.

Figure I-7-15 Transition Effects

(2) To see more transition effects, click the More button.

(3) Change the settings of the transition effects such as the duration of the transition between the previous slide and the current slide, and the sound effects. If you want the next slide to appear either when you click the mouse or automatically after the number of seconds you enter—whichever comes first—select both the On Mouse Click and the After check boxes.

7.3.2 Animate Objects

You can animate the text, pictures, shapes, tables, SmartArt graphics, and other objects in your Microsoft PowerPoint presentation to give them visual effects.

Note: In PowerPoint, animations are not the same as transitions. A transition animates the way one slide changes to the next.

There are four different types of animation effects:

- Entrance effects make an object fade gradually into focus, fly onto the slide from an edge, or bounce into view.
- Exit effects include making an object fly off of the slide, disappear from view, or spiral off of the slide.
- Emphasis effects include making an object shrink or grow in size, change color, or spin on its center.
- You can use Motion Paths to make an object move up or down, left or right, or in a star or circular pattern (among other effects). You also can draw your own motion path.

The following steps show us how to add animation to an object.

(1) Select the object or text on the slide that you want to animate.

(2) On the Animations tab (ref. Figure I-7-16), in the Animation group, click the More button, and then select the animation effect that you want.

Figure I-7-16 Animations tab

(3) You can set the effect options, timing, or order for an animation.

7.3.3 Hyperlink

In PowerPoint, a hyperlink can be a connection from one slide to another slide in the same presentation (such as a hyperlink to a custom show) or to a slide in another presentation, an e-mail address, a Web page, or a file.

You can create a hyperlink from text or from an object, such as a picture, graph, shape, or WordArt.

On the Insert tab, in the Links group, click Hyperlink, we can create a hyperlink to a slide in a different presentation (Existing File or Web Page), or a hyperlink to a page or file on the Web (Existing File or Web Page), or a hyperlink to a slide in the same presentation (Place in This Document), or a hyperlink to a new file (Create New Document), or a hyperlink to an e-mail address, as shown in Figure I-7-17.

7.3.4 Action Button

An action button is a ready-made button that you can insert into your presentation and define hyperlinks for. Action buttons contain shapes and commonly understood symbols for going to next, previous, first, and last slides, and for playing movies or sounds.

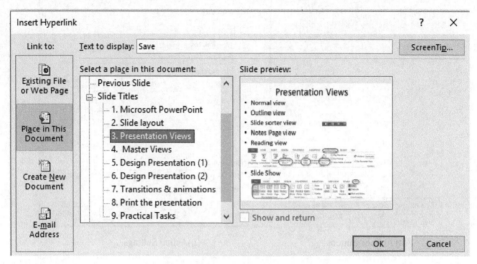

Figure I-7-17　Insert Hyperlink

The following steps show us how to insert an action button.

(1) On the Insert tab, in the Illustrations group, click the arrow under Shapes, and then click the More button.

(2) Under Action Buttons (ref. Figure I-7-18(a)), click the button that you want to add.

(3) Click a location on the slide, and then drag to draw the shape for the button.

(4) In the Action Settings dialog box (ref. Figure I-7-18(b)), do one of the following:

- To choose the behavior of the action button when you click it, click the Mouse Click tab.
- To choose the behavior of the action button when you move the pointer over it, click the Mouse Over tab.

(5) To choose what will happen when you click or move the pointer over the action button, do one of the following:

- If you don't want anything to happen, click None.
- To create a hyperlink, click Hyperlink to, and then select the destination for the hyperlink.
- To run a program, click Run program, click Browse, and then locate the program that you want to run.
- To run a macro, click Run macro, and then select the macro that you want to run.
- If you want the shape that you chose as an action button to perform an action, click Object action, and then select the action that you want it to perform.
- To play a sound, select the Play sound check box, and then select the sound that you want to play.

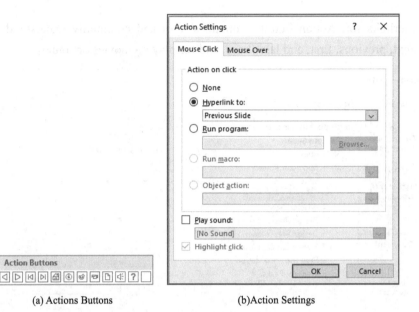

(a) Actions Buttons   (b)Action Settings

Figure I-7-18   Insert an action button

7.3.5   Insert a Sound and/or Video Clip on a Slide

7.3.5.1   Add Audio to Your Slide Show

You can add audio, such as music, narration, sound bites, and more to your presentations to add emphasis.

The following steps show us how to add and play audio clips in a presentation.

(1) On the Insert tab, in the Media group, click the arrow under Audio (ref. Figure I-7-19(a)).

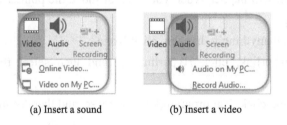

(a) Insert a sound   (b) Insert a video

Figure I-7-19   Insert a sound and/or video clip on a slide

(2) To add a sound from your local drive or a network share, click Audio on my PC, locate the audio clip that you want, and then click Insert. When you add music or other sounds to a slide, an audio icon as well as the Audio Tools appear on the tab (ref. Figure I-7-20).

(3) On the Playback tab in the Audio Tools, click Play to play the sound. You can also set the other options for the audio on the Playback tab.

(4) In Slide Show view, just click the audio icon to play the sound.

Figure I-7-20　Audio Tools

7.3.5.2　Add Video to Your Slide Show

To prevent possible problems with links, it is a good idea to copy the video into the same folder as your presentation before you add the video to your presentation.

The following steps show us how to add video to your slide show.

(1) On the Insert tab, in the Media group, click the arrow under Video (ref. Figure I-7-19(b)).

(2) To add a video from your local drive or a network share, click Video on my PC, locate the video clip that you want, and then click Insert. When you add video to a slide, a video-related icon as well as the Video Tools appear on the tab (ref. Figure I-7-21).

(3) On the Playback tab in the Video Tools, click Play to play the video. You can also set the other options for the video on the Playback tab.

(4) In Slide Show view, just click the video icon to play the video.

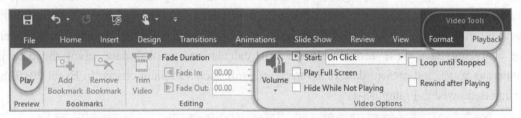

Figure I-7-21　Video Tools

(5) You can also insert online video by clicking Online Video.

Text 8
Introduction to Spreadsheets

8.1 Introduction to Excel

8.1.1 Definition of a Spreadsheets

A spreadsheet is a computer application that simulates a paper, accounting worksheet. It displays multiple cells that together make up a grid consisting of rows and columns, each cell containing alphanumeric text, numeric values or formulas. A formula defines how the content of that cell is to be calculated from the contents of any other cell (or combination of cells) each time any cell is updated. Spreadsheets are frequently used for financial information because of their ability to re-calculate the entire sheet automatically after a change to a single cell is made.

Excel is a spreadsheet program in the Microsoft Office system. Users can use Excel to create and format workbooks (a collection of spreadsheets) in order to analyze data and make more informed business decisions. Specifically, users can use Excel to track data, build models for analyzing data, write formulas to perform calculations on that data, pivot the data in numerous ways, and present data in a variety of professional looking charts. In Excel 2010 the design is showed in Figure I-8-1.

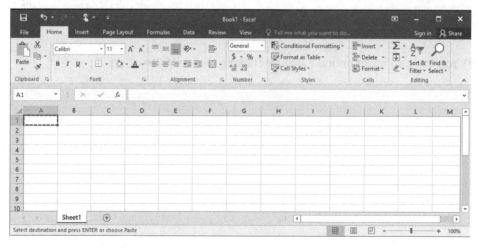

Figure I-8-1 Excel Window

Features of spreadsheets
- Workbook: contains one or more related worksheets. By default, each workbook contains three worksheets.
- Worksheet (or sheet): a single, two-dimensional array of data.
- Rows: identified by numbers.
- Columns: identified by letters.
- Cells: intersection of column and row.

8.1.2 Contents of Cells

Either a numeric value, a formula, text or simply blank.

8.1.2.1 Numerical Values

- A number, e.g.: 7, 3.14, etc.;
- Displayed right justified within the cell by default;
- May contain math operation, e.g.: =6–3+7 is displayed as 10.

8.1.2.2 Numeric Formula

Content is not what is displayed; Formula may refer to values in other cells.

E.g.,   Contents of B6:   =   B3 + B2

i.e., values in B2 and B3 are added, results displayed in B6.

Order of calculations		
1	brackets ()	left to right
2	* /	left to right
3	+ –	left to right

8.1.2.3 Text (or Labels)

- Anything else, though usually a heading;
- Treated as a sequence of individual characters;
- Displayed left justified within the cell by default;
- To enter numbers as text, enter the single quote before the number.

E.g.:   '1234

8.1.2.4 Blank

Mathematical value of zero.

8.1.3 Specifying a Range of Cells

Figure I-8-2 shows the range of Excel cells: A1:A5, A7, C1:F1, and C3:G7, which contains

5+1+4+5*5 = 35 cells.

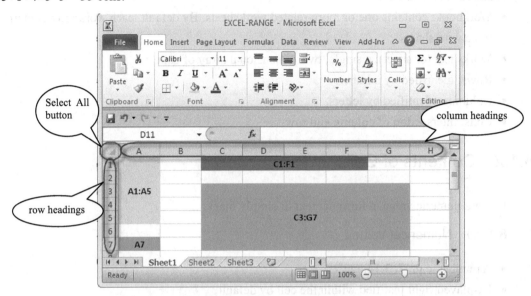

Figure I-8-2   Range of Excel Cells

Table I-8-1 shows how to select a cell or a range in Excel. To cancel a selection of cells, click any cell on the worksheet.

Table I-8-1   Select a Cell or a Range in Excel

To select	Do this
A single cell	Click the cell, or press the arrow keys to move to the cell
A range of cells	Click the first cell in the range, and then drag to the last cell, or hold down Shift while you press the arrow keys to extend the selection. You can also select the first cell in the range, and then press F8 to extend the selection by using the arrow keys. To stop extending the selection, press F8 again
A large range of cells	Click the first cell in the range, and then hold down Shift while you click the last cell in the range. You can scroll to make the last cell visible
All cells on a worksheet	Click the Select All button or press Ctrl+A
Nonadjacent cells or cell ranges	Select the first cell or range of cells, and then hold down Ctrl while you select the other cells or ranges
An entire row or column	Click the row or column heading
Adjacent rows or columns	Drag across the row or column headings. Or select the first row or column; then hold down Shift while you select the last row or column
Nonadjacent rows or columns	Click the column or row heading of the first row or column in your selection; then hold down Ctrl while you click the column or row headings of other rows or columns that you want to add to the selection

To select	Do this
The first or last cell in a row or column	Select a cell in the row or column, and then press Ctrl+Arrow key (RIGHT ARROW or LEFT ARROW for rows, UP ARROW or DOWN ARROW for columns)
The first or last cell on a worksheet or in a Microsoft Office Excel table	Press Ctrl+Home to select the first cell on the worksheet or in an Excel list. Press Ctrl+End to select the last cell on the worksheet or in an Excel list that contains data or formatting
Cells to the last used cell on the worksheet (lower-right corner)	Select the first cell, and then press Ctrl+Shift+End to extend the selection of cells to the last used cell on the worksheet (lower-right corner)
Cells to the beginning of the worksheet	Select the first cell, and then press Ctrl+Shift+Home to extend the selection of cells to the beginning of the worksheet
More or fewer cells than the active selection	Hold down Shift while you click the last cell that you want to include in the new selection. The rectangular range between the active cell and the cell that you click becomes the new selection

8.2 Spreadsheet Formulas and Functions

Formulas are equations that can perform calculations, return information, manipulate the contents of other cells, test conditions, and more. A formula typically has the format:

=*expression*

where the *expression* consists of:

- values, such as 2, 9.14 or 6.67E-11;
- references to other cells, such as A1 for a single cell or B1:B3 for a range;
- arithmetic operators, such as +, −, *, /, and others;
- relational operators, such as >=, <, and others;
- functions, such as SUM(), AVERAGE(), and many others.

When a cell contains a formula, it often contains references to other cells. Such a cell reference is a type of variable. Its value is the value of the referenced cell or some derivation of it. If that cell in turn references other cells, the value depends on the values of those. References can be relative (e.g., A1, or B1:B3), absolute (e.g., $A$1, or $B$1:$B$3) or mixed row-wise or column-wise absolute/relative (e.g., $A1 is column-wise absolute and A$1 is row-wise absolute). Function Key F4 is used to toggle between absolute and relative referencing.

Table I-8-2 shows some examples of formulas and their descriptions.

Table I-8-2   Examples of Formulas

Formula	Description
=5+2*3	Adds 5 to the product of 2 times 3
=SQRT(A1)	Uses the SQRT function to return the square root of the value in A1
=TODAY()	Returns the current date
=IF(A1>0, "Plus", "Minus")	Tests the cell A1 to determine if it contains a value greater than 0. If the result of the test is true, the text "Plus" appears in the cell; if the result is false, the text "Minus" appears

8.2.1 Some Common Functions

Microsoft Excel provides a vast library of built-in worksheet functions. Some common functions are:

(1) =SUM(range)

Adds all the numbers in a range of cells. (Remember text and blank cells have a value of 0)

(2) =AVERAGE(range)

Returns the average (arithmetic mean) of the arguments.

(3) =MIN(range)

Returns the smallest number in a set of values.

(4) =MAX(range)

Returns the largest value in a set of values.

(5) =TODAY()

Returns the serial number of the current date.

(6) =NOW()

Returns the serial number of the current date and time.

(7) =DATE(year,month,day)

Returns the sequential serial number that represents a particular date.

(8) =YEAR(serial_number)

Returns the year corresponding to a date. The year is returned as an integer in the range 1900~9999.

(9) =IF(logical_test,value_if_true,value_if_false)

Returns one value if a condition you specify evaluates to TRUE and another value if it evaluates to FALSE. Use IF to conduct conditional tests on values and formulas.

(10) =VLOOKUP(lookup_value,table_array,col_index_num,range_lookup)

Searches for a value in the leftmost column of a table, and then returns a value in the same row from a column you specify in the table. Use VLOOKUP instead of HLOOKUP when your comparison values are located in a column to the left of the data you want to find. The V in VLOOKUP stands for "Vertical".

8.2.2 Common Formula Patterns

E.g., Figure I-8-3 shows the profit-loss projections spreadsheet. Figure I-8-4 shows the regular driver's income spreadsheet.

1) Constant Increases

Assume salary increases by $500 per year.

Each salary depends on the initial salary in B8.

What is the formula in C8?

Hint: = B8 +$E$3

	A	B	C	D	E
1	Profit-Loss Projections over 4 years:				
2					
3	Salary constant increase per annum =				$ 500
4	Sales percentage increase per annum =				12%
5					
6	Income	2004	2005	2006	2007
7		----------	----------	----------	----------
8	Salary	$11,500			
9	Sales	$22,000			

Figure I-8-3   Profit-Loss Projections Spreadsheet

	A	B	C
1	Regular Driver's income		$1,100
2	bonus for a race win		$600
3			
4	Driver	Race Win	Income
5	Gary Scelzi	Yes	
6	John Force	No	

Figure I-8-4   Regular Driver's Income Spreadsheet

2) Percentage Increases

Assume projected sales will increase by 12% from the previous year, beginning at 22 000 in 2004.

What is the formula for C9?

Hint: = B9*(1 + $E$4)

8.2.3 Copying Formulas—Absolute vs. Relative References

(1) Copy command automatically adjusts cell references in formula as it copies.

(2) Absolute referencing is used for key values.

E.g., If the formula in C8 is copied across to D8, the formula will become

= C8 + $E$3

and when the formula in C9 is copied to D9, the formula will become

= C9 * (1 + $E$4)

8.2.4 Standard Spreadsheet Functions

8.2.4.1 IF Function

The IF function returns one value if a condition users specify evaluates to TRUE, and another value if that condition evaluates to FALSE. For example, the formula =IF(A1>10,"Over 10","10 or less") returns "Over 10" if A1 is greater than 10, and "10 or less" if A1 is less than or equal to 10.

Syntax:

=IF(logical_test,value_if_true,value_if_false)

The IF function syntax has the following arguments.

- logical_test: required. Any value or expression that can be evaluated to TRUE or FALSE. This argument can use any logical comparison operator.
- value_if_true: optional. The value that users want to be returned if the logical_test argument evaluates to TRUE.

- value_if_false: optional. The value that users want to be returned if the logical_test argument evaluates to FALSE.

If function is a useful function when there is a choice between two alternatives. Allowable logical comparisons includes:

=, >, <, <=, >=, <> (not equal)

E.g., in Figure I-8-4, a driver's income is either $1 100 or $1 100+$600 depending on whether there is a "No" or "Yes" in the "Race Win" column.

What is the formula in C5?

Hint: =IF(B5="Yes", $C$1+$C$2, $C$1)

What is another possible formula in C5?

Hint: =IF(B5="No", $C$1, $C$1+$C$2)

Now this formula can be copied down column C for each player listed.

8.2.4.2　VLOOKUP Function

VLOOKUP function can be used to search the first column of a range of cells, and then return a value from any cell on the same row of the range.

Syntax:

=VLOOKUP(LookupValue, TableRange, ColNum, [NotExact])

· The VLOOKUP function looks DOWN the first column of TableRange for the last row with a value <= LookupValue, then returns the value in ColNum column across within TableRange. If NotExact is FALSE, VLOOKUP only finds exact matches. If NotExact is TRUE or omitted and an exact match is not found then the largest value less than or equal to LookupValue is matched, as described above. The VLOOKUP function syntax has the following arguments.

(1) LookupValue: required. The value to search in the first column of the TableRange. The LookupValue argument can be a value or a reference. If the value you supply for the LookupValue argument is smaller than the smallest value in the first column of the TableRange argument, VLOOKUP returns the #N/A error value.

(2) TableRange: required. The range of cells that contains the data. The values in the first column of TableRange are the values searched by LookupValue. These values can be text, numbers, or logical values. Uppercase and lowercase text are equivalent.

(3) ColNum: required. The column number in the TableRange argument from which the matching value must be returned. A ColNum argument of 1 returns the value in the first column in TableRange; a ColNum of 2 returns the value in the second column in TableRange, and so on. If the ColNum argument is:

- Less than 1, VLOOKUP returns the #VALUE! error value.
- Greater than the number of columns in TableRange, VLOOKUP returns the #REF! error value.

(4) NotExact: optional. A logical value that specifies whether you want VLOOKUP to find

an exact match or an approximate match.
- If NotExact is either TRUE or is omitted (in such case, the values in the first column of TableRange must be placed in ascending sort order), an exact or approximate match is returned. If an exact match is not found, the next largest value that is less than LookupValue is returned.
- If the NotExact argument is FALSE (in such case, the values in the first column of TableRange do not need to be sorted), VLOOKUP will find only an exact match. If there are two or more values in the first column of TableRange that match the LookupValue, the first value found is used. If an exact match is not found, the error value #N/A is returned.

Example 1:

Figure I-8-5 shows an example of excel spreadsheet.

(1) =VLOOKUP(17, $A$1:$C$4, 2)

In the first example, =VLOOKUP searches down the first column of the specified block (column A), looking for the largest number less than or equal to 17. It stops at cell A2, then moves across the specified number of columns 2. It stops at cell B2 and returns the value 56. To search horizontally through a table, use HLOOKUP function.

(2) Find the values of the following:

=VLOOKUP(5, $A$1:$C$4, 1)
=VLOOKUP(50, $A$1:$C$4, 3)
=VLOOKUP(1, $A$1:$C$4, 2)
=VLOOKUP(18, $A$1:$C$4, 9)
=VLOOKUP("18", $A$1:$C$4, 3)

Example 2:

Figure I-8-6 shows another example of excel spreadsheet. For VLOOKUP to work properly with text fields, it is very important to include FALSE in the NotExact field in the VLOOKUP.

	A	B	C
1	4	34	hello
2	10	56	out
3	20	145	there
4	40	6	7/18/2010

Figure I-8-5 An Example of Excel Spreadsheet(1)

	A	B	C
1	Sunday	34	hello
2	Monday	56	out
3	Tuesday	145	there
4	Wednesday	6	7/27/2010
5	Thursday	62	how
6	Friday	290	are
7	Saturday	77	you

Figure I-8-6 An Example of Excel Spreadsheet(2)

(1) =VLOOKUP("Tuesday", $A$1:$C$7,2, FALSE)

Which means that only an exact match will do, so Excel will keep looking down the list until an exact match is found.

(2) What would be the right answer now?

=VLOOKUP("Tuesday", $A$1:$C$7,2,FALSE)

=VLOOKUP("Thursday", $A$1:$C$7,2, FALSE)

=VLOOKUP("Saturday", $A$1:$C$7,3,FALSE)

Example 3:

Suppose that in your business, you give a bonus to staff if they work more than a certain number of hours per week. The bonus table is given as the following in Figure I-8-7.

How would you set up the above table in your spreadsheet so that the bonus is given automatically by looking up this table and the hours worked by a staff member? (Ref. Figure I-8-8)

What is the VLOOKUP formula in cell C8?

=VLOOKUP(B8,$A$2:$B$5,2)

Then copy this formula down column C for each staff listed.

Hours worked	Bonus
0-20	$0
21-30	$50
31-45	$80
>45	$100

Figure I-8-7 An Example of Excel Spreadsheet(3)

	A	B	C
1	Hours worked	Bonus	
2	0	$0	
3	21	$50	
4	31	$80	
5	45	$100	
6			
7	Staff	Hours worked	Bonus
8	Mary	34	$80
9	John	57	$100
10	Tom	48	$100
11	Jenny	23	$50
12	Robert	16	$0

Figure I-8-8 An Example of Excel Spreadsheet(4)

8.2.5 Useful Financial Functions

8.2.5.1 PMT Function

Syntax:

= PMT(Rate, Nper, Pv, [Fv], [Type])

The PMT function calculates the fully amortized periodic payment needed to repay a loan with a principal of Pv dollars at Rate percent per period over Nper periods. The PMT function syntax has the following arguments.

(1) Rate: required. The interest rate for the loan.

(2) Nper: required. The total number of payments for the loan.

(3) Pv: required. The present value, or the total amount that a series of future payments is worth now; also known as the principal.

(4) Fv: optional. The future value, or a cash balance you want to attain after the last payment is made.

(5) Type: optional. The number 0 (zero) or 1 and indicates when payments are due.
- 0 or omitted: at the end of the period.
- 1: at the beginning of the period.

Rate must correlate with the unit used for Nper. If payments are monthly, Rate must equal the annual rate divided by 12.

E.g., to calculate a monthly payment (paid on the last day of the month) for a three-year loan of $10 000 at an annual 15% interest rate, enter:
=PMT(0.15/12,3*12,10000)

The monthly payment for a 3 year, $10 000 loan at 15% is
$ 346.65

8.2.5.2 FV Function

Syntax:
= FV(Rate,Nper,Pmt,[Pv],[Type])

The FV function returns the future value of an investment where Pmt is invested for Nper periods at the rate of Rate per period. The FV function syntax has the following arguments:

(1) Rate: required. The interest rate per period.

(2) Nper: required. The total number of payment periods in an annuity.

(3) Pmt: required. The payment made each period; it cannot change over the life of the annuity. Typically, Pmt contains principal and interest but no other fees or taxes. If Pmt is omitted, you must include the Pv argument.

(4) Pv: optional. The present value, or the lump-sum amount that a series of future payments is worth right now.

(5) Type: optional. The number 0 or 1 and indicates when payments are due.
- 0 or omitted: at the end of the period.
- 1: at the beginning of the period.

E.g., Assume you want to set aside $500 at the end of each year in a savings account that earns 15% annually. To determine what the account will be worth at the end of six years, enter this formula:
=FV(15%,6,–500)

In 6 years, the yearly payments of $500 will be worth: $4 376.87.

8.3 Spreadsheet Charts

8.3.1 Spreadsheet Chart Elements

Charts are used to display series of numeric data in a graphical format to make it easier to understand large quantities of data and the relationship between different series of data. A chart has many elements (ref. Figure I-8-9). Some of these elements are displayed by default, others can be added as needed. Users can change the display of the chart elements by moving them to other locations in the chart, resizing them, or by changing the format. Users can also remove chart elements.

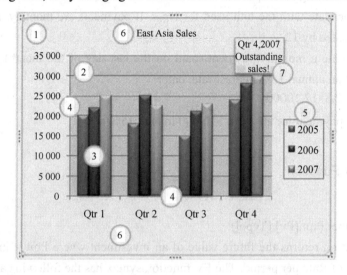

Figure I-8-9　Spreadsheet Chart Elements

① The chart area: the entire chart and all its elements.

② The plot area: in a 2-D chart, the area bounded by the axes, including all data series; in a 3-D chart, the area bounded by the axes, including the data series, category names, tick-mark labels, and axis titles.

③ The data points of the data series that are plotted in the chart: individual values plotted in a chart and represented by bars, columns, lines, pie or doughnut slices, dots, and various other shapes called data markers; data markers of the same color constitute a data series.

④ The horizontal (category) and vertical (value) axis along which the data is plotted in the chart.

⑤ The legend: a box that identifies the patterns or colors that are assigned to the data series or categories in a chart.

⑥ Chart title and axis title.

⑦ A data label: identifies the details of a data point in a data series. A data label provides additional information about a data marker, which represents a single data point or value that originates from a datasheet cell.

8.3.2　Creating Spreadsheet Charts

Steps to construct a basic spreadsheet chart.

(1) On the worksheet, arrange the data that you want to plot in a chart.

(2) Select the cells that contain the data that you want to use for the chart.

(3) Click the **Insert** tab; in the Charts / Tours / Sparklines group (ref. Figure I-8-10), do one of the following.

- Click the chart type, and then click a chart subtype.
- To see all available chart types, click ![icon] in the **Charts** group to launch the **Insert Chart** dialog box (ref. Figure I-8-11), and then scan and choose the chart type and its corresponding subtype.

Part I　Knowledge Points（第一部分　知识点）

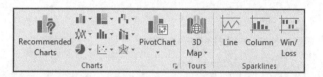

Figure I-8-10　Chart Types in the Charts & Tours & Sparklines Group

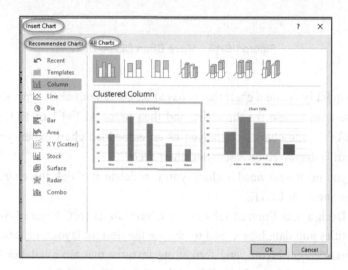

Figure I-8-11　Insert Chart Dialog Box

Tip: A ScreenTip displays the chart type name when you rest the mouse pointer over any chart type or chart subtype.

(4) By default, the chart is placed on the worksheet as an embedded chart. If you want to place the chart in a separate chart sheet, you can change its location by doing the following:

- Click anywhere in the embedded chart to activate it. This displays the **Chart Tools**, adding the **Design** and **Format** tabs (ref. Figure I-8-12).

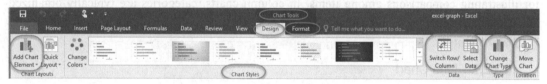

Figure I-8-12　Chart Tools

- On the **Design** tab, in the **Location** group, click **Move Chart** (ref. Figure I-8-13) to open the **Move Chart** dialog box.
- Under **Choose where you want the chart to be placed**, do one of the following:
  ■ To display the chart in a chart sheet, click **New sheet**.

Tip: If you want to replace the suggested name for the chart, you can type a new name in the **New sheet** box.

  ■ To display the chart as an embedded chart in a worksheet, click **Object in**, and then click a worksheet in the Object in box.

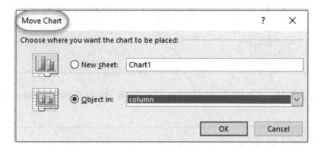

Figure I-8-13 Move Chart Dialog Box

Notes:

- To quickly create a chart that is based on the default chart type, select the data that you want to use for the chart, and then press ALT+F1 or F11. When you press ALT+F1, the chart is displayed as an embedded chart; when you press F11, the chart is displayed on a separate chart sheet.
- If you no longer need a chart, you can delete it. Click the chart to select it, and then press DELETE.

(5) Use the **Design** and **Format** tabs in the **Chart Tools** (ref. Figure I-8-12) to add chart elements such as titles and data labels, and to change the design, layout, or format of your chart. You can also access design, layout, and formatting features that are available for specific chart elements (such as chart axes or the legend) by right-clicking those chart elements in the chart.

Tip: If you don't see the **Chart Tools**, click anywhere inside the chart to activate it.

Note: Since the chart is linked to the workbook data, any subsequent changes made to the workbook are automatically reflected in the chart.

8.3.2.1 Column Chart

Data that is arranged in columns or rows on a worksheet can be plotted in a column chart (ref. Figure I-8-14). Column charts are useful for showing data changes over a period of time or for illustrating comparisons among items. In column charts, categories are typically organized along the horizontal axis and values along the vertical axis. The column charts are usually vertically oriented.

8.3.2.2 Line Chart

Data that is arranged in columns or rows on a worksheet can be plotted in a line chart (ref. Figure I-8-15). Line charts can display continuous data over time, set against a common scale, and are therefore ideal for showing trends in data at equal intervals. In a line chart, category data is distributed evenly along the horizontal axis, and all value data is distributed evenly along the vertical axis.

8.3.2.3 Pie Chart

Data that is arranged in one column or row only on a worksheet can be plotted in a pie chart. Pie charts show the size of items in one data series, proportional to the sum of the items. The data points in a pie chart are displayed as a percentage of the whole pie. For example: Figure

I-8-16 shows the hours worked for each staff.

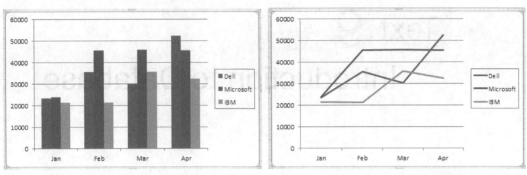

Figure I-8-14   Column Chart   Figure I-8-15   Line Chart

8.3.2.4   Bar Chart

A bar chart or bar graph is a chart with rectangular bars of lengths usually proportional to the magnitudes or frequencies of what they represent. Bar charts illustrate comparisons among individual items. The bars are usually horizontally oriented (ref. Figure I-8-17).

 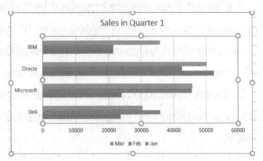

Figure I-8-16   Pie Chart   Figure I-8-17   Bar Chart

8.3.2.5   XY Graph

Data that is arranged in columns and rows on a worksheet can be plotted in an xy (scatter) chart (ref. Figure I-8-18). Scatter charts show the relationships among the numeric values in several data series, or plot two groups of numbers as one series of xy coordinates.

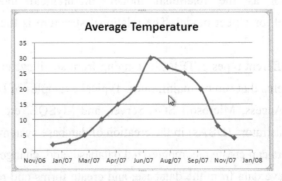

Figure I-8-18   xy Graph

Text 9
Introduction to Database

9.1 Database

Database, often abbreviated DB, is a collection of information organized in such a way that a computer program can quickly select desired pieces of data.

Traditional databases are organized by fields, records, and files. A field is a single piece of information; a record is one complete set of fields; and a file is a collection of records.

For example, a telephone book is analogous to a file. It contains a list of records, each of which consists of three fields: name, address, and telephone number.

To access information from a database, users need a database management system (DBMS), which is a collection of programs that enables users to enter, organize, and select data in a database.

9.2 Database Management Systems

A Database Management Systems (DBMS) is a collection of programs that enables users to store, modify, and extract information from a database. DBMSs may use any of a variety of database models, such as the relational model, hierarchical model, network model, multidimensional model, or object model. The relational structure is the most commonly used today.

There are many different types of DBMSs, ranging from small systems that run on personal computers to huge systems that run on mainframes. Typical examples of DBMSs include Oracle, IBM DB2, Microsoft Access, Microsoft SQL Server, and MySQL, etc. DBMSs are typically used by database administrators (DBAs) in the creation of database systems.

With a DBMS users can create a computerized database; add, change, and delete data in the database; sort and retrieve data from the database; and create forms and reports from the data in the database.

9.2.1 Relational Database Hierarchy

A relational database is a database that conforms to the relational model, and refers to a database's data and schema (the database's structure of how that data is arranged). The term relational database was originally defined and coined by E.F. Codd.

- A DB consists of a number of files—File1, File2, File3…
- Each file consists of many records—Record1, Record2…
- Records have a number of fields—Field1, Field2…
- Fields contain characters, etc—Paul, Thomas…

E.g., University Database:
- File1 is for students;
- File2 for student examination results;
- File3 for academic staff details;
- File4 for academic staff pay details.

Within each file every record must be of the same field structure.

1) Data Files (Data Tables)

A database includes a group of related data files (data tables). A data file is a collection of related records stored on a disk such as a hard disk, CD-ROM or DVD-ROM.

2) Records

A collection of fields is called a record. A record is a complete set of information. Records are composed of fields, each of which contains one item of information. A set of records constitutes a file. Each member record in the data file contains different data. In relational database management systems, records are called tuples.

For example, a personnel file might contain records that have four fields: an employee ID field, a name field, a department field and an email address field.

3) Fields

A field is a space allocated for a particular item of information. Fields are defined by Name, Type and Length.

4) Primary Key

In relational database design, a primary key or a unique key or key field is a candidate key to uniquely identify each row (record) in a table. A unique key or primary key comprises a single column or set of columns. No two distinct rows in a table can have the same value (or combination of values) in those columns. Depending on its design, a table may have arbitrarily many unique keys but at most one primary key. A primary key must always have a value.

5) Foreign Key

In relational database design, a foreign key is a referential constraint between two tables. A foreign key contains values that correspond to values in the primary key of another table. For example (ref. Figure I-9-1), you might have an Orders table in which each order has a customer

ID number that corresponds to a record in a Customers table. The customer ID field is a foreign key of the Orders table.

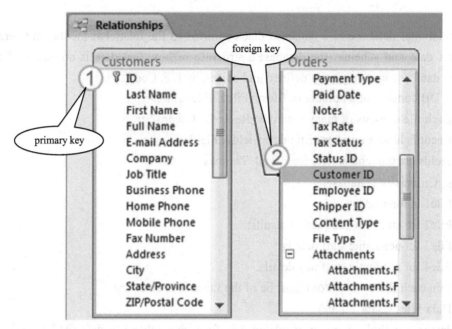

Figure I-9-1   Table Relationship

9.2.2 Examples of a Database File

STUDENT FILE—a collection of cards

Record = one student's details. Each record has a unique identifier.

1) Operations on Student Cards

(1) Maintenance: Add records, modify records, and delete records.

(2) Information Production: Inquiries, reports.

2) Ordering of Cards

Arrange Student Cards in Student ID or Student Name order to facilitate retrieval.

Student ID is the KEYFIELD.

3) Index

Index means a list of keys (or keywords), each of which identifies a unique record. Index makes it faster to find specific records and to sort records by the index field—that is, the field used to identify each record.

NAME INDEX is a separate card stack containing only index field(s) data with cross reference into main data file. Need a unique field to identify each card.

E.g., STUDENT ID

Now to access students via their ID number, create a separate ID INDEX.

Using indexes we can:
- Quickly find a student record knowing either STUDENT ID or STUDENT NAME;
- List students in either STUDENT ID or NAME sequence.

9.2.3 Data Integrity

Data integrity ensures that data is "whole" or complete, that the condition in which data is identically maintained during any operation (such as transfer, storage, and retrieval), and that the preservation of data for its intended use, or, relative to specified operations, the a priori expectation of data quality. Put simply, data integrity is the assurance that data is consistent and correct.

9.2.4 Data Redundancy

Data redundancy means duplicated data. It will slow system down because of maintenance problems. To prevent redundancy in database tables, database normalization should be done to prevent redundancy and any other problems that might affect the performance of the database.

9.3 Introduction to Microsoft Office Access

Microsoft Office Access, previously known as Microsoft Access, is a pseudo-relational database management system from Microsoft that combines the relational Microsoft Jet Database Engine with a graphical user interface and software-development tools. It is a member of the Microsoft Office suite of applications, included in the Professional and higher editions or sold separately.

9.3.1 Access Data Files

All features related with the one database are contained in the one file which has an extension: .accdb (in previous version:.mdb).

9.3.2 Field Data Types

The data type specifics the kind of data a field can contain and how the field is used.
(1) AutoNumber. unique number. Number type that is automatically incremented.
(2) Text (also called alphanumeric): letters, numbers, or special characters.
(3) Number: numbers only, can include decimal places.

(4) Date/Time: month, day, year, and time.

(5) Yes/No (also called Boolean): can only contain Yes or No, or True or False, values. Used for conditions.

(6) Currency: number type data formatted for currency.

(7) OLE Object: linked to an object external to Access.

(8) Hyperlink: an address which contains the path to an object, document, web page, or other destination.

(9) Lookup Wizard: creates a field that allows you to choose a value from another table or from a list of values.

9.3.3 Creating a Database File

(1) Open Access, click **Blank desktop database**, or on the **File** tab, click **New**, and then click **Blank desktop database**.

(2) In the Blank desktop database dialog, type a file name in the **File Name** box. To change the location of the file from the default, click **Browse** () for a location to put your database, browse to the new location, and then click **OK**.

(3) Click **Create**. Access creates the database with an empty table named **Table1**, and then opens Table1 in Datasheet view (ref. Figure I-9-2). The cursor is placed in the first empty cell in the **Click to Add** column.

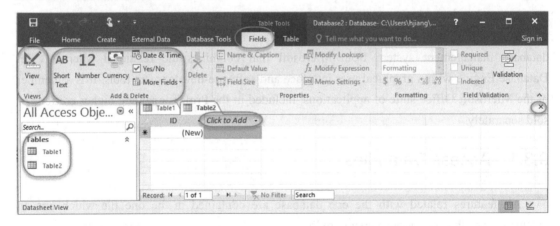

Figure I-9-2 Create a Database With an Empty Table Named Table1

(4) Begin typing to add data, or you can paste data from another source. Note:
- If you do not want to enter data in Table1 at this time, click Close . If you made any changes to the table, Access prompts you to save the changes. Click **Yes** to save your changes, click **No** to discard them, or click **Cancel** to leave the table open.
- If you close Table1 without saving it at least once, Access deletes the entire table, even if you have entered data in it.

9.3.4　Create a new table

9.3.4.1　Create a table, starting in Datasheet view

In Datasheet view, you can enter data immediately and let Access build the table structure behind the scenes. Field names are assigned numerically (Field1, Field2, and so on), and Access automatically sets each field's data type, based on the data you enter.

(1) On the **Create** tab, in the **Tables** group, click **Table**. Access creates the table and selects the first empty cell in the **Click to Add** column.

(2) On the **Fields** tab, in the **Add & Delete** group, click the type of field that you want to add. If you don't see the type that you want, click **More Fields** (ref. Figure I-9-2). Access displays a list of commonly used field types. Click the field type that you want, and Access adds the new field to the datasheet at the insertion point. You can move the field by dragging it. When you drag a field in a datasheet, a vertical insertion bar appears where the field will be placed.

(3) To add data, begin typing in the first empty cell, or paste data from another source.

(4) To rename a column (field), double-click the column heading, and then type the new name.

(5) To move a column, click its heading to select the column, and then drag the column to the location that you want. You can also select multiple contiguous columns and then drag them to a new location all at once. To select multiple contiguous columns, click the column header of the first column, and then, while holding down Shift, click the column header of the last column.

9.3.4.2　Create a table, starting in Design view

In Design view, you first create the table structure. You then switch to Datasheet view to enter data, or enter data by using some other method, such as pasting, or importing.

(1) On the **Create** tab, in the **Tables** group, click **Table Design**.

(2) For each field in your table, type a name in the **Field Name** column, and then select a data type from the **Data Type** list.

(3) If you want, you can type a description for each field in the **Description** column (ref. Figure I-9-3). The description is then displayed on the status bar when the cursor is located in that field in **Datasheet** view. The description is also used as the status bar text for any controls in a form or report that you create by dragging the field from the Field List pane, and for any controls that are created for that field when you use the Form Wizard or Report Wizard.

(4) After you have added all of your fields, on the **File** tab, click **Save** to save the table.

(5) You can begin typing data in the table at any time by switching to **Datasheet** view and clicking in the first empty cell. You can also paste data from another source.

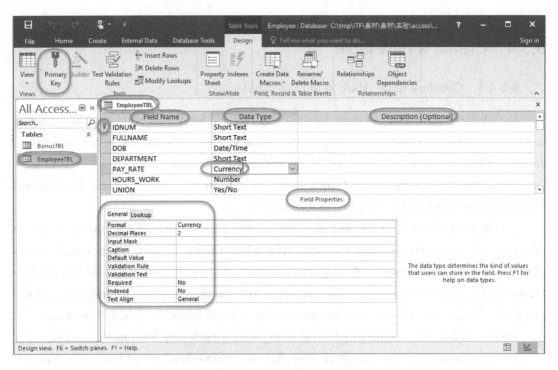

Figure I-9-3　Create a New Field in Design View

9.3.4.3　Set a table's primary key

You should specify a primary key for a table to gain the following benefits.

- Access automatically creates an index for the primary key, which can help improve database performance.
- Access makes sure that every record has a value in the primary key field.
- Access makes sure that each value in the key field is unique. Unique values are crucial, because otherwise there is no way to reliably distinguish a particular record from other records.

When you create a new table in **Datasheet** view, Access automatically creates a primary key for you and assigns it a field name of ID and the AutoNumber data type.

In **Design** view, you can change or remove the primary key, or set the primary key for a table that doesn't already have one. Use the following procedure:

- Select the table whose primary key you want to set or change.
- On the **Home** tab, in the **Views** group, click **View**, and then click **Design View**.
- In the table design grid, select the field or fields that you want to use as the primary key. To select one field, click the row selector for the field that you want. To select more than one field, hold down **Ctrl**, and then click the row selector for each field.
- On the **Design** tab, in the **Tools** group, click **Primary Key**. A key indicator appears to the left of the field or fields that you specify as the primary key.

9.3.4.4 Set a table's properties

You can set properties that apply to an entire table or to entire records.

(1) Select the table whose properties you want to set.

(2) On the **Home** tab, in the **Views** group, click **View**, and then click **Design View**.

(3) On the **Design** tab, in the **Show/Hide** group, click **Property Sheet** . The table property sheet is shown.

(4) On the property sheet, click the **General** tab.

(5) Click the box to the left of the property that you want to set, and then enter a setting for the property (ref. Figure I-9-4).

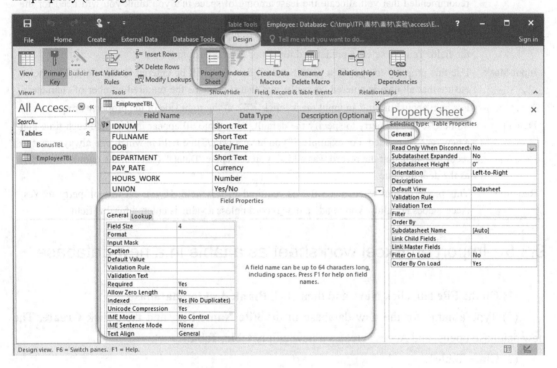

Figure I-9-4 Set a Table's Properties

(6) To save your changes, press **Ctrl+S**.

9.3.4.5 Set a field's properties

Regardless of how you created your table, it is a good idea to examine and set field properties. While some properties are available in Datasheet view, some properties can only be set in Design view. To switch to Design view, right-click the table in the **Navigation Pane** and then click **Design View**. To see a field's properties, click the field in the design grid. The properties are displayed below the design grid, under **Field Properties** (ref. Figure I-9-3 & Figure I-9-4).

To see a description of each field property, click the property and read the description in the box next to the property list under **Field Properties**. You can get more detailed information by clicking the Help button.

The following table (Table I-9-1) describes some of the field properties that are commonly adjusted.

Table I-9-1 Common Field Properties

Property	Description
Field Size	For Text fields, this property sets the maximum number of characters that can be stored in the field. The maximum is 255. For Number fields, this property sets the type of number that will be stored (Long Integer, Double, and so on). For the most efficient data storage, it is recommended that you allocate the least amount of space that you think you will need for the data. You can adjust the value upwards later, if your needs change
Format	This property sets how the data is displayed. It does not affect the actual data as it is stored in the field. You can select a predefined format or enter a custom format
Input Mask	Use this property to specify a pattern for all data that will be entered in this field. This helps ensure that all data is entered correctly, and that it contains the required number of characters. For help about building an input mask, click ⋯ at the right side of the property box
Default	Value Use this property to specify the default value that will appear in this field each time that a new record is added. For example, if you have a Date/Time field in which you always want to record the date that the record was added, you can enter "Date()" (without the quotation marks) as the default value
Required	This property sets whether a value is required in this field. If you set this property to Yes, Access does not allow you to add a new record unless a value is entered for this field

9.3.5 Import an Excel worksheet as a table in a new database

(1) On the **File** tab, click **New**, and then click **Blank desktop database**.

(2) Type a name for the new database in the **File Name** box, and then click **Create**. The new database opens, and Access creates a new empty table, Table1.

(3) Close Table1.

(4) On the **External Data** tab, in the **Import & Link** group, click **Excel**.

(5) In the **Get External Data-Excel Spreadsheet** dialog box, click **Browse**.

(6) Use the **File Open** dialog box to locate your file.

(7) Select the file, and then click **Open**.

(8) In the **Get External Data-Excel Spreadsheet** dialog box, ensure that the **Import the source data into a new table in the current database** option is selected.

(9) Click **OK**. The **Import Spreadsheet Wizard** starts, and asks you a few questions about your data.

(10) Follow the instructions, clicking **Next** or **Back** to navigate through the pages. On the last page of the wizard, click **Finish**. Access imports the data into a new table, and then displays it under All Tables in the **Navigation Pane**.

9.4 Access Query Design

A query is a request for data results, for action on data, or for both. You can use a query to answer a simple question, to perform calculations, to combine data from different tables, or even to add, change, or delete table data. Queries that you use to retrieve data from a table or to make calculations are called select queries. Queries that add, change, or delete data are called action queries.

(1) Open an existing Access database.

(2) On the **Create** tab, in the **Queries** group, click **Query Design**.

(3) In the **Show Table** dialog box (ref. Figure I-9-5), on the **Tables** tab, double-click the tables you want to show.

(4) Close the **Show Table** dialog box.

(5) The query design screen (ref. Figure I-9-6) provides a mechanism for selecting specific data from datafile(s) by:

- Displaying only selected fields (can also change the order of fields);
- Only records that match given criteria;
- Reorder records;
- Can also link across multiple files.

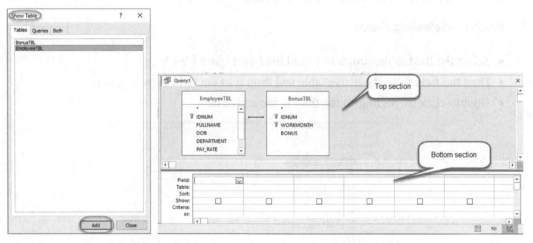

Figure I-9-5   Show Table   Figure I-9-6   Access Query Screen

9.4.1 Top Section

Top section shows the tables to be used in the query.

9.4.1.1 Show Tables

(1) On the **Home** tab, in the **Views** group, click **View**, and then click **Design View**.

(2) In the **Query Setup** group, click **Show Table** 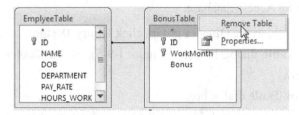, or right-click the blank area of the top section, then click **Show Table** command in the corresponding shortcut menu.

(3) In the **Show Table** dialog box, double-click the tables you want to add, and then click **Close**.

9.4.1.2 Remove Tables

Right-click the table showed in the top section, and then click **Remove Table** command in the corresponding shortcut menu (ref. Figure I-9-7).

Figure I-9-7 Remove a Table

9.4.2 Bottom Section

Bottom section contains fields and criteria used in the query.

9.4.2.1 Selecting Fields

- Select the field in the dropdown field list (ref. Figure I-9-8);
- Drag the field from the active table and drop it in the field view skeleton;
- Double-click (to highlight) the field in the active table.

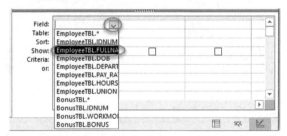

Figure I-9-8 Select the Field in the Dropdown Field List

9.4.2.2 Selecting Records

Enter criteria into bottom section of the query design screen.

1) Character field

Data is entered into the file skeleton as is or encased in quotes (ref. Figure I-9-9).

E.g., retrieve all employees who are in Forward department.

Figure I-9-9　Character Field

2) Numeric field

E.g., Retrieve all records with PAY_RATE = $8.75 (ref. Figure I-9-10).

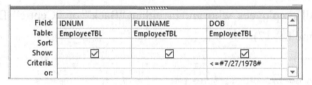

Figure I-9-10　Numeric Field

3) Date Fields

E.g., retrieve all employees born on or before 27th July 1978 (ref. Figure I-9-11).

Figure I-9-11　Date Field

4) Logical Fields

Enter　Yes　or　No

E.g., retrieve all employees who are UNION members (ref. Figure I-9-12).

Figure I-9-12　Logical Field

5) Multiple Criteria

(1) Condition1 AND Condition2:

Enter conditions on the same line.

E.g.1, DEPARTMENT= "Utility"　AND　PAY_RATE > $7　(ref. Figure I-9-13).

Figure I-9-13　Multiple Criteria (AND Condition)(1)

E.g.2, PAY_RATE between $8.5 and $9   (ref. Figure I-9-14 or Figure I-9-15).

Field:	IDNUM	FULLNAME	PAY_RATE	PAY_RATE
Table:	EmployeeTBL	EmployeeTBL	EmployeeTBL	EmployeeTBL
Sort:				
Show:	☑	☑	☑	☑
Criteria:			>=8.5	<=9
or:				

Figure I-9-14   Multiple Criteria (AND Condition)(2)

or

Field:	IDNUM	FULLNAME	PAY_RATE
Table:	EmployeeTBL	EmployeeTBL	EmployeeTBL
Sort:			
Show:	☑	☑	☑
Criteria:			>=8.5 and <=9
or:			

Figure I-9-15   Multiple Criteria (AND Condition)(3)

(2) Condition1 OR Condition2:

Enter conditions on different lines in criteria skeleton.

E.g., retrieve all employees who are in either "Back" or "Centre" DEPARTMENT (ref. Figure I-9-16 or Figure I-9-17).

Field:	IDNUM	FULLNAME	DEPARTMENT	DEPARTMENT
Table:	EmployeeTBL	EmployeeTBL	EmployeeTBL	EmployeeTBL
Sort:				
Show:	☑	☑	☑	☑
Criteria:			"Back"	
or:				"Centre"

Figure I-9-16   Multiple Criteria (OR Condition)(1)

Field:	IDNUM	FULLNAME	DEPARTMENT
Table:	EmployeeTBL	EmployeeTBL	EmployeeTBL
Sort:			
Show:	☑	☑	☑
Criteria:			"Back" Or "Centre"
or:			

Figure I-9-17   Multiple Criteria (OR Condition)(2)

9.4.2.3 Ordering of Records

Sort the output ascending/descending on character, numeric & date fields.

E.g.1, sort all employees in ascending order according to NAME. Show the information of FULLNAME, DOB, and DEPARTMENT (ref. Figure I-9-18).

Field:	FULLNAME	DOB	DEPARTMENT
Table:	EmployeeTBL	EmployeeTBL	EmployeeTBL
Sort:	Ascending		
Show:	☑	☑	☑
Criteria:			
or:			

Figure I-9-18   Ordering of Records(1)

E.g.2, retrieve all employees who have a PAY_RATE over $8 and who are also UNION members. Display the records in descending PAY_RATE order within each DEPARTMENT. Show the information of FULLNAME, DEPARTMENT, and PAY_RATE (ref. Figure I-9-19).

Field:	FULLNAME	DEPARTMENT	PAY_RATE	UNION
Table:	EmployeeTBL	EmployeeTBL	EmployeeTBL	EmployeeTBL
Sort:		Ascending	Descending	
Show:	☑	☑	☑	☐
Criteria:			>8	Yes
or:				

Figure I-9-19   Ordering of Records(2)

9.4.2.4   Showing Records

Tick the SHOW box if you want the field to be displayed (ref. Figure I-9-19, FULLNAME, DEPARTMENT, and PAY_RATE will be displayed whereas UNION will not be displayed).

9.4.2.5   Calculation Fields

E.g., retrieve all employees whose Salary exceed $250. Here the Salary = PAY_RATE * HOURS_WORK.
In the Field row of a new column enter
Salary: PAY_RATE * HOURS_WORK
Or right-click the mouse on the Field row of a new column and choose the **Build** option on the menu, an Expression Builder window will be opened. Double-click PAY_RATE field name, then click * operator, finally double-click HOURS_WORK field name, the calculated expression will be displayed automatically in the **Expression Builder** window (ref. Figure I-9-20). Click **OK** button, close **Expression Builder** window. In the query design view, rename "Expr1" as "Salary" (ref. Figure I-9-21).

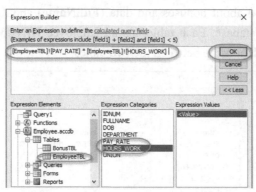

Figure I-9-20   Expression Builder Window

Field:	EmployeeTBL.*	Salary: [EmployeeTBL]![PAY_RATE]*[EmployeeTBL]![HOURS_WORK]
Table:	EmployeeTBL	
Sort:		
Show:	☑	☑
Criteria:		>250
or:		

Figure I-9-21   Calculation Field

9.4.2.6　Grouping Data

Totals queries partition data into groups and calculate summary information for each group. Each group in a totals query is mutually exclusive.

E.g., find the number of employee within each department. Show the information of DEPARTMENT, and Head Count (ref. Figure I-9-22(a) or Figure I-9-22(b)).

(a)　Method 1　　　　　　　　　　(b)　Method 2

Figure I-9-22　Grouping Data

Hint:

On the **Design** tab, in the **Show/Hide** group, click **Totals** Σ. In the bottom section of query design screen, an additional row will appear on the query design grid with the word **Total** to the left. Note: to begin with, each column in the query will have the words "**Group By**" in this **Total** row.

9.4.2.7　Multiple Data Tables Query

Tables in a well-designed database bear logical relationships to each other. These relationships exist on the basis of fields that the tables have in common. When you want to review data from related tables, you use a select query.

E.g., retrieve the bonuses of employees in each month (ref. Figure I-9-23). Data (**WorkMonth** and **Bonus**) about bonuses and data (**FullName**) about employees are stored in two tables in the same database. Each table has an employee **IDNUM** field, which forms the basis of a one-to-many relationship between the two tables.

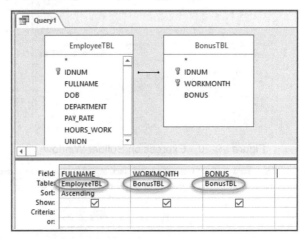

Figure I-9-23　Multiple Data Tables Query

9.5 Access Report Generator

9.5.1 Access Reports Overview

Reports are an important part of any computer business application. Features which are important to obtain meaningful reports:

- Presentation must be acceptable, proper page breaks with headings on each page.
- Must be able to print the report in a variety of orders, e.g. alphabetical.
- Calculate totals of numeric columns (fields).
- Group like rows together, and print subtotals for each group, e.g. subtotal of sales for each customer.

The Access report looks like Figure I-9-24.

```
Date: 2007-7-18                          Page: 1
            Royal Charities Pty. Ltd.
          Previous Month Collection Amounts

        NAME            AMOUNT
        ------------------------------
        JOE             60
        FRED            50
        JOE             30
        MARY            100
        MARY            70
        JOE             80
        ------------------------------
                        390
```

Figure I-9-24 An Example of Access Report

A report can be created by using the **Report** command or the **Report Wizard** or the **Blank Report** command in the **Reports** group of the **Create** tab.

9.5.2 Format of an Access Report

In Access, the design of a report is divided into sections. You can view your report in **Design view** to see its sections (ref. Figure I-9-25).

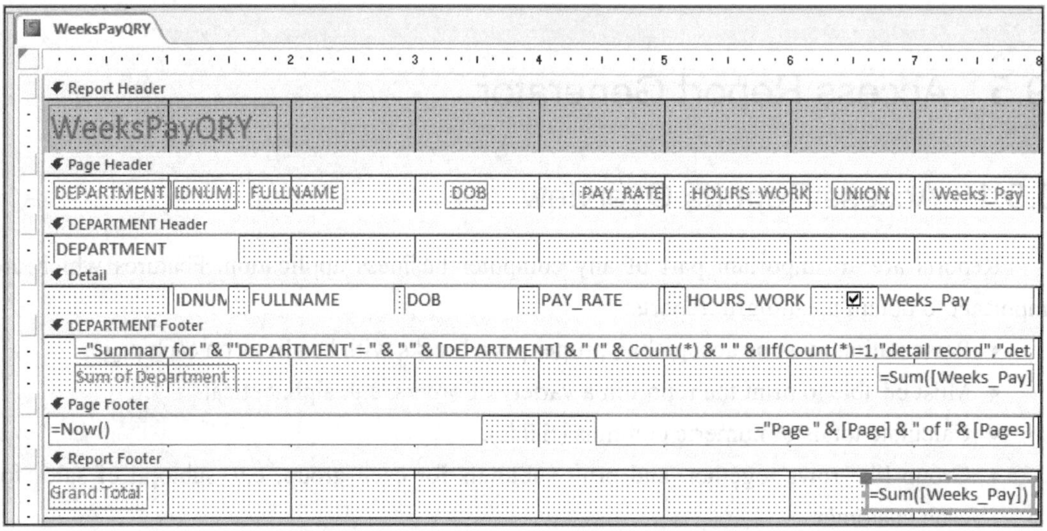

Figure I-9-25   Format of an Access Report

9.5.2.1   Report Header

This section is printed just once, at the beginning of the report. Use the report header for information that might normally appear on a cover page, such as a logo, a title, or a date. When you place a calculated control that uses the Sum aggregate function in the report header, the sum calculated is for the entire report. The report header is printed before the page header.

9.5.2.2   Page Header

This section is printed at the top of every page. For example, use a page header to repeat the report title on every page.

9.5.2.3   Group Header

This section is printed at the beginning of each new group of records. Use the group header to print the group name. For example, in a report that is grouped by product, use the group header to print the product name. When you place a calculated control that uses the Sum aggregate function in the group header, the sum is for the current group.

E.g.1, the following report is grouped by NAME. All records with the same name are grouped together (ref. Figure I-9-26).

NAME	AMOUNT
FRED	50
Subtotal	**50**
JOE	60
JOE	30
JOE	80
Subtotal	**170**
MARY	100
MARY	70
Subtotal	**170**
Total	**390**

Figure I-9-26   All Records are Grouped by NAME

E.g.2, the following report (ref. Figure I-9-27) is grouped by DEPARTMENT.

DEPARTMENT	ID	NAME	DOB	PAY_RATE	HOURS_WORK	UNION	Weeks_Pay
Back							
	1011	Graham, Ben	1/25/1978	$6.00	30	☐	$180.00
	1013	Harley, Tom	7/8/1979	$9.00	35	☐	$315.00
	1030	Scarlett, Matthew	9/10/1979	$8.60	40	☑	$344.00
Sum of Department							$839.00
Centre							
	1018	Kilpatrick, Glen	9/17/1973	$8.60	20	☑	$172.00
	1022	Clarke, David	7/19/1980	$9.10	32	☑	$291.20
	1026	Riccardi, Peter	12/17/1972	$6.75	30	☑	$202.50
Sum of Department							$665.70

Figure I-9-27   All Records are Grouped by DEPARTMENT

The department footer is SUM(Weeks_Pay). If we want to count how many employees are in each department, the footer will be COUNT(ID) or COUNT(NAME).

9.5.2.4 Detail

This section is printed once for every row in the record source. This is where you place the controls that make up the main body of the report.

9.5.2.5 Group Footer

This section is printed at the end of each group of records. Use a group footer to print summary information for a group.

9.5.2.6 Page Footer

This section is printed at the end of every page. Use a page footer to print page numbers or per-page information.

9.5.2.7 Report Footer

This section is printed just once, at the end of the report. Use the report footer to print report totals or other summary information for the entire report.

Text 10
Data Communications and Networks

10.1 Data Communications

Computer communications describes the transmission of data from one computer to another, or from one device to another. A communication device, therefore, is any machine that assists data transmission. For example, modems, hubs, and routers are all communication devices. Communication software refers to programs that make it possible to transmit data.

10.1.1 Data Communications Components

SENDING UNIT →	TRANSMISSION CHANNEL →	RECEIVING UNIT
Computer Terminal	Telephone Line	Computer Terminal
+	Radio Waves	+
Software	Microwaves	Software
	Co-axial Cable	
	Fibre-optic Cable	

10.1.2 MODEM

For any communication using telephone system a MODEM is essential to convert computer data to signals that can be carried by the telephone system.

A modem (modulator-demodulator, ref. Figure I-10-1) is a device or program that enables a computer to transmit data over, for example, telephone or cable lines. Computer information is stored digitally, whereas information transmitted over telephone lines is transmitted in the form of analog waves. A modem converts a digital signal to an analog signal and vice versa.

(1) Modem Features:
- Auto-answer;
- Auto-dial;

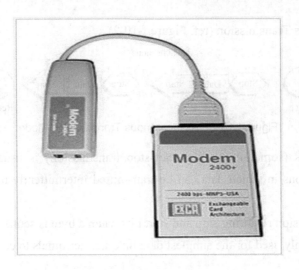

Figure I-10-1   A 2400bps Modem for a Laptop

- Automatic disconnect;
- Error correction;
- Data compression.

(2) Modem Data Speeds: Units are bits per second (bps).

10.1.3   Communication Software

Communication software is used to provide remote access to systems and exchange files and messages in text, audio and/or video formats between different computers or user IDs. Communication software consists of programs that:

- Help users establish a connection to another computer or network;
- Manage the transmission of data, instructions, and information;
- Provide all interfaces for users to communicate with one another.

The first two are system software and the third is application software. Examples of application software for communications are: E-mail, FTP, Web browser, newsgroup/message boards, chat rooms, instant messaging, video conferencing, and video telephone calls.

10.1.4   Data Transfer

Data is usually transferred in PACKETS made up of a set number of bits. Sending and receiving devices must be in synchronization for successful data transfer.

There are two methods of data transfer: asynchronous transmission and synchronous transmission.

1) Asynchronous Transmission (ref. Figure I-10-2)

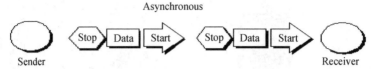

Figure I-10-2　Asynchronous Transmission Mode

- Asynchronous (sometimes called start-stop transmission) is usually used to describe communications in which data can be transmitted intermittently rather than in a steady stream.
- Data transmission requiring stop and start bits when a byte is sent.
- Most commonly used for the simplest data link, i.e.: terminals to computers.

E.g.,

Start	1 char	stop
1 bit	8 bits	1 bit

Send 1000 chars:

Data = $1000 \times 8 = 8000$bits

Extra = $1000 \times 2 = 2000$bits

Total = 10 000bits

2) Synchronous Transmission (ref. Figure I-10-3)

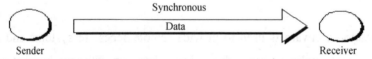

Figure I-10-3　Synchronous Transmission Mode

- Occurring at regular intervals;
- No start and stop bits used;
- No pause between sending characters;
- Involves sending large blocks of characters;
- Each block has special synchronous characters and control information which are sent at the beginning of these blocks. Each block finishes with some check sum information for error checking.

E.g.,

SYN	CI	Data	CS
3 bytes	2 bytes		5 bytes

Send 1000 chars:

Data = $1000 \times 8 = 8000$bits

Extra = $10 \times 8 = 80$bits

Total = 8080bits

10.1.5　Protocol

A protocol is a formal description of digital message formats and the rules for exchanging messages in or between computing systems and in telecommunications. Protocols may include signaling, authentication and error detection and correction capabilities. A protocol describes the syntax, semantics, and synchronization of communication and may be implemented in hardware or software, or both. The protocol determines the following:

- The type of error checking to be used;
- Data compression method, if any;
- How the sending device will indicate that it has finished sending a message;
- How the receiving device will indicate that it has received a message.

The TCP/IP (Transmission Control Protocol/Internet Protocol) is the suite of communications protocols used to connect hosts (e.g., computers) on the Internet and permit any computer to communicate with the Internet. Prominent members of TCP/IP include:

1) Transmission Control Protocol (TCP)

The Transmission Control Protocol (TCP) is one of the main protocols in TCP/IP networks. Whereas the IP protocol deals only with packets, TCP enables two hosts to establish a connection and exchange streams of data. TCP guarantees delivery of data and also guarantees that packets will be delivered in the same order in which they were sent. TCP is the protocol that major Internet applications rely on, applications such as the World Wide Web, E-mail, and file transfer. Other applications, which do not require reliable data stream service, may use the User Datagram Protocol (UDP) which provides a datagram service that emphasizes reduced latency over reliability.

2) User Datagram Protocol (UDP)

The User Datagram Protocol (UDP) is one of the core members of the Internet Protocol Suite, the set of network protocols used for the Internet. With UDP, computer applications can send messages (datagrams) to other hosts on an IP network without requiring prior communications to set up special transmission channels or data paths. It's a connectionless protocol that used primarily for broadcasting messages over a network.

3) Internet Protocol (IP)

The Internet Protocol (IP) is the principal communications protocol used for relaying datagrams (packets) across an internetwork using the Internet Protocol Suite. Responsible for routing packets across network boundaries, it is the primary protocol that establishes the Internet.

4) Internet Control Message Protocol (ICMP)

The Internet Control Message Protocol (ICMP) is one of the core protocols of the Internet Protocol Suite. It is an extension to the Internet Protocol (IP). ICMP supports packets containing error, control, and informational messages. The PING command, for example, uses ICMP to test an Internet connection.

5) Hypertext Transfer Protocol (HTTP)

The Hypertext Transfer Protocol (HTTP) is the underlying protocol used by the World Wide Web. HTTP defines how messages are formatted and transmitted, and what actions Web servers and browsers should take in response to various commands. For example, when the user enters a URL in a browser, this actually sends an HTTP command to the Web server directing it to fetch and transmit the requested Web page.

6) Post Office Protocol (POP3)

The Post Office Protocol (POP) is a protocol used to retrieve E-mail from a mail server. Most E-mail applications (sometimes called an e-mail client) use the POP protocol, although some can use the newer IMAP (Internet Message Access Protocol). The POP protocol has been developed through several versions, with version 3 (POP3) being the current standard. POP3 is used for most webmail services such as Gmail and Yahoo! Mail.

7) Internet Message Access Protocol (IMAP)

The Internet Message Access Protocol (IMAP) is one of the two most prevalent Internet standard protocols for E-mail retrieval, the other being the Post Office Protocol (POP). Virtually all modern E-mail clients and mail servers support both protocols as a means of transferring E-mail messages from a server.

8) File Transfer Protocol (FTP)

File Transfer Protocol (FTP) is a standard network protocol used exchanging files over the Internet. FTP works in the same way as HTTP for transferring Web pages from a server to a user's browser and SMTP for transferring electronic mail across the Internet in that, like these technologies, FTP uses the Internet's TCP/IP protocols to enable data transfer.

FTP is most commonly used to download a file from a server using the Internet or to upload a file to a server (e.g., uploading a Web page file to a server).

10.1.6 Direction of Data Communications

- SIMPLEX transmissions.
- HALF-DUPLEX transmissions.
- FULL-DUPLEX transmissions.

1) SIMPLEX transmissions—one direction only

Simplex refers to one-way communications where one party is the transmitter and the other is the receiver. An example of simplex communications (ref. Figure I-10-4) is a simple radio or a television, which you can receive data from stations but can't transmit data.

2) HALF-DUPLEX transmissions—one direction at a time

Half-duplex refers to the transmission of data in both directions, but only one direction at a time (not simultaneously). For example, a walkie-talkie (ref. Figure I-10-5) is a half-duplex device because only one party can talk at a time.

Figure I-10-4   SIMPLEX Transmission

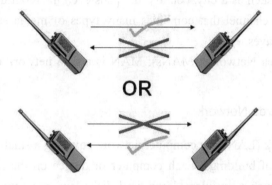

Figure I-10-5   HALF-DUPLEX Transmission

3) FULL-DUPLEX transmissions—both directions at once

Full-duplex refers to the transmission of data in two directions simultaneously. For example, a telephone (ref. Figure I-10-6) is a full-duplex device because both parties can talk at once.

Figure I-10-6   FULL-DUPLEX Transmission

10.2 Networks

A computer network, often simply referred to as a network, is a collection of computers and devices interconnected by communications channels that facilitate communications among users and allows users to share resources, such as hardware, software, data, and information.

- A number of computers or terminals connected together make up a NETWORK.
- Each computer is known as a NODE.
- Computers and devices that allocate resources for a network are called SERVERS.
- The shape of the network is its TOPOLOGY.

10.2.1 Types of Computer Networks

There are many types of computer networks, including:
- Local Area Networks (LANs): LAN is a network that connects computers and devices in a limited geographical area such as home, school, computer laboratory, office building, or closely positioned group of buildings.
- Wide Area Networks (WANs): WAN is a computer network that covers a large geographic area such as a city, country, or spans even intercontinental distances, using a communications channel that combines many types of media such as telephone lines, cables, and air waves.
- Metropolitan Area Networks (MANs): MAN is a data network designed for a town or city.

10.2.1.1 Local Area Network

A local area network (LAN) is a computer network covering a small geographic area, like a home, office, or group of buildings. Each computer or device on the network, called a node, often shares resources such as printers, large hard disks, and programs. Often, the nodes are connected via cables. LAN consists of:
- Hardware—PC + additional communication board in the computer;
- A software package, e.g., Windows 7;
- Cabling joining the computers and other peripheral devices together;
- Allows sharing of software, hardware (server, printer) and information;
- Server workstation has large hard disk with application programs and data that others can access;
- LAN's normally within the same building or campus;
- Cabling is usually twisted pair, co-axial, or optical fiber.

10.2.1.2 Wide Area Network

A Wide Area Network (WAN) is a computer network that spans a relatively large geographical area. Typically, a WAN consists of two or more LANs. Computers connected to a wide-area network are often connected through public networks, such as the telephone system. They can also be connected through leased lines or satellites.
- Uses communications channel that combines many type of media such as telephone lines, cables and radio waves.
- The Internet is the world's largest WAN.

10.2.1.3 Internet

The Internet is a global system of interconnected computer networks that use the standard

Internet Protocol Suite (TCP/IP) to serve billions of users worldwide. It is a network of networks that consists of millions of private, public, academic, business, and government networks, of local to global scope, that are linked by a broad array of electronic and optical networking technologies. The Internet carries a vast range of information resources and services, such as the inter-linked hypertext documents of the World Wide Web (WWW) and the infrastructure to support electronic mail.

10.2.1.4 Intranet

A network based on TCP/IP protocols (an internet) belonging to an organization, usually a corporation, accessible only by the organization's members, employees, or others with authorization. An intranet's Web sites look and act just like any other Web sites, but the firewall surrounding an intranet fends off unauthorized access.

Like the Internet itself, intranets are used to share information. Secure intranets are now the fastest-growing segment of the Internet because they are much less expensive to build and manage than private networks based on proprietary protocols.

10.2.1.5 Extranet

An extranet is a private network that uses Internet protocols, network connectivity, and possibly the public telecommunication system to securely share part of an organization's information or operations with suppliers, vendors, partners, customers or other businesses. Whereas an intranet resides behind a firewall and is accessible only to people who are members of the same company or organization, an extranet provides various levels of accessibility to outsiders. Users can access an extranet only if they have a valid username and password, and their identity determines which parts of the extranet they can view. Extranets are becoming a very popular means for business partners to exchange information.

10.2.2 Network Topologies

Network topology is the layout pattern of interconnections of the various elements (links, nodes, etc.) of a computer network. How different nodes in a network are connected to each other and how they communicate are determined by the network's topology. Network topologies may be physical or logical. Physical topology means the physical design of a network including the devices, location and cable installation. Logical topology refers to how data is actually transferred in a network as opposed to its physical design. Common topologies include a ring, star, bus, line, tree, and fully connected, etc. (ref. Figure I-10-7).

10.2.2.1 Ring Topology

A ring network is a network topology in which each node connects to exactly two other nodes, forming a single continuous pathway for signals through each node—a ring. Data travels from node to node, with each node along the way handling every packet. Ring topologies are relatively expensive and difficult to install, but they offer high bandwidth and can span large distances.

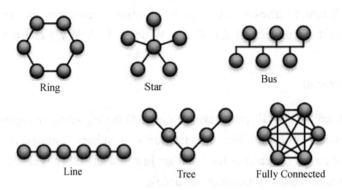

Figure I-10-7 Network Topologies

- Incoming data read from token.
- Outgoing data attached to empty token.
- Communication is usually one-way around the ring, i.e. no collisions.
- One node going down disables the network.
- Wiring is twisted pair or coaxial cable.
- Need a Token Ring interface card.

10.2.2.2 Star Topology

All of the computers and devices (nodes) on the network are connected to a central hub. The central device that provides a common connection point for nodes on the network is called the hub. Star networks are relatively easy to install and manage, but bottlenecks can occur because all data must pass through the hub.

On a star network, if one node fails, only that node is affected. The other nodes continue to operate normally. If the hub fails, however, the entire network is inoperable until the hub is repaired.

10.2.2.3 Bus Topology

All devices are connected to a shared communications cable, called the bus or backbone. Bus networks are relatively inexpensive and easy to install for small networks. Ethernet systems use a bus topology.

10.2.3 Network Communication Technologies

10.2.3.1 Ethernet

Ethernet is a local-area network (LAN) architecture developed by Xerox Corporation in cooperation with DEC and Intel in 1976. Ethernet uses a bus or star topology and supports data transfer rates of 10 Mbps. The Ethernet specification served as the basis for the IEEE 802.3 standard, which specifies the physical and lower software layers. Ethernet uses the CSMA/CD access method to handle simultaneous demands. It is one of the most widely implemented LAN standards.

A newer version of Ethernet, called 100Base-T (or Fast Ethernet), supports data transfer rates of 100 Mbps. And the newest version, Gigabit Ethernet supports data rates of 1 gigabit (1 000 megabits) per second.

10.2.3.2 Token Ring

Token ring (ref. Figure I-10-8) is a type of computer network in which all the computers are arranged (schematically) in a circle. A token, which is a special bit pattern, travels around the circle. To send a message, a computer catches the token, attaches a message to it, and then lets it continue to travel around the network.

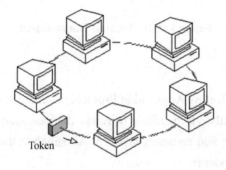

Figure I-10-8   Token Ring Network

10.2.3.3   IEEE 802.11

IEEE 802.11 and IEEE 802.11x refers to a family of specifications developed by the IEEE for wireless LAN technology. IEEE 802.11 specifies an over-the-air interface between a wireless client and a base station or between two wireless clients.

10.2.3.4   Bluetooth

Bluetooth is an open wireless technology standard for exchanging data over short distances (using short wavelength radio transmissions) from mobile phones, laptops, PCs, printers, digital cameras, and video game consoles, creating wireless personal area networks (PANs) with high levels of security.

10.2.3.5   IrDA

IrDA (Infrared Data Association) is a group of device manufacturers that developed a standard for transmitting data via infrared light waves. Increasingly, computers and other devices (such as mobilephones, PDAs, etc.) come with IrDA ports. This enables users to transfer data from one device to another without any cables (ref. Figure I-10-9). IrDA ports support roughly the same transmission rates as traditional parallel ports. The only restriction on their use is that the two devices must be within a few feet of each other and there must be a clear line of sight between them.

IrDA interfaces are used in medical instrumentation, test and measurement equipment, palmtop computers, mobile phones, and laptop computers (most laptops and phones also offer Bluetooth but it is now becoming more common for Bluetooth to simply replace IrDA in new versions of products).

Figure I-10-9 Infrared Transmission

10.2.3.6 WAP

WAP is short for the Wireless Application Protocol, a secure specification that allows users to access information instantly via handheld wireless devices such as mobile phones, pagers, two-way radios, smartphones and communicators. WAP supports most wireless networks and is supported by all operating systems.

10.2.3.7 Wireless Network

Wireless network (ref. Figure I-10-10) refers to any type of computer network that is wireless, and is commonly associated with a telecommunications network whose interconnections between nodes are implemented without the use of wires. Wireless telecommunications networks are generally implemented with some type of remote information transmission system that uses electromagnetic waves, such as radio waves and/or microwaves.

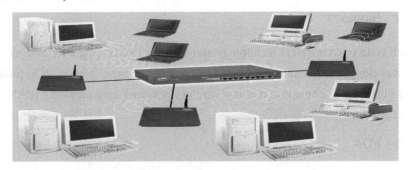

Figure I-10-10 Wireless Network Architecture

10.2.4 Connecting Networks

10.2.4.1 Repeater and Hub

A repeater (ref. Figure I-10-11) is a network device used to regenerate or replicate a signal. Repeaters are used in transmission systems to regenerate analog or digital signals distorted by transmission loss. Analog repeaters frequently can only amplify the signal while digital repeaters can reconstruct a signal to near its original quality.

In a data network, a repeater can relay messages between subnetworks that use different protocols or cable types. Hubs can operate as repeaters by relaying messages to all connected computers. A repeater cannot do the intelligent routing performed by bridges and routers.

A hub (ref. Figure I-10-12) is a central device that provides a common connection point for nodes on a network. Hubs are commonly used to connect segments of a LAN. A hub contains multiple ports. When a packet arrives at one port, it is copied to the other ports so that all segments of the LAN can see all packets.

Figure I-10-11 Wireless Repeater

Figure I-10-12 Four-Port Ethernet Hub

10.2.4.2 Bridge and Switch

A bridge is a device that connects two local-area networks (LANs), or two segments of the same LAN that use the same protocol, such as Ethernet or Token-Ring.

A network switch or switching hub is a computer networking device that filters and forwards packets between LAN segments. LANs that use switches to join segments are called switched LANs or, in the case of Ethernet networks, switched Ethernet LANs.

10.2.4.3 Router

A router is an electronic device that interconnects two or more computer networks, and selectively interchanges packets of data between them. Routers are located at gateways, the places where two or more networks connect. Routers use headers and forwarding tables to determine the best path for forwarding the packets, and they use protocols such as ICMP (Internet Control Message Protocol, an extension to the Internet Protocol (IP)) to communicate with each other and configure the best route between any two hosts. Figure I-10-13 shows Cisco CRS-1 Carrier routing system.

Figure I-10-13 Cisco CRS-1 Carrier Routing System

10.2.4.4 Gateway

A gateway is a node on a network that serves as an entrance to another network. In enterprises, the gateway is the computer that routes the traffic from a workstation to the outside network that is serving the Web pages. In homes, the gateway is the ISP that connects the user to the Internet. A gateway can also be a computer system located on earth that switches data signals and voice signals between satellites and terrestrial networks.

10.2.5 Network Architecture

10.2.5.1 Client-Server

Client-server model is a kind of network architecture in which one or more computers act as a server, and the other computers on the network, called clients, request services from the server (ref. Figure I-10-14). Servers are powerful computers or processes dedicated to managing disk drives (file servers), printers (print servers), or network traffic (network servers). Clients are PCs or workstations on which users run applications. Clients rely on servers for resources, such as files, devices, and even processing power.

Client-server architectures are sometimes called two-tier architectures.

10.2.5.2 Peer-to-Peer

Peer-to-Peer, Abbreviated P2P, is a type of network in which each workstation has equivalent capabilities and responsibilities (ref. Figure I-10-15). Peers are both suppliers and consumers of resources, in contrast to the traditional client-server architecture where only servers supply, and clients consume.

10.2.6 Communication Channel

A communication channel is the transmission media on which data, instructions, or information travel, in either analog or digital form, depending on the type of communication channel. Two examples of communication channels are cable television lines and telephone lines.

Figure I-10-14   Client-Server Network Architecture   Figure I-10-15   Peer-to-Peer Network Architecture

10.2.6.1   Basic Concepts

1) Bandwidth

Bandwidth, network bandwidth, data bandwidth or digital bandwidth, is a bit rate measure of available or consumed data communication resources. For digital devices, the bandwidth is usually expressed in bits per second (bps) or bytes per second. For analog devices, the bandwidth is expressed in cycles per second, or Hertz (Hz).

2) Latency

Latency is the amount of time it takes a packet to travel from source to destination. Together, latency and bandwidth define the speed and capacity of a network.

3) Baseband

Baseband is the original band of frequencies of a signal before it is modulated for transmission at a higher frequency. Baseband also means a type of data transmission in which digital or analog data is sent over a single unmultiplexed channel, such as an Ethernet LAN. Baseband transmission use TDM (Time Division Multiplexing) to send simultaneous bits of data along the full bandwidth of the transmission channel.

4) Broadband

Broadband is used to describe a type of data transmission in which a single medium (wire) can carry several channels at once. Cable TV, for example, uses broadband transmission. In contrast, baseband transmission allows only one signal at a time.

Most communications between computers, including the majority of Local Area Networks, use baseband communications. An exception is B-ISDN networks, which employ broadband transmission.

5) Transmission Media

A transmission medium (plural transmission media) is the material or substance (solid, liquid, gas, or plasma) capable of carrying one or more signals in a communication channel.

Transmission media are one of two types: physical or wireless.

10.2.6.2 Physical Transmission Media

Physical transmission media use wire, cable, and other tangible materials to send communications signals. Examples of physical transmission media include twisted-pair cable, coaxial cable, and fiber optic cable.

1) Twisted-pair Cable

Twisted-pair cable is a type of cable that consists of two independently insulated wires twisted around one another. The use of two wires twisted together helps to reduce crosstalk and electromagnetic induction. While twisted-pair cable is used by older telephone networks and is the least expensive type of local-area network (LAN) cable, most networks contain some twisted-pair cabling at some point along the network. Other types of cables used for LANs include coaxial cables and fiber optic cables. Figure I-10-16 shows unshielded twisted pair cables with different twist rates.

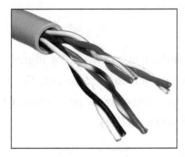

Figure I-10-16 Twisted Pair Cables

2) Coaxial Cable

Coaxial cable (ref. Figure I-10-17), or coax, is an electrical cable consisting of a round conducting wire, surrounded by an insulating spacer, surrounded by a cylindrical conducting sheath, usually surrounded by a final insulating layer (jacket). It is used as a high-frequency transmission line to carry a high-frequency or broadband signal. Because the electromagnetic field carrying the signal exists (ideally) only in the space between the inner and outer conductors, it cannot interfere with or suffer interference from external electromagnetic fields.

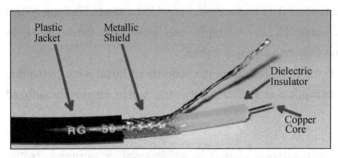 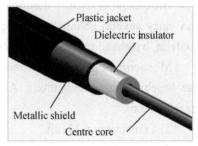

Figure I-10-17 Coaxial Cable and its Structure

3) Fiber-optic Cable

An optical fiber cable is a cable consists of dozens or hundreds of thin strands (optical fibers) of glass or plastic that use light to transmit signals (ref. Figure I-10-18). The optical fiber elements are typically individually coated with plastic layers and contained in a protective tube suitable for the environment where the cable will be deployed. In practical fibers, the cladding is usually coated with a tough resin buffer layer, which may be further surrounded by a jacket layer, usually plastic. These layers add strength to the fiber but do not contribute to its optical wave guide properties. Rigid fiber assemblies sometimes put light-absorbing ("dark") glass between the fibers, to prevent light that leaks out of one fiber from entering another. This reduces cross-talk between the fibers, or reduces flare in fiber bundle imaging applications.

 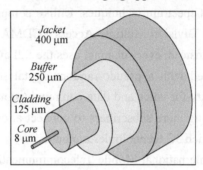

Figure I-10-18   Optical Fibers and Structure of a Typical Single-Mode Optical Fiber

10.2.6.3   Wireless Transmission Media

Examples of wireless transmission media include infrared, radio, microwave, and communication satellite.

1) Infrared

Abbreviated as IR, infrared is a wave of light that in the area beyond the visible part of the color spectrum. While it is invisible to human eye infrared is often used to enhance visibility when using night vision devices.

2) Broadcast Radio

Broadcast radio is type of one-way wireless transmission medium that distributes radio signals through the air over long distances such as between cities, regions, and countries and short distances such as within an office or home.

3) Cellular Radio

A cellular radio is a form of broadcast radio that is used widely for mobile communications, especially wireless modems and cellular telephones.

- Cellular Telephone: A cellular telephone or mobile telephone is an electronic device used for full duplex two-way radio telecommunications over a cellular network of base stations known as cell sites. Mobile phones connect to a wireless communications network through radio wave or satellite transmissions. In addition to the standard voice function of a telephone, modern mobile phones also support many additional services and accessories, such as SMS (Short Message Service), Web browsing, E-mail, Internet access, gaming, Bluetooth and infrared short range wireless communication, camera, MMS (Multimedia Message Service), MP3 player, radio and GPS. Low-end mobile phones are often referred to as feature phones, whereas high-end mobile phones that offer more advanced computing ability are referred to as smartphones.

- GSM: Short for Global System for Mobile Communications, one of the leading digital cellular systems. GSM uses narrowband TDMA (Time Division Multiple Access), which allows eight simultaneous calls on the same radio frequency.
- CDMA: Short for Code-Division Multiple Access, a digital cellular technology that uses spread-spectrum techniques. Unlike competing systems, such as GSM, that use TDMA (Time Division Multiple Access), CDMA does not assign a specific frequency to each user. Instead, every channel uses the full available spectrum. Individual conversations are encoded with a pseudo-random digital sequence. CDMA consistently provides better capacity for voice and data communications than other commercial mobile technologies, allowing more subscribers to connect at any given time, and it is the common platform on which 3G technologies are built.
- 3G: International Mobile Telecommunications-2000 (IMT-2000), better known as 3G or 3rd Generation (analog cellular was the first generation, GSM the second, and CDMA the 2.5G), is a generation of standards for mobile phones and mobile telecommunications services. Application services include wide-area wireless voice telephone, mobile Internet access, video calls and mobile TV, all in a mobile environment. Compared to the older 2G and 2.5G standards, a 3G system must provide peak data rates of at least 200 kbit/s according to the IMT-2000 specification. Recent 3G releases, often denoted 3.5G and 3.75G, also provide mobile broadband access of several Mbit/s to laptop computers and smartphones. 3G will work over wireless air interfaces such as GSM, TDMA and CDMA.
- 4G: 4G stands for the fourth generation of cellular wireless standards. It is a successor to 3G and 2G families of standards. Speed requirements for 4G service set the peak download speed at 100 Mbit/s for high mobility communication (such as from trains and cars) and 1 Gbit/s for low mobility communication (such as pedestrians and stationary users). A 4G system is expected to provide a comprehensive and secure all-IP based mobile broadband solution to smartphones, laptop computer wireless modems and other mobile devices. Facilities such as ultra-broadband Internet access, IP telephony, gaming services, and streamed multimedia may be provided to users.
- 5G: 5th. generation mobile networks or 5th generation wireless systems, are the proposed next telecommunications standards beyond the current 4G/IMT-Advanced standards. Rather than faster peak Internet connection speeds, 5G planning aims at higher capacity than current 4G, allowing higher number of mobile broadband users per area unit, and

allowing consumption of higher or unlimited data quantities in gigabyte per month and user. This would make it feasible for a large portion of the population to stream high-definition media many hours per day with their mobile devices, when out of reach of Wi-Fi hotspots. 5G research and development also aims at improved support of machine to machine communication, also known as the Internet of things, aiming at lower cost, lower battery consumption and lower latency than 4G equipment.

4) Microwaves

Microwaves are electromagnetic waves with wavelengths ranging from as long as one meter to as short as one millimeter, or equivalently, with frequencies between 300MHz (0.3 GHz) and 300GHz.

5) Communication Satellite

A communication satellite (sometimes abbreviated to COMSAT) is an artificial satellite stationed in space for the purposes of telecommunications. Modern communication satellites use geostationary orbits, Molniya orbits or low (polar and non-polar) Earth orbits. Communications satellites are commonly used for mobile phone signals, weather tracking, or broadcasting television programs. Communications satellites are artificial satellites that relay receive signals from an earth station and then retransmit the signal to other earth stations (ref. Figure I-10-19).

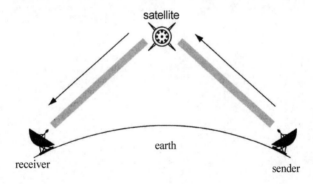

Figure I-10-19 Satellite Communication

10.2.7 Data Processing

Computer data processing is any process that uses a computer program to enter data and summarise, analyse or otherwise convert data into usable information. The process may be automated and run on a computer. It involves recording, analysing, sorting, summarising, calculating, disseminating and storing data. Because data is most useful when well-presented and actually informative, data-processing systems are often referred to as information systems.

10.2.7.1 Centralized Data Processing

- Places everything-processing, hardware and software—in one central location.

- Very inefficient system.
- Data has to be physically transported to the computer.
- Processed material has to be delivered to users.

10.2.7.2 Distributed Data Processing

- Centralized mainframe computer system is linked with minis and micros at external locations.
- Users have control over processing local data.
- Central mainframe handles applications involving whole company.

Text 11
Introduction to WWW

11.1 Introduction to WWW

The World Wide Web, abbreviated as WWW and commonly known as the Web, is a system of interlinked hypertext documents accessed via the Internet. With a web browser, one can view web pages that may contain text, images, audios, videos, and other multimedia and navigate between them via hyperlinks.

11.1.1 Web Browsers

Web browsers, often referred to just as browsers, are software applications used to locate and display Web pages on the World Wide Web. Browsers are able to display Web pages due to an underlying Web protocol called HyperText Transfer Protocol (HTTP). HTTP defines how messages are formatted and transmitted, and what actions Web servers and browsers should take in response to various commands. It is what allows Web clients and Web servers to communicate with each other. When you enter a Web address (URL) in your browser, this actually sends an HTTP command to the Web server directing it to fetch and transmit the requested Web page and display the information in your browser. All Web servers serving Web sites and pages support the HTTP protocol.

The two most popular browsers are Microsoft Internet Explorer and Firefox. Both of these are graphical browsers, which mean that they can display graphics as well as text. Most modern browsers can present multimedia information, including sound and video, though they require plug-ins for some formats.

Although browsers are primarily intended to access the World Wide Web, they can also be used to access information provided by Web servers in private networks or files in file systems. Some browsers can also be used to save information resources to file systems.

11.1.2 Uniform Resource Locator

Uniform Resource Locator (URL) is the global address of documents and other resources on the World Wide Web. The first part of URL address is called a protocol identifier and it indicates what protocol to use, and the second part is called a resource name and it specifies the IP address or the domain name where the resource is located. The protocol identifier and the resource name are separated by a colon and two forward slashes.

For example, the two URLs below point to two different files at the domain www.cc.ecnu.edu.cn. The first specifies a Web page that should be fetched using the HTTP protocol; the second specifies an executable file that should be fetched using the FTP protocol:

http://www.cc.ecnu.edu.cn/index.html

ftp://ftp.cc.ecnu.edu.cn/exercises.exe

11.1.3 Domain Names

A domain name is a name that identifies one or more IP addresses. For example, the domain name microsoft.com represents about a dozen IP addresses. Domain names are used in URLs to identify particular servers. For example, in the URL http://www.microsoft.com/index.html, the domain name is www.microsoft.com.

Every domain name has a suffix that indicates which top level domain (TLD) it belongs to. There are only a limited number of such domains. For example:

- gov—government agencies;
- edu—educational institutions;
- org—organizations (nonprofit);
- mil—military;
- com—commercial business;
- net—network organizations;
- int—international;
- cn—China;
- ca—Canada;
- jp—Japan.

Because the Internet is based on IP addresses, not domain names, every Web server requires a Domain Name System (DNS) server to translate domain names into IP addresses, and vice versa. For example, the domain name www.baidu.com might be translated to 119.75.217.56.

11.1.4 Web Servers

A web server is a computer program that delivers (serves) content, such as web pages, using the Hypertext Transfer Protocol (HTTP), over the World Wide Web. Every Web server has an IP address and possibly a domain name. For example, if the user enters the URL http://www.cc.ecnu.edu.cn/index.html in the browser, this sends a request to the server whose domain name is www.cc.ecnu.edu.cn. The server then fetches the page named index.html and sends it to the browser.

Any computer can be turned into a Web server by installing server software and connecting the machine to the Internet. There are many Web server software applications, including public domain software from NCSA and Apache, and commercial packages from Microsoft, Netscape and others.

11.2 Introduction to HTML

The World Wide Web is a system of Internet servers that support especially electronic documents. The documents are formatted in a markup language called HTML (Hyper Text Markup Language) that supports links to other documents, as well as graphics, audio, and video files. A web site is a collection of related Web pages and associated items, such as documents and pictures, stored on a web server.

11.2.1 HTML

HTML is the authoring language used to create text files with styles and links for use with World Wide Web browsers, such as Internet Explorer or Firefox.

HTML defines the structure and layout of a Web document by using a variety of tags and attributes. The correct structure for an HTML document starts with <HTML><HEAD>(enter here what document is about)</HEAD><BODY> and ends with </BODY></HTML>. All the information you'd like to include in your Web page fits in between the <BODY> and </BODY> tags.

E.g.,

```
<HTML>
<HEAD>
    <TITLE>  IT Fundamentals  </TITLE>
</HEAD>
<BODY>
Everything you need to know about <B> IT Fundamentals </B> is on the Web Page
</BODY>
</HTML>
```

11.2.2 Requirements

11.2.2.1 Authoring Tools

We need a text editor to write the HTML code. The most simple one is Notepad, which can be loaded by selecting Start → All programs → Accessories → NOTEPAD.

Today, however, there are more sophisticated authoring tools available which do not require the author to know any HTML at all. You simply paint the page as you compose a document in a word processor. Two popular applications of this type are Dreamweaver and FrontPage, which allow you to type up a Web page in a WYSIWYG (What You See Is What You Get) environment, without dealing with or editing any HTML codes directly. You simply enter in the contents of your page, using conventional formatting styles such as variable font sizing and text alignment, and these tools transparently translate them into the corresponding HTML codes.

11.2.2.2　Browsers

We need a browser to test the code to see if the web page is successful or not. We will be using either Internet Explorer or Firefox.

11.2.2.3　Servers

We need to be connected to a server to test links to other sites and to make the web pages visible to the world.

11.2.3　Tags

HTML is composed of tagged elements that instruct Web browsers to perform a defined task.

Tags tell the browser how to display the information provided. Tags often (usually) come in pairs, a beginning tag and an ending tag. Each tag is enclosed on the <and> symbols.

E.g. 1:　　<HTML> & </HTML>, where
　<HTML>　　indicates the beginning of an HTML file, and
　</HTML>　　indicates the end of an HTML file.
E.g. 2:　　<B> This text will be bold </B>
The /indicates the end of the tag. Tags are not case sensitive.

11.2.4　Basic HTML Document Structure

```
<HTML>
<HEAD>
    <TITLE> Greatwall Football Club</TITLE>
</HEAD>
<BODY>
    All the news about the sports
</BODY>
</HTML>
```

<HEAD> defines the HEAD section which contains the pair of <TITLE> tags that provide the title for the page.

The entire contents of the web page will be enclosed within the <BODY> tags.

11.2.5　Some HTML Tags

These tags would be used in the <BODY> section of the document.

- <B>bold</B>
- <I>italic</I>
- <U>underline</U>
- <CENTER> Center text horizontally </CENTER>
- <EM> Emphasis </EM>
- <STRONG> Strong emphasis </STRONG>
-
 Break to new line
- <P> Paragraph </P>
- <HR> Horizontal line
- <PRE> preformatted text </PRE>

White space is not ignored in the <PRE> tag.

Other markup tags can be embedded into the preformatted text.

E.g., try to draw the following web page:

```
NAME        AGE         POSITION
Billy       30          CHF
Gary        35          Full Forward
```

11.2.6 Heading Tags

Heading tags produce six kinds of text sizes.
<H1> Largest Heading </H1>
<H6> Smallest Heading </H6>

11.2.7 Font Size and Colors

11.2.7.1 Pages Font Size

Include the following in the <HEAD> section.
<BASEFONT size = 7>
largest = 7   default = 3

11.2.7.2 Pages Font Color, Background Color and Background Image

BGCOLOR="red" TEXT="blue"   or BACKGROUND="imagefile" TEXT="blue"
This is included in the opening <BODY> tag.
I.e.:   <BODY> is replaced by
    <BODY BGCOLOR = "blue" TEXT = "yellow" > or
<BODY BACKGROUND = "bg.jpg" TEXT="orange" >
E.g.,

```
<HTML>
<HEAD>
    <BASEFONT size = 7>
    <TITLE> Greatwall Football Club</TITLE>
</HEAD>
<BODY BGCOLOR = "blue"  TEXT = "yellow">
    Welcome to Expos 2010, Shanghai!
</BODY>
</HTML>
```

11.2.7.3 Words Style, Size, and Color

<FONT FACE = "ARIAL"> Welcome to Shanghai </FONT>

<FONT SIZE = 7> Go Shopping </FONT>

<FONT COLOR = "blue"> Greatwall </FONT>

E.g.,

```
<HTML>
<HEAD>
    <TITLE>Words Style, Color and Size</TITLE>
    <BASEFONT size = 4>
</HEAD>
<BODY  bgcolor="pink" text="blue">
   <p align="center"><FONT COLOR = "green">East China Normal University</FONT> </p>
   <HR>
   words <FONT FACE = "Arial" SIZE = 7 COLOR = "red">style,color & size</FONT> <FONT
   FACE = "Symbol" SIZE = 6 COLOR = "purple">style,color & size</FONT> ,Get it?
</BODY>
</HTML>
```

will look like Figure I-11-1.

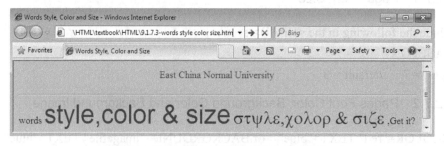

Figure I-11-1　Words Style, Size, and Color

11.2.8　Steps to Create a Web Page

(1) Get into Notepad;

(2) Type in the text;
(3) Save as filename.html or filename.htm;
(4) Get into File Explorer / Windows Explorer;
(5) Double click on filename.html or filename.htm, enjoy your web page;
(6) Go back to notepad—make changes, and **Save**;
(7) **Refresh** in web browser (e.g., Microsoft Edge / Internet Explorer), enjoy your web page;
(8) Repeat steps(6)and(7)to develop your web page.

11.3 Constructing Web Pages (1)—List, Image, Anchor

11.3.1 List

There are two kinds of lists:
(1) Unordered list:
  <UL> and </UL>
(2) Ordered list:
<OL> and </OL>
Example 1:

```
Some of my favorite food includes:
<UL>
   <LI> Chocolate
   <LI> Ice Cream
   <LI> Pizza
</UL>
```

will look like Figure I-11-2.

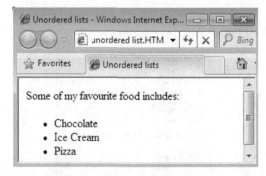

Figure I-11-2   Unordered List

Example 2:

```
Some of my favorite food includes:
<OL>
```

```
    <LI> Chocolate
    <LI> Ice Cream
    <LI> Pizza
</OL>
```

will look like Figure I-11-3.

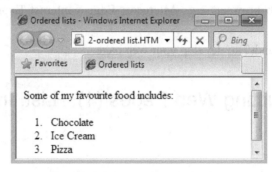

Figure I-11-3　Ordered List

Example 3:

```
Some of my favorite food includes:
<OL>
<LI> Desserts
    <UL>
        <LI> Ice cream
        <LI> Chocolate cake
    </UL>
<LI> Main Meals
    <UL>
        <LI> Pizza
        <LI> Pie
    </UL>
</OL>
```

will look like Figure I-11-4.

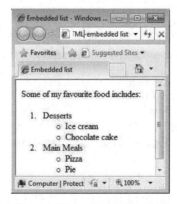

Figure I-11-4　Embedded List

11.3.2 Adding Image

11.3.2.1 Image File Formats

The file formats most commonly used on the Internet are GIF (Graphical Interchange Format) and JPG (Joint Photographic Experts Group).

11.3.2.2 The Image Tag

<IMG SRC= "...">

- The <IMG> tag inserts a graphical image into the document's normal text flow.
- An image can be placed inline with text anywhere on a page.
- The default alignment of an image is with the baseline of the surrounding text.
- The image tag has no closing tag.

E.g., <IMG SRC="myphoto.gif ">

11.3.2.3 Image Attributes

(1) SRC= "url"

(2) ALIGN=   TOP | MIDDLE | BOTTOM | LEFT | RIGHT

E.g., <IMG SRC="myphoto.gif " ALIGN=RIGHT>

(3) WIDTH = n--------image width in pixels

   HEIGHT = n-------image height in pixels

The width and height attributes are not required but are recommended to help the browser render the page faster. The height and width attributes can also be used to scale an image in the horizontal and vertical dimensions.

(4) WIDTH= n% HEIGHT= n %

where n is the width given as a percentage of the available browser window width.

(5) ALT—Alternative text

ALT defines a short description of the image (ref. Figure I-11-5).

E.g., <IMG SRC="snoopy.gif "   ALIGN=RIGHT   ALT="My Favorite Pet">

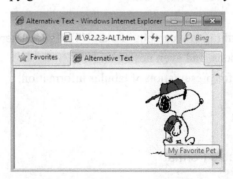

Figure I-11-5 Alternative Text

Note: Only the src=... attribute is required, the rest (align, width, height, ALT, etc.) are optional.

11.3.3 Anchor Tag

11.3.3.1 Anchor Tag Pair

Links are accomplished using the anchor tag pair, <A> and </A>.
The format to link to another web page looks like this:
<A HREF="URL"> Name of hypertext</A>
E.g.1, <A HREF="http://www.ecnu.edu.cn">Our University</A>
links to the ECNU web site when "Our University" is clicked.
E.g.2, <A HREF="specials.html">My favorite foods</A>
links to a web page called specials.html when "My favorite foods" is clicked.
E.g.3, <A HREF="mailto:Melissa@hotmail.com">Please contact me</A>
links to Melissa's email when "Please contact me" is clicked.

11.3.3.2 Naming a Target Anchor

Target anchors are used to tag locations within HTML documents as hypertext targets. This is especially useful if the HTML document is a large one and you want the user to be able to navigate to selected parts of the webpage quickly.
E.g., <A NAME=summary></A>
Other documents can navigate to that location using a hypertext reference anchor, i.e.,
<A HREF="#summary">My report</A>
takes you to the name anchor "summary" in the current document. Target anchors are useful for large web pages.

11.4 Constructing Web Pages (2)—Table

11.4.1 Table

Tables can contain anything you can put into an HTML document.
E.g., text, graphics, videos, and even nested tables.
Tables are very useful for presentation of tabular information.

11.4.2 Table Tags

11.4.2.1 General Table Format

The general format of a table looks like this:

```
<TABLE>                                    ←start of table definition
  <CAPTION> caption contents </CAPTION>    ←caption definition
  <TR>                                     ←start of first row definition
    <TH> cell contents </TH>               ←first cell in row 1 (a head)
    ⋮
    <TH> cell contents </TH>               ← last cell in row 1 (a head)
  </TR>                                    ←end of first row definition
  <TR>                                     ←start of second row definition
    <TD> cell contents </TD>               ←first cell in row 2
    ⋮
    <TD> cell contents </TD>               ←last cell in row 2
  </TR>                                    ←end of second row definition
    ⋮
  <TR>                                     ←start of last row definition
    <TD> cell contents </TD>               ←first cell in last row
    ⋮
    <TD> cell contents </TD>               ←last cell in last row
  </TR>                                    ←end of last row definition
</TABLE>                                   ←end of table definition
```

11.4.2.2 Table Elements

Table Elements lists as follow:

Element	Description
<TABLE>...</TABLE>	Defines a table in HTML. If the BORDER attribute is present, your browser displays the table with a border
<CAPTION>...</CAPTION>	Define the caption for the title of the table. The default position of the title is centered at the top of the table. The attribute ALIGN= BOTTOM can be used to position the caption below the table. NOTE: Any kind of markup tag can be used in the caption
<TR>...</TR>	Specifies a table row within a table. You may define attributes for the entire row. ALIGN (LEFT, CENTER, RIGHT) and /or VALIGN (TOP, MIDDLE, BOTTOM). See Table Attributes for more information
<TH>...</TH>	Defines a table header cell. By default the text in this cell is bold and centered. Table header cells may contain other attributes to determine the characteristics of the cell and/or its contents
<TD>...</TD>	Defines a table data cell. By default the text in this cell is aligned left and centered vertically. Table data cells may contain other attributes to determine the characteristics of the cell and/or its contents

(1) The <TABLE> and </TABLE> tags are container tags and therefore must surround the entire table definition.

(2) The first item inside the table is the CAPTION, (optional).

(3) You can have any number of rows defined by the <TR> and </TR> tags.

(4) Within a row you can have any number of cells defined by the <TH>...</TH> or <TD>...</TD> tags.

(5) Each row of a table is formatted independently of the rows above and below it.

A simple example:

```
<HTML>
  <TABLE>
  <CAPTION> JUST 9 CELLS </CAPTION>
  <TR>
      <TH> First Column </TH>
      <TH> Second Column </TH>
      <TH> Third Column </TH>
  </TR>
  <TR>
      <TD>This is cell 1.</TD>
      <TD>This is cell 2.</TD>
      <TD>This is cell 3.</TD>
  </TR>
  <TR>
      <TD>This is cell 4.</TD>
      <TD>This is cell 5.</TD>
      <TD>This is cell 6.</TD>
  </TR>
  </TABLE>
</HTML>
```

looks like Figure I-11-6.

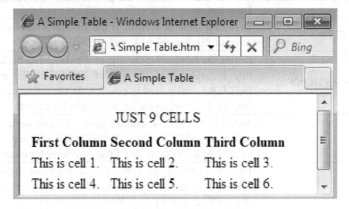

Figure I-11-6　A Simple Table

11.4.2.3　Table Attributes

Table Attributes lists as follow:

NOTE: Attribute defined within <TH>…</TH> or <TD>…</TD> cells override the default alignment set in a <TR>…</TR>

Attribute	Description
WIDTH= n|%	Specifies the width of the table—in pixels or as a percentage of document width

	Continued
NOTE: Attribute defined within <TH>...</TH> or <TD>...</TD> cells override the default alignment set in a <TR>...</TR>	
HEIGHT= n|%	Specifies the height of the table
BORDER=n	Specifies the border width of the table. Tip: Set border=0 to display tables with no borders
CELLPADDING=n|%	Specifies the space between the cell's border and its contents
CELLSPACING=n|%	Specifies the space between cells
ALIGN = LEFT|CENTER|RIGHT	Horizontal alignment of a cell
VALIGN = TOP|MIDDLE|BOTTOM	Vertical alignment of a cell
COLSPAN=n	Sets the number (n) of columns a cell spans (default=1)
ROWSPAN=n	Sets the number (n) of rows a cell spans (default=1)
BGCOLOR=RGB(r,g,b)|#rrggbb|colorname	Specifies the background color of the table
BORDERCOLOR=RGB(r,g,b)|#rrggbb|colorname	Specifies the border color of the table
NOWRAP	Turn off word wrapping within a cell

Note:

a <TD> bgcolor will override a <TR> bgcolor and a <TR> bgcolor will override a <TABLE> bgcolor.

E.g.1,

```
<HTML>
   <TABLE WIDTH=75% BGCOLOR=pink BORDER=3 CELLPADDING=10>
   <CAPTION> JUST 4 CELLS </CAPTION>
   <TR>
      <TD WIDTH=30% COLSPAN=3 ALIGN="center">Famous Athletes</TD>
   </TR>
   <TR>
      <TD WIDTH=20% BGCOLOR=cyan>Liu Xiang</TD>
      <TD>Yao Ming</TD>
      <TD WIDTH=50% BGCOLOR=yellow>Yi Jianlian</TD>
   </TR>
   </TABLE>
</HTML>
```

will give you the following web page showed in Figure I-11-7.

E.g.2,

Back to the previous example, replace the following line:

<TD> Yao Ming </TD>

with

<TD ALIGN="center"><A HREF="http://www.yaoblog.org"> Yao Ming </A></TD>

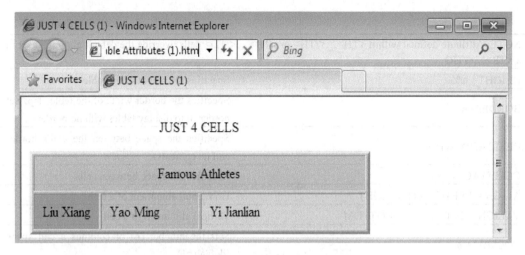

Figure I-11-7　Set Table Attributes (1)

which uses an anchor to navigate to another location and we get the following web page showed in Figure I-11-8.

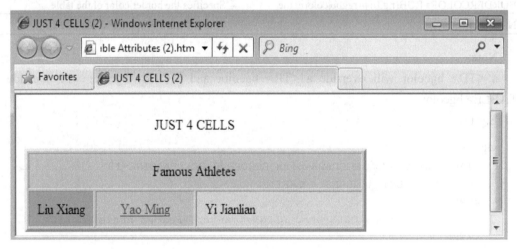

Figure I-11-8　Set Table Attributes (2)

11.4.3　Tables Incorporating an Inline Image

Example1:

```
<TABLE BORDER=3 WIDTH=100 HEIGHT=75>
<TR>
  <TD ALIGN="left" VALIGN="middle"><IMG SRC="baby.gif" ALT="baby's
  portrait"> </TD>
</TR>
</TABLE>
```

The result is showed in Figure I-11-9.

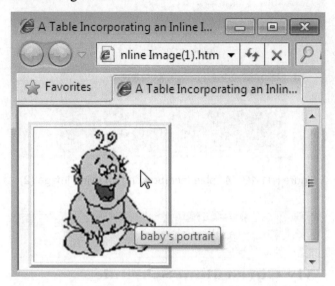

Figure I-11-9   A Table Incorporating an Inline Image(1)

Example2:

```
<TABLE BORDER=3 WIDTH=80% ALIGN ="center" BGCOLOR="orange" CELLSPACING=5>
   <CAPTION><EM>My Interests and Hobbies</EM></CAPTION>
   <TR>
      <TH COLSPAN=2 BGCOLOR="gray">Sports</TH>
      <TH BGCOLOR="olive">Main Hobbies </TH>
   </TR>
   <TR>
      <TD ALIGN ="right">Swimming</TD>
      <TD ALIGN="center">Soccer</TD>
      <TD ALIGN="left">Stamp Collecting</TD>
   </TR>
   <TR>
      <TD>Badminton</TD>
      <TD>Football</TD>
      <TD ALIGN="CENTER"><IMG SRC="dancing.gif"></TD>
   </TR>
</TABLE>
```

What would the web page look like? (Ref. Figure I-11-10)

Example3:

How would you write the code to create this table showed in Figure I-11-11?

Reference answer:

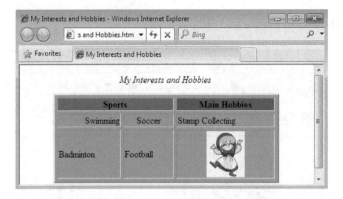

Figure I-11-10　A Table Incorporating an Inline Image (2)

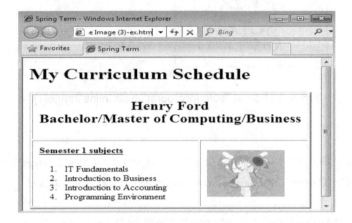

Figure I-11-11　A Table Incorporating an Inline Image (3)

```
<HTML>
<HEAD>
  <TITLE>Spring Term</TITLE>
</HEAD>
<BODY>
   <H1>My Curriculum Schedule</H1>
   <TABLE BORDER=3 CELLPADDING=10>
    <TR>
     <TD COLSPAN=2><H2 ALIGN="center"> Henry Ford <BR>
        Bachelor/Master of Computing/Business</h2>
     </TD>
    </TR>
    <TR>
     <TD><U><B>Semester 1 subjects</U></B><BR>
      <OL>
         <LI>IT Fundamentals
         <LI>Introduction to Business
```

```
            <LI>Introduction to Accounting
            <LI>Programming Environment
         </OL>
      </TD>
      <TD><IMG SRC="happy.gif" ></TD>
    </TR>
  </TABLE>
</BODY>
</HTML>
```

Text 12 Computers and Society

12.1 E-Business

12.1.1 E-Business Basics

E-Business (eBusiness), or Electronic Business, is the administration of conducting business via computer networks. It can be defined as the application of information and communication technologies (ICT) in support of all the activities of business, which includes the buying and selling of goods and services, along with providing technical or customer support through the Internet.

E-commerce is a subset of an overall e-business strategy. E-commerce is the process of buying, transferring, or exchanging goods and services via the Internet.

12.1.2 E-Business Models

12.1.2.1 Business-to-business (B2B)

Business-to-business (B2B, sometimes also called Business-to-Business) is a marketing strategy which involves the transaction of goods and services between businesses, such as between a manufacturer and a wholesaler, or between a wholesaler and a retailer.

12.1.2.2 Business-to-consumer (B2C)

Business-to-consumer (B2C, sometimes also called Business-to-Customer) is the exchange of goods and services from a business to a consumer. An example of a B2C transaction would be a person buying a pair of shoes from a retailer.

12.1.2.3 Consumer-to-business (C2B)

Consumer-to-business (C2B, sometimes also called Customer-to-Business) is an e-business model in which consumers (individuals) offer goods and services to companies and the companies pay them.

12.1.2.4 Consumer-to-consumer (C2C)

Consumer-to-consumer (C2C, sometimes also called Customer-to-Customer) involves the electronically-facilitated transactions between consumers through some third party. A common example is the online auction, in which a consumer posts an item for sale and other consumers bid to purchase it; the third party generally charges a flat fee or commission. The sites are only intermediaries, just there to match consumers. They do not have to check quality of the products being offered.

12.1.3 Electronic Shopping Carts

An electronic shopping cart is an electronic commerce tool (software or service) that is the user-interface for the customer to shop at online stores. It allows users to place items in a "shopping basket", and the cart remembers these items for a predetermined length of time. Extra features such as different color or size options, quantity of order, and matching item links can be found linked from the shopping cart. Once a shopper inputs his shipping address, taxes and shipping costs can also be tallied from within the shopping cart. For the merchant, the shopping cart also provides important information, which is often transparent to the shopper including a cart number to track the order.

12.2 Electronic Data Interchange

Electronic Data Interchange (EDI) transfers business documents (including purchase orders, invoices, shipping notices, catalogues, etc.) between different companies using networks, such as VANs (Value Added Networks) or the Internet. As more and more companies get connected to the Internet, EDI is becoming increasingly important as an easy mechanism for companies to buy, sell, and trade information.

(1) Problems caused by EDI are:
- Defining formats;
- Security;
- Available technology.

(2) Why use EDI?
- Save time and money;
- Improve customer service;
- End repetition;
- Expand customer base.

12.3 E-mail

Electronic mail, commonly called email or e-mail, is a store-and-forward method of composing, sending, storing, and receiving messages over electronic communication systems.

An email message consists of two components, the message header, and the message body, which is the email's content. The message header contains control information, including, minimally, an originator's email address and one or more recipient addresses. Usually additional information is added, such as a subject header field.

Email was possibly the first real commercial use on the Internet, because email is:
- Cheap;
- Fast;
- Convenient;
- One email can be sent to many places;
- Use of attachments.

12.4 Instant Messaging

Instant messaging (IM) is a form of real-time direct text-based communication between two or more people using personal computers or other devices, along with shared software clients. The user's text is conveyed over a network, such as the Internet. More advanced instant messaging software clients also allow enhanced modes of communication, such as live voice or video calling. Most widely used services include WeChat, AOL Instant Messenger, Windows Live Messenger, Yahoo! Messenger, and Skype, etc.

Instant messaging falls under the umbrella term online chat, as it is a real-time text-based networked communication system, but is distinct in that it is based on clients that facilitate connections between specified known users (often using "Buddy List", "Friend List" or "Contact List"), whereas online "chat" also includes web-based applications that allow communication between (often anonymous) users in a multi-user environment.

12.5 New Technologies and Patterns

12.5.1 Big Data

Big data is a term for data sets that are so large or complex that traditional data processing applications are inadequate to deal with them. Challenges include analysis, capture, data curation, search, sharing, storage, transfer, visualization, querying, updating and information privacy. The term "big data" often refers simply to the use of predictive analytics, user behavior analytics, or certain other advanced data analytics methods that extract value from data.

Big data can be described by the following characteristics:

- Volume

The quantity of generated and stored data. The size of the data determines the value and potential insight—and whether it can actually be considered big data or not.

- Variety

The type and nature of the data. This helps people who analyze it to effectively use the resulting insight.

- Velocity

In this context, the speed at which the data is generated and processed to meet the demands and challenges that lie in the path of growth and development.

- Variability

Inconsistency of the data set can hamper processes to handle and manage it.

- Veracity

The quality of captured data can vary greatly, affecting accurate analysis.

- Value

The goal, the outcome, the prioritization and the overall value and relevance.

12.5.2　Internet Plus

Internet Plus, similar to Information Superhighway and Industry 4.0, is proposed by China's prime minister Li Keqiang in his Government Work Report on March 5, 2015 so as to keep pace with the Information Trend. According to China's official website, "Internet plus" was on the list of significant economic keywords in 2015 and is one of the newest expressions of the two sessions (National People's Congress of the People's Republic of China and Chinese People's Political Consultative Conference) of the year.

Internet Plus refers to the application of the internet and other information technology in conventional industries. It is an incomplete equation where various internets (mobile Internet, cloud computing, big data or Internet of Things) can be added to other fields, fostering new industries and business development in China.

12.5.3　Cloud Computing

Cloud computing is a type of Internet-based computing that provides shared computer processing resources and data to computers and other devices on demand. It is a model for enabling ubiquitous, on-demand network access to a shared pool of configurable computing resources (e.g., computer networks, servers, storage, applications and services), which can be rapidly provisioned and released with minimal management effort.

Cloud computing and storage solutions provide users and enterprises with various capabilities to store and process their data in third-party data centers that may be located far from the user—ranging in distance from across a city to across the world. Cloud computing relies on sharing of resources to achieve coherence and economy of scale, similar to a utility (like the electricity grid) over an electricity network.

12.5.4 The Internet of Things

The Internet of things (IoT) is the internetworking of physical devices, vehicles (also referred to as "connected devices" and "smart devices"), buildings and other items—embedded with electronics, software, sensors, actuators, and network connectivity that enable these objects to collect and exchange data. In 2013 the Global Standards Initiative on Internet of Things (IoT-GSI) defined the IoT as "the infrastructure of the information society." The IoT allows objects to be sensed and/or controlled remotely across existing network infrastructure, creating opportunities for more direct integration of the physical world into computer-based systems, and resulting in improved efficiency, accuracy and economic benefit. When IoT is augmented with sensors and actuators, the technology becomes an instance of the more general class of cyber-physical systems, which also encompasses technologies such as smart grids, smart homes, intelligent transportation and smart cities. Each thing is uniquely identifiable through its embedded computing system but is able to interoperate within the existing Internet infrastructure. Experts estimate that the IoT will consist of almost 50 billion objects by 2020.

Typically, IoT is expected to offer advanced connectivity of devices, systems, and services that goes beyond machine-to-machine (M2M) communications and covers a variety of protocols, domains, and applications. The interconnection of these embedded devices (including smart objects), is expected to usher in automation in nearly all fields, while also enabling advanced applications like a smart grid, and expanding to the areas such as smart cities.

12.5.5 Mobile Web

The mobile Web refers to the use of browser-based Internet services from handheld mobile devices, such as smartphones or feature phones, through a mobile or other wireless network.

Traditionally, access to the World Wide Web has been via fixed-line services on laptops and desktop computers. However, the Web is becoming more accessible by portable and wireless devices.

An early 2010 ITU (International Telecommunication Union) report said that with the current growth rates, web access by people on the go — via laptops and smart mobile devices – is likely to exceed web access from desktop computers within the next five years.

The distinction between mobile Web applications and native applications is anticipated to become increasingly blurred, as mobile browsers gain direct access to the hardware of mobile devices (including accelerometers and GPS chips), and the speed and abilities of browser-based applications improve. Persistent storage and access to sophisticated user interface graphics functions may further reduce the need for the development of platform-specific native applications.

12.5.6 Artificial Intelligence

Artificial intelligence (AI) is intelligence exhibited by machines. In computer science, an ideal "intelligent" machine is a flexible rational agent that perceives its environment and takes actions that maximize its chance of success at some goal. Colloquially, the term "artificial intelligence" is applied when a machine mimics "cognitive" functions that humans associate with other human minds, such as "learning" and "problem solving". As machines become increasingly capable, mental facilities once thought to require intelligence are removed from the definition. For example, optical character recognition is no longer perceived as an exemplar of "artificial intelligence", having become a routine technology. Capabilities currently classified as AI include successfully understanding human speech, competing at a high level in strategic game systems (such as Chess and Go), self-driving cars, and interpreting complex data. AI is also considered a danger to humanity if it progresses unabatedly. AI research is divided into subfields that focus on specific problems or on specific approaches or on the use of a particular tool or towards satisfying particular applications.

The central problems (or goals) of AI research include reasoning, knowledge, planning, learning, natural language processing (communication), perception and the ability to move and manipulate objects. General intelligence is among the field's long-term goals. Approaches include statistical methods, computational intelligence, soft computing (e.g. machine learning), and traditional symbolic AI. Many tools are used in AI, including versions of search and mathematical optimization, logic, methods based on probability and economics. The AI field draws upon computer science, mathematics, psychology, linguistics, philosophy, neuroscience and artificial psychology.

12.5.7 Business Intelligence

Business intelligence (BI) is a philosophy which includes the strategies, processes, applications, data, products, technologies and technical architectures used to support the collection, analysis, presentation and dissemination of business information. BI technologies are capable of handling large amounts of structured and sometimes unstructured data to help identify, develop and otherwise create new strategic business opportunities. The goal of BI is to allow for the easy interpretation of these large volumes of data. Identifying new opportunities and implementing an effective strategy based on insights can provide businesses with a competitive market advantage and long-term stability.

BI technologies provide historical, current and predictive views of business operations. Common functions of business intelligence technologies are reporting, online analytical processing, analytics, data mining, process mining, complex event processing, business performance management, benchmarking, text mining, predictive analytics and prescriptive analytics.

BI can be used to support a wide range of business decisions ranging from operational to strategic. Basic operating decisions include product positioning or pricing. Strategic business decisions include priorities, goals and directions at the broadest level. In all cases, BI is most effective when it combines data derived from the market in which a company operates (external data) with data from company sources internal to the business such as financial and operations data (internal data). When combined, external and internal data can provide a more complete picture which, in effect, creates an "intelligence" that cannot be derived by any singular set of data. Amongst myriad uses, BI tools empower organisations to gain insight into new markets, assess demand and suitability of products and services for different market segments and gauge the impact of marketing efforts.

12.5.8　Deep Learning

Deep learning (DL, also known as deep structured learning, hierarchical learning or deep machine learning) is a branch of machine learning based on a set of algorithms that attempt to model high level abstractions in data by using a deep graph with multiple processing layers, composed of multiple linear and non-linear transformations.

Deep learning is part of a broader family of machine learning methods based on learning representations of data. An observation (e.g., an image) can be represented in many ways such as a vector of intensity values per pixel, or in a more abstract way as a set of edges, regions of particular shape, etc. Some representations are better than others at simplifying the learning task (e.g., face recognition or facial expression recognition). One of the promises of deep learning is replacing handcrafted features with efficient algorithms for unsupervised or semi-supervised feature learning and hierarchical feature extraction.

Research in this area attempts to make better representations and create models to learn these representations from large-scale unlabeled data. Some of the representations are inspired by advances in neuroscience and are loosely based on interpretation of information processing and communication patterns in a nervous system, such as neural coding which attempts to define a relationship between various stimuli and associated neuronal responses in the brain.

Various deep learning architectures such as deep neural networks, convolutional deep neural networks, deep belief networks and recurrent neural networks have been applied to fields like computer vision, automatic speech recognition, natural language processing, audio recognition and bioinformatics where they have been shown to produce state-of-the-art results on various tasks.

Deep learning has been characterized as a buzzword, or a rebranding of neural networks.

12.5.9 Docker

Docker is an open-source project that automates the deployment of Linux applications inside software containers.

Docker containers wrap up a piece of software in a complete filesystem that contains everything it needs to run: code, runtime, system tools, system libraries—anything you can install on a server. This guarantees that it will always run the same, regardless of the environment it is running in.

Docker provides an additional layer of abstraction and automation of operating-system-level virtualization on Linux. Docker uses the resource isolation features of the Linux kernel such as Cgroups and kernel namespaces, and a union-capable file system such as aufs and others to allow independent "containers" to run within a single Linux instance, avoiding the overhead of starting and maintaining virtual machines.

12.5.10 3D Printing

3D printing (3DP), also known as additive manufacturing (AM), refers to processes used to synthesize a three-dimensional object in which successive layers of material are formed under computer control to create an object. Objects can be of almost any shape or geometry and are produced from digital model data 3D model or another electronic data source such as an Additive Manufacturing File (AMF) file.

The term 3D printing has its origin sense, 3D printing in reference to a process that deposits a binder material onto a powder bed with inkjet printer heads layer by layer. More recently, the term is being used in popular vernacular to encompass a wider variety of additive manufacturing techniques.

12.6 Social Issues

12.6.1 Computer Crime

Computer crime, cybercrime, e-crime, hi-tech crime or electronic crime, refers to criminal activity where a computer or network is the tool, target, or place of a crime. Computer crime or cybercrime also includes traditional crimes, such as fraud, theft, blackmail, forgery, and embezzlement, in which computers or networks are used to facilitate the illegal activities.

12.6.1.1 Computer Security Risks

A computer security risk is any event or action that could cause a loss of or damage to computer hardware, software, data, information, or processing capability. Computer security risks include:

1) System Failure

System failure is the prolonged malfunction of a computer.

2) Information Theft

Information theft is a kind of computer security risk that occurs when someone steals personal or confidential information.

3) Software Theft

Software theft is a kind of computer security risk that occurs when someone steals software media, intentionally erases programs, or illegally copies a program.

4) Hardware Theft

Hardware theft is the act of stealing computer equipments.

5) Computer Viruses, Worms & Trojan horses

(1) Computer Virus: A computer virus is a program or piece of code that is loaded onto users' computer without users' knowledge and runs against users' wishes. Viruses can also replicate themselves. All computer viruses are manmade. A simple virus that can make a copy of itself over and over again is relatively easy to produce. Even such a simple virus is dangerous because it will quickly use all available memory and bring the system to a halt. An even more dangerous type of virus is one capable of transmitting itself across networks and bypassing security systems.

(2) Worm Virus: Worm virus is a program or algorithm that replicates itself over a computer network and usually performs malicious actions, such as using up the computer's resources and possibly shutting the system or network down.

(3) Trojan Horse Virus: Trojan horse virus is a malicious program that looks like a legitimate program but when executed, can erase files and damage hard disks. Trojan horses do not replicate themselves but they can be just as destructive. Some Trojans are designed to be more annoying than malicious (like changing the desktop, adding silly active desktop icons) or they can cause serious damage by deleting files and destroying information on the computer system. Trojans are also known to create a backdoor on the computer that gives malicious users access to the computer system, possibly allowing confidential or personal information to be compromised. Unlike viruses and worms, Trojans do not reproduce by infecting other files nor do they self-replicate.

6) Hacker

A hacker is a person who breaks into computers and computer networks, either for profit or motivated by the challenge. A hacker is usually a computer enthusiast, i.e. a person who enjoys learning programming languages and computer systems and can often be considered an expert on the subject(s).

(1) White Hat Hacker: A white hat hacker breaks security for non-malicious reasons, for instance testing their own security system. Often, this type of "white hat" hacker is called an ethical hacker.

(2) Black Hat Hacker: A black hat hacker is someone who breaks computer security without authorization or uses technology (usually a computer, phone system or network) for vandalism, credit card fraud, identity theft, piracy, or other types of illegal activity.

(3) Hacktivist: Formed by combining "hack" with "activist," a hacktivist is a hacker who utilizes technology to announce a social, ideological, religious, or political message. Unlike a malicious hacker, who may disrupt a system for financial gain or out of a desire to cause harm, the hacktivist performs the same kinds of disruptive actions (such as a DoS attack) in order to draw attention to a cause. For the hacktivist, it is an Internet-enabled way to practice civil disobedience and protest.

(4) Script Kiddie: A script kiddie, script bunny, or script kitty, is a non-expert who breaks into computer systems by using pre-packaged automated tools written by others, usually with little understanding of the underlying concept—hence the term script (i.e. a prearranged plan or set of activities) kiddie (i.e. kid, child—an individual lacking knowledge and experience, immature).

(5) Cyberterrorist: A cyberterrorist is a hacker who utilizes the Internet as a weapon to conduct terrorist activities, including acts of deliberate, large-scale disruption of computer networks, especially of personal computers attached to the Internet.

(6) Cracker: Cracker is someone who accesses a computer or network illegal with the intent of destroying data, stealing information, or other malicious action.

12.6.1.2 Internet Security Risks

On the Internet, where no central administrator is present, the security risk is even greater.

1) Denial of Service Attacks (DoS attack)

A denial-of-service attack (DoS attack) or distributed denial-of-service attack (DDoS attack) is a type of attack on a network that is designed to bring the network to its knees by flooding it with useless traffic. DoS attacks are implemented by either forcing the targeted computer(s) to reset, or consuming its resources so that it can no longer provide its intended service or obstructing the communication media between the intended users and the victim so that they can no longer communicate adequately.

Hackers use DoS attacks to prevent legitimate uses of computer network resources. DoS attacks attempt to flood a network, disrupt connections between two computers, prevent an individual from accessing a service or disrupt service to a specific system or person. Those on the receiving end of a DoS attack may lose valuable resources, such as their E-mail services, Internet access or their Web server. Some DoS attacks may eat up all the bandwidth or even use up all of a system resource, such as server memory.

2) Digital Certificates

A digital certificate (also known as or identity certificate or a public key certificate) is an attachment to an electronic message used for security purposes. The most common use of a digital certificate is to verify that a user sending a message is who he or she claims to be, and to provide the receiver with the means to encode a reply.

3) Digital Signature

Digital signature is a digital code that can be attached to an electronically transmitted message that uniquely identifies the sender. Like a written signature, the purpose of a digital signature is to guarantee that the individual sending the message really is who he or she claims to be. Digital signatures are especially important for electronic commerce and are a key component of most authentication schemes. To be effective, digital signatures must be unforgeable. There are a number of different encryption techniques to guarantee this level of security.

12.6.2 Security

A computer system owner may choose to build a number of controls and safeguards into their system to protect the system and its data against accidental or deliberate damage.

Access requirements:
- Physical—ID card;
- Software—password, voice recognition, finger print reader, laser eye scan, etc.;
- Copy protecting software;
- Auditor checks of the system;
- Cryptographing of communicated data (encryption);
- Development of a disaster recovery plan.

12.6.2.1 User ID

User ID (UID, or user name) is the unique combination of characters, such as letters of the alphabet and/or numbers that identifies a specific user.

12.6.2.2 Password

A password is a secret word or string of characters that is used for authentication, to prove identity or gain access to a resource (a file, computer, or program). The password should be kept secret from those not allowed access.

User names and passwords are commonly used by people during a log in process that controls access to protected computer operating systems, mobile phones, cable TV decoders, automated teller machines (ATMs), etc. A typical computer user may require passwords for many purposes: logging in to computer accounts, retrieving E-mail from servers, accessing programs, databases, networks, web sites, and even reading the morning newspaper online.

12.6.2.3 Biometric Device

A biometric device authenticates a person's identity by translating a personal characteristic into a digital code that is compared with a digital code stored in the computer verifying a physical or behavioral characteristic.

Examples of biometric devices and systems include fingerprint scanners, hand geometry systems, face recognition systems, voice verification systems, signature verification systems, and iris recognition systems.

12.6.2.4 Encryption

Encryption is the translation of data into a secret code. Encryption is the most effective way to achieve data security. To read an encrypted file, you must have access to a secret key or password that enables you to decrypt it. Unencrypted data is called plain text (or plaintext); encrypted data is referred to as cipher text (or ciphertext).

12.6.2.5 Firewall

A firewall (ref. Figure I-12-1) is a part of a computer system or network that is designed to prevent unauthorized access while permitting authorized communications. It is a device or set of devices that is configured to permit or deny network transmissions based upon a set of rules and other criteria.

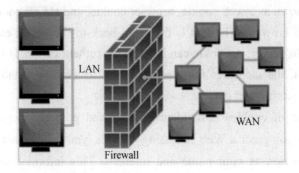

Figure I-12-1   An Illustration of Where a Firewall Would be Located in a Network

Firewalls can be implemented in either hardware or software, or a combination of both. Firewalls are frequently used to prevent unauthorized Internet users from accessing private networks connected to the Internet, especially intranets. All messages entering or leaving the intranet pass through the firewall, which inspects each message and blocks those that do not meet the specified security criteria.

There are several types of firewall techniques.

- Packet filter: Packet filtering inspects each packet passing through the network and

accepts or rejects it based on user-defined rules. Although difficult to configure, it is fairly effective and mostly transparent to its users. It is susceptible to IP spoofing.
- Application gateway: Applies security mechanisms to specific applications, such as FTP and Telnet servers. This is very effective, but can impose a performance degradation.
- Circuit-level gateway: Applies security mechanisms when a TCP or UDP connection is established. Once the connection has been made, packets can flow between the hosts without further checking.
- Proxy server: Intercepts all messages entering and leaving the network. The proxy server effectively hides the true network addresses.

12.6.3 Privacy

Information privacy is the right of individuals and companies to deny or restrict the collection and use of information about them. Privacy is sometimes related to anonymity, the wish to remain unnoticed or unidentified in the public realm. When something is private to a person, it usually means there is something within them that is considered inherently special or personally sensitive. The degree to which private information is exposed therefore depends on how the public will receive this information, which differs between places and over time. Privacy is broader than security and includes the concepts of appropriate use and protection of information.

12.6.3.1 Cookies

A cookie, also known as a web cookie, browser cookie, and HTTP cookie, is a piece of text stored by a user's web browser. The text is then sent back to the server each time the browser requests a page from the server. A cookie can be used for authentication, storing site preferences, shopping cart contents, the identifier for a server-based session, or anything else that can be accomplished through storing text data.

The main purpose of cookies is to identify users and possibly prepare customized Web pages for them. When you enter a Web site using cookies, you may be asked to fill out a form providing such information as your name and interests. This information is packaged into a cookie and sent to your Web browser which stores it for later use. The next time you go to the same Web site, your browser will send the cookie to the Web server. The server can use this information to present you with custom Web pages. So, for example, instead of seeing just a generic welcome page you might see a welcome page with your name on it.

Cookies have some important implications on the privacy and anonymity of Web users. While cookies are sent only to the server setting them or the server in the same Internet domain, a Web page may contain images or other components stored on servers in other domains.

Besides privacy concerns, cookies also have some technical drawbacks. In particular, they do not always accurately identify users, they can be used for security attacks.

12.6.3.2 Spyware

Spyware is a type of malware that is installed surreptitiously on a personal computer to intercept or take partial control over the user's interaction with the computer, without the user's informed consent.

Spyware applications are typically bundled as a hidden component of freeware or shareware programs that can be downloaded from the Internet. Once installed, the spyware monitors user activity on the Internet and transmits that information in the background to someone else. Spyware can also gather information about e-mail addresses and even passwords and credit card numbers.

Because spyware exists as independent executable programs, they have the ability to monitor keystrokes, scan files on the hard drive, snoop other applications, such as chat programs or word processors, install other spyware programs, read cookies, change the default home page on the Web browser, consistently relaying this information back to the spyware author who will either use it for advertising/marketing purposes or sell the information to another party.

Cookies are not considered spyware because people know they exist.

12.6.3.3 Spam

Spam is the unsolicited E-mail message or newsgroups posting sent to many recipients or newsgroups at once. Spam is Internet junk mail.

12.6.3.4 Backdoor

A backdoor is also called a trapdoor. It is a method of bypassing normal authentication, securing remote access to a program, online service or an entire computer system, while attempting to remain undetected. The backdoor is written by the programmer who creates the code for the program. It is often only known by the programmer. A backdoor is a potential security risk.

12.6.3.5 Content Filtering

Content filtering is the process of restricting access to certain material on the Web. It is most widely used on the internet to filter email and web access.

1) Internet Content Rating Association (ICRA)

Internet Content Rating Association (ICRA) is an international non-profit organization with offices in the United States and the United Kingdom. Its mission is to help users find the content they want, to trust what they find and to filter out what they don't want for themselves or for their children. ICRA also acts as a forum through which both policy and technical infrastructure are defined to help shape the way that the World Wide Web and content distribution channels work.

2) Web Filtering Software

Web filtering software, content-control software or censorware, is a program that restricts access to specified websites. Some also filter sites that use specific words. Others allow users to filter E-mail messages, chat rooms, and programs.

12.6.4 Computer Ethics

Computer ethics is a branch of practical philosophy which deals with how computing professionals should make decisions regarding professional and social conduct. With the growth of the Internet, privacy issues as well as concerns regarding computing technologies such as spyware and web browser cookies have called into question ethical behavior in technology.

12.6.4.1 Intellectual Property (IP) Rights

Intellectual property (IP) refers to any property that is created using original thought. Common types of intellectual property include copyrights, trademarks, patents, industrial design rights and trade secrets in some jurisdictions. Unlike tangible property, rights are not extinguished when the property is destroyed. Intellectual property rights are the rights to which creators are entitled for their work.

12.6.4.2 Copyright

Copyright is the legal right granted to an author, composer, playwright, publisher, or distributor to exclusive publication, production, sale, or distribution of a literary, musical, dramatic, or artistic work.

12.6.4.3 Plagiarism

Plagiarism is the wrongful appropriation, close imitation, or purloining and publication, of another author's language, thoughts, ideas, or expressions, and the representation of them as one's own original work.

Plagiarism is now considered academic dishonesty and a breach of journalistic ethics, subject to sanctions like expulsion and other severe career damage.

Plagiarism is not a crime but is disapproved more on the grounds of moral offence.

12.6.4.4 Piracy

Software piracy is the unauthorized use, duplication, distribution, or sale of copyrighted computer software. Most retail programs are licensed for use at just one computer site or for use by only one user at any time. By buying the software, the buyer becomes a licensed user rather than an owner. Users are allowed to make copies of the program for backup purposes, but it is against the law to give copies to friends and colleagues.

Some common types of software piracy include counterfeit software, OEM (Original Equipment Manufacturer) unbundling, softlifting, hard disk loading, and Internet software piracy.

Part II Reference Version

（第二部分 参考译文*）

---

*本部分译文仅作为理解时参考使用，并非逐词逐句的翻译。

课文 1 信息技术简介

1.1 计算机文化

计算机文化是指有效地使用计算机及其技术的知识和能力，也可指一个人使用计算机程序以及与计算机相关的其他应用程序的舒适程度，还可指对计算机如何运作的了解程度。

"计算机文化"这个术语源于 1972 年阿特·鲁赫曼（Art Luehrmann）发表的一篇会议论文"Should the computer teach the student, or vice-versa?"。计算机文化一般定义为使用计算机和应用软件的技能，要具备这种技能就需要学习基本的计算机知识，理解计算机的硬件、软件和它们的使用方法。计算机文化涵盖的内容随着计算机的发展而变化。例如，美国有些高校要求的最低技能包括文件管理（File Manipulation）、文字处理（Word Processing）、电子表格（Spreadsheet）、演示文稿（PowerPoint）等软件的使用技能。

1.2 信息系统

信息系统是信息技术以及使用该技术来支持协作、管理和决策的系统。通常用"信息系统"这个词时，特指依赖于计算机技术的信息系统。信息系统一般包含五个部分：人、规范、软件、硬件和数据。
- 人：使用计算机的最终用户；
- 规范：计算机操作的规则或指南；
- 软件：为计算机硬件提供运行指令；
- 硬件：处理数据以产生信息的设备，包括键盘、鼠标、显示器、系统部件以及其他设备；
- 数据：原始的、未加工的事实，包括文本、数字、图像、音频和视频等。

1.3 信息技术

信息是数据处理的结果，为用户提供知识。信息技术（IT）是用于研究、设计、开发、实现、支持或管理计算机信息系统所采用的各种技术总称。简而言之，信息技术是指应用计算机科学和通信技术对信息进行安全的转换、存储、保护、处理、传输和检索的技术。信息技术的应用包括计算机硬件和软件、网络和通信技术、应用软件开发工具等。

1.4 信息通信技术

信息通信技术（ICT）是信息技术的拓展，它是电信服务、信息服务、IT 服务及应用的有机结合，强调统一通信的作用以及电信通信（电话线、无线信号）、计算机、企业软件、中间件、存储、音视频系统等的集成，以帮助用户更有效地访问、存储、传输和操纵信息。

在过去的几十年中，信息通信技术为社会提供了大量新的通信能力。例如，位于不同国家的人们可以使用诸如即时消息、IP 语音（VoIP）和视频会议等技术实时通信。

现代信息和通信技术已经创建了一个"地球村"，人们可以在世界范围相互通信。

1.5 什么是计算机

计算机是一种电子设备，它是在存储在其内存中的指令控制之下工作，可以接收数据、根据指定的规则处理数据、输出结果、存储结果。

1.6 计算机组成

计算机的硬件组成包括系统部件、输入设备、输出设备、存储设备和通信设备。系统部件主要包括 CPU 和内存，输入设备将用户的数据和指令输入计算机，输出设备将计算机中的信息传递（输出）给用户，存储设备存储数据和指令，通信设备在一个或多个计算机间发送和接收数据、指令和信息。

1.6.1 硬件

硬件是构成计算机的电子电路和机械部件，具体包括系统部件、输入设备、输出设备、存储设备和通信设备。

1.6.1.1 系统部件

系统部件也被称为基本部件。它是个人计算机的主要部件，通常有一个金属或塑料（少数情况）外壳，内含主板、电源、散热风扇、内置磁盘驱动器、内存模块和扩展卡（例如视频卡、网卡），但不包括键盘、显示器等外部设备。

1.6.1.2 输入设备

输入设备是将数据或指令输入计算机中的硬件组件。常用的输入设备：键盘、鼠标、话筒（麦克风）、扫描仪、数码相机、计算机摄像头等。

1.6.1.3 输出设备

输出设备是将计算机中的信息传递（输出）给用户的硬件组件。常用的输出设备：显示器、打印机、扬声器（喇叭）等。

1.6.1.4 存储设备

存储设备用于存储数据，一般指大容量存储设备，如硬盘、软盘、光盘、U 盘、磁带等。

1.6.1.5 通信设备

通信设备是在一个计算机内或多个计算机间发送和接收数据、指令和信息的硬件组件，如调制解调器。

1.6.1.6 总线

系统部件、输入设备、输出设备、存储设备和通信设备通过总线互连。总线是计算机各部件之间传输数据和指令的物理通道，它以一种通用的方式为各部件提供数据传送和逻辑控制功能。

1）数据总线

数据总线在 CPU 与 RAM 之间传输需要处理或存储的数据。

2）地址总线

地址总线用来指定在内存中存储的数据的地址。

3）控制总线

控制总线在 CPU 与其他设备之间传输控制信息。

1.6.2 软件

计算机软件，或软件，是计算机程序和相关文档的集合，是控制计算机的数据和指令的集合。软件一般分为系统软件和应用软件。

1.6.2.1 系统软件

系统软件包括管理和控制硬件以使应用软件执行任务的程序。系统软件包括操作系统和实用程序。

1）操作系统

操作系统是计算机中最重要的程序。为了运行其他程序，每台通用计算机必须装有操作系统。操作系统的主要功能包括管理输入输出设备、管理磁盘上的文件和目录、管理磁盘驱动器和打印机等外围设备。

2）实用程序

实用程序（也称为服务程序、工具软件）是系统软件的一个组成部分，用来帮助用户更好地控制、管理和使用计算机硬件、操作系统和应用程序，执行备份和恢复、磁盘清理、磁盘碎片整理、计算机杀毒等任务。

1.6.2.2 应用软件

应用软件（也称为最终用户程序、终端用户程序）是指针对使用者的某种应用目的而开发的软件。它可以是一个特定的程序，如图像浏览器。也可以是一组功能联系紧密，可以互相协作的程序的集合，如微软的 Office 软件。也可以是一个由众多独立程序组成的庞大的软件系统，如数据库管理系统。

较常见的应用软件有数据库管理程序（如 Microsoft Access、Oracle、Microsoft SQL Server、IBM DB2）、文字处理软件（如 Microsoft Word）、电子表格软件（如 Microsoft Excel）和媒体播放软件（如 MediaPlayer、RealPlayer、暴风影音）等。

系统软件是应用软件的基础，如果没有操作系统和实用程序，应用软件将无法运行。

1.7 计算机的分类

- 巨型机；
- 大型机；
- 中型机；
- 小型机；
- 个人计算机。

1.7.1 巨型机

巨型机，又称超级计算机，指计算能力（尤其是计算速度）为世界顶尖的电子计算机。现有的超级计算机运算速度大都可以达到每秒万亿次以上。超级计算机是最快、最强劲、最贵的计算机。

1.7.2 大型机

大型机，又称大型计算机、大型主机、主机等，是从 IBM System/360 开始的一系列计算机及与其兼容或同等级的计算机，主要用于大量数据和关键项目的计算，例如银行金融交易及数据处理、人口普查、企业资源规划等等，可以同时支持成百上千用户的使用。

1.7.3 中型机

中型机是指功能介于大型计算机和个人计算机之间的计算机。中型机曾被称为小型机。

1.7.4 小型机

小型机是指运行原理类似于个人计算机和服务器，但性能及用途又与它们有所不同的一种高性能计算机。它是20世纪70年代由DEC（数字设备公司）公司首先开发的一种高性能计算产品。小型机的规模和能力介于工作站和大型机之间。

1.7.5 个人计算机

个人计算机（PC），或称个人电脑，是在大小、性能以及价位等多个方面适合个人使用，并由最终用户直接操控的计算机的统称。它与批处理计算机或分时系统等一般同时由多人操控的大型机相对。从桌面计算机、笔记本电脑到平板电脑等都属于个人计算机的范畴。个人计算机提供输入、处理、输出、存储等功能，并包含处理器、内存、输入设备、输出设备和存储设备。

1.7.5.1 工作站

工作站是一种高端的个人计算机，它是为了单用户使用并提供比个人计算机更加强大的性能，尤其是在图形处理能力、任务并行方面的能力。工作站通常连接局域网并装有多用户操作系统，常用于执行诸如计算机辅助设计、绘图建模、科学和工程计算、图像处理、动画和电影的视觉效果等任务。

工作站通常比较昂贵，价格一般是标准PC的数倍。

1.7.5.2 桌面计算机

桌面计算机（或称台式计算机、台式电脑）是指在办公室或家中办公桌上使用的计算机，不同于笔记本电脑或PDA等便携式电脑。台式电脑也被称为微型计算机、微机。

1.7.5.3 笔记本电脑

笔记本电脑（也称为手提电脑或膝上电脑）是一种小型（小到可以放置在膝上）、可以方便携带的个人电脑。当前的发展趋势是体积越来越小，重量越来越轻，而功能却越发强大。为了缩小体积，笔记本电脑通常拥有液晶显示器（液晶屏），现在新式的有触摸屏。除了键盘以外，有些还装有触控板（touchpad）或触控点（pointing stick）作为定位设备（pointing device）。

1.7.5.4 平板电脑

平板电脑是一种小型的、扁平式的、方便携带的个人电脑，以触摸屏作为基本的输入设备。其触摸屏允许用户通过触控笔或数字笔或手指而不是传统的键盘或鼠标来进行计算机操作。2010年，苹果公司通过推广iPad产品带动了平板电脑市场的发展。

1.7.5.5 手持电脑

手持电脑（也称为掌上电脑、手持设备、移动设备、移动手持设备、口袋电脑）是一个口袋大小的计算设备，通常有一个带触摸输入和/或微型键盘的显示屏幕。智能手机和PDA、电子书是深受欢迎的手持设备。

智能手机是一种移动电话，提供了比基本功能的手机更先进的计算能力和连接性。一个例子是诺基亚9210是第一台彩色屏幕的通信手机，是第一个真正的带有开放操作系统的智能手机。另一个例子是苹果公司推出的iPhone。

个人数码助理（PDA）是集计算、电话、传真、Internet和网络功能于一体的手持设备。PDA通常采用触控笔作为输入设备，而存储卡作为外部存储介质。在无线传输方面，大多数PDA具有红外和蓝牙接口，以保证无线传输的便利性。许多PDA还具备Wi-Fi连接以及GPS全球卫星定位系统。

电子书是指将文字、图片、声音、影像等内容数字化的出版物和植入或下载数字化文字、图片、声音、影像等内容的集存储和显示终端于一体的手持阅读器。代表人们所阅读的数字化出版物，区别于以纸张为载体的传统出版物。电子书通过数码方式记录在以光、电、磁为介质的设备中，必须借助于特定的设备来读取、复制和传输。

课文 2 系统部件

系统部件也被称为计算机基本部件,是个人计算机的主要部件,通常有一个金属或(少数情况)塑料外壳,包括机箱、微处理器、内存、总线和端口,但不包括键盘、显示器等其他外部设备。

1)主板

主板(又称主机板、系统板、逻辑板、母板、底板等),是微机的主电路板。典型的主板能提供一系列接合点。一般主板上接有 CPU、BIOS、内存、存储器接口、并行口、串行口、扩展槽以及控制显示屏、键盘和磁盘等标准外设的控制器。主板上最重要的构成组件是芯片组(Chipset)。这些芯片组为主板提供一个通用平台供不同设备连接,控制不同设备的沟通。它亦包含对不同扩充插槽的支持。一些高档主板还集成红外通信技术、蓝牙和 802.11(Wi-Fi)等功能。

2)计算机芯片

计算机芯片是一种嵌入集成电路的半导体材料(通常是硅),一般包含成百上千的电子部件(晶体管)。计算机就是由许多置于印刷电路板上的各类芯片构成。例如,CPU 芯片(又称为微处理器)就包含一个完整的处理单元,而内存芯片则包含空白的内存。

系统部件的两个主要组成是 CPU 和内存。

2.1 中央处理器

中央处理器(CPU),简称为处理器,是电子计算机的主要设备之一,其功能主要是解释计算机指令以及处理计算机中的数据。CPU、内存和输入/输出设备是计算机的三大核心部件。

由集成电路制造的 CPU,20 世纪 70 年代以前,本来是由多个独立单元构成,后来发展出微处理器 CPU 复杂的电路可以做成单一微小功能强大的单元。

CPU 包含三个部件:控制单元、算术逻辑单元和寄存器。

2.1.1 控制单元

控制单元用于从内存中提取、解释并执行指令。控制单元负责指挥 CPU 工作,控制单元发出控制信号,通过该信号对 CPU 各部件加以控制:

- 指导和协调计算机中的大部分指令的操作;

- 解释每条指令并对动作初始化。

2.1.2 算术逻辑单元

算术逻辑单元是 CPU 的执行单元，是所有 CPU 的核心组成部分，其主要功能是进行二进制的算术运算（加、减、乘等）和逻辑运算（与、或、非、异或等）。

2.1.3 寄存器

寄存器是 CPU 中有限存储容量的高速存储部件，可用来暂存指令、数据和地址。所有数据在处理前必须先存储在寄存器中。

寄存器主要有以下三种。
- 累加器：存储 ALU 运算所产生的中间结果。
- 程序计数器：也称为指令指针，存储即将要执行的下一条指令的地址。
- 指令寄存器：存储正在执行的指令。

2.1.4 CPU 的特性

2.1.4.1 指令周期

指令周期，也称为机器周期或读取-执行周期，是指 CPU 从内存中获取一条指令到执行此条指令所经历的步骤。每台计算机的指令集可能不同，因此指令周期也不同，但一般有以下四个阶段。

1）指令获取

CPU 内有程序计数器（PC），它储存了下一条将要执行的指令的地址。处理器按 PC 储存的地址，到内存中获取指令的内容，PC 加 1，经数据总线将指令存入指令寄存器（IR）中。

2）指令解码

指令译码器解释 IR 内的指令。如果指令有一个间接地址，则有效地址和所需数据从主存中读取，并存储于数据寄存器中。

3）指令执行

控制单元将译码信息作为控制信号序列传送给 CPU 相应的功能单元来执行指令，如读取寄存器中的值、将寄存器中的值传送给 ALU 以执行算术或逻辑运算、将结果写回寄存器等。ALU 接收到相应的信息后，会反馈一个条件信号给控制单元。

4）结果存储

操作所产生的结果可以存储在主存，或者发送到输出设备。根据 ALU 反馈的信息，程序计数器将更新为下一指令执行的地址。

此周期将重复。其中，步骤 1 和 2 称为取指周期。这些步骤对每个指令而言都是相同的。取指周期从指令字获取操作码和操作数并进行译码。步骤 3 和 4 称为执行周期。这些步骤因指令而异。执行周期首先是内存处理，数据在 CPU 和输入/输出模块之间传输；其次是数据处理阶段，对数据进行算术运算以及逻辑运算；然后是集中变更阶段，执行诸如跳转等操作序列；最后是对所有步骤进行综合。

2.1.4.2　流水线

流水线，亦称管线，是现代计算机处理器中必不可少的部分，是指将计算机指令处理过程拆分为多个步骤，并通过多个硬件处理单元并行执行来加快指令执行速度。CPU 在第一条指令完成之前即开始执行第二条指令。

2.1.4.3　RISC 和 CISC

指令集是指 CPU 可以执行的指令集合。CPU 的指令集可以是 RISC 或 CISC。

1）RISC（精简指令集计算）

精简指令集是计算机 CPU 的一种设计模式。这种设计思路对指令数目和寻址方式都做了精简，使其实现更容易，指令并行执行程度更好，编译器的效率更高。许多工作站和个人计算机采用了 RISC 方式。RISC 的主要特点：

- 指令数目少、程序较长；
- 每条指令都采用标准字长、执行时间短；
- 比 CISC 快。

2）CISC（复杂指令集计算）

复杂指令集是一种微处理器指令集架构，每个指令可执行若干低阶操作，诸如从内存读取、储存和计算操作，全部集于单一指令中。CISC 的特点：

- 指令数目多而复杂；
- 程序较短；
- 每条指令字长并不相等。

2.1.4.4　系统时钟

系统时钟是一块小型石英晶体电路，是 CPU 控制计算机操作的计时工具。时钟速度，也称为时钟频率，用于衡量微处理器执行指令的速度。每台计算机包含一个内部时钟，用于调控指令的执行速度、同步计算机各部件的执行。CPU 每执行一条指令，需要固定的时钟周期。时钟速度越快，CPU 每秒就可以执行更多的指令。时钟速度以兆赫兹（MHz）或吉赫兹（GHz）表示。总之，系统时钟：

- 同步计算机所有操作的时间；
- 时钟每个滴答（每个时钟周期）→执行一条指令（以前的计算机）；
- 当今的 CPU 每个时钟周期可以执行多条指令；
- 时钟速度的衡量单位为赫兹（即若干次周期每秒）。

2.1.4.5　字长

字是计算机数据的基本存储单位。字长（字的大小或字的长度）是指计算机在单位时间可以处理的二进制位的多少。不同的计算机字长也会不同。对于 16 位 CPU 的计算机，一个字是 16 位（2 字节）。大型机的字长可以高达 64 位（8 字节）。某些计算机和编程语言会区分 Shortwords 和 Longwords。Shortword 通常为 2 个字节长，而 Longword 是 4 个字节。

2.1.4.6 计算机的速度

计算机的速度主要根据以下参数来衡量：
- 字长；
- 总线大小；
- 时钟频率。

2.1.4.7 散热器和热导管

所谓的主动性散热、被动性散热，主要是以是否需要额外的驱动能源执行散热来分别，散热器是最典型的被动性散热元件，除此之外热导管（Heat Pipe）也是近年来日益普及与推崇的被动性散热元件，而主动式散热元件则有散热风扇（用马达、电力驱动）、水冷循环等。

因此，无风扇的散热器称为"被动性散热元件"，而有风扇的散热器称为"主动性散热元件"。

散热器以导热性佳、质轻、易加工的金属（多为铝或铜）贴附于发热表面，并采用薄而大块的鳍片状结构，以复合的热交换模式来散热。散热器是附在发热设备上的一层良好导热介质，有时在导热介质的基础上还会加上风扇等来起到加快散热的效果。

热导管（或称热管）是一种具有快速均温特性的特殊材料，其中空的金属管体使其具有质轻的特点，而其快速均温的特性则使其具有优异的热超导性能，是现今电子产品散热装置中最普遍高效的导热（非散热）元件。

2.1.4.8 并行处理

并行处理（或称并行计算、平行计算）是多个 CPU 同时执行一个程序的计算模式。理想的情况下，并行处理可以使得程序更快运行，因为有多个 CPU 同时运行程序。而实际上，往往很难将一个程序合理分割让不同的 CPU 执行不同的部分而不会彼此干扰。

2.2 内存

内存（又称为主存）是计算机内部的存储区域，是一种利用半导体技术做成的电子设备，用来存储处理器将要执行的指令、指令所需要的数据以及处理的结果（信息）。内存通常由主板或其他电路板上的一个或更多芯片构成。内存一般用来存储运算时的数据，一般速度很快。大多数内存（RAM）只是临时存储数据和指令，也有一小部分内存（ROM）是非易失性存储器。

2.2.1 计算机内存单元

计算机中的所有数据，数字、字符、控制信号、地址等都采用二进制的方式编码。二进制是逢二进位的进位制，0、1 是基本算符。

一个 bit（位）就是一个二进制位（0 或 1），bit 是 binary digit（二进制数比特）的缩写。bit 是计算机可以处理的数据的最小单位。

（1）Byte（字节）是内存的基本存储单位。一个字节由八位二进制组成（如 01010110 表示大写字母 T）。

- 存储器的每一个存储单元都有一个地址。
- 地址是标识字节在内存中位置的唯一编号。

（2）
- 1KB = 1024B = 2^{10} Bytes
- 1MB =1024KB = 2^{20} Bytes
- 1GB =1024MB = 2^{30} Bytes
- 1TB =1024GB = 2^{40} Bytes

（3）以下是三种最常用的二进制编码系统。
- ASCII（美国信息交换标准代码）是一个 8 位字符的编码方案，包括 128 个字符（请参见附录 A）：33 个非打印控制字符（现在大多已过时），一般用于控制文本和空格；95 个打印字符。ASCII 码是现今最通用的单字节编码系统。
- EBCDIC（扩展二进制编码的十进制交换码）是一个 8 位字符编码方案，由 IBM 开发。
- Unicode（统一码、万国码、单一码、标准万国码）是 Unicode 协会开发的一种字符编码标准，可以代表几乎世界上所有的书面语言。Unicode 字符有多种表示形式，包括 UTF-8（8 位）、UTF-16（16 位）和 UTF-32（32 位）。大多数 Windows 系统使用 UTF-16 形式。

2.2.2 内存的分类

系统部件包含两类内存：易失性内存和非易失性内存。

2.2.2.1 易失性内存

易失性内存是指当电源关掉后所存储的数据便会消失的存储器。所有的 RAM（除了用于 BIOS 的 CMOS）都是易失性内存。

2.2.2.2 非易失性内存

非易失性内存是指当电源关掉后所存储的数据不会消失的存储器。非易失性内存通常用于长期存储。ROM 是非易失性内存。

辅助存储器（硬盘、软盘、磁带和光盘等）都是非易失性存储器。

2.2.3 随机存储器

随机存取存储器（RAM）又称作"随机存储器"，是与 CPU 直接交换数据的内部存储器，主存主要是指 RAM。RAM 可以随时读写，而且速度很快，但当电源关闭时不能保留数据，通常作为操作系统或其他正在运行中的程序的临时数据存储媒介。主存是计算机内部最主要的内存，用来加载各式各样的程序与数据以供 CPU 直接使用。

RAM 可以分为动态随机存取存储器（DRAM）和静态随机存取存储器（SRAM）两大类。DRAM 需要定时刷新（数千次每秒），但由于具有较低的单位容量价格，扩展性也不错，所以被大量地采用作为系统的主存。而 SRAM 不需要刷新，因此具有快速访问的优点，但生产成本较昂贵，一个典型的应用是高速缓存。

2.2.4 高速缓存

高速缓存（Cache），其原始意义是指访问速度比一般 RAM 快的一种特殊 RAM。个人计算机中通常有两种类型的 Cache：内存高速缓存和硬盘高速缓存。

2.2.4.1 内存高速缓存

一般而言，内存高速缓存不像系统主存那样使用 DRAM 技术，而使用昂贵但较快速的 SRAM 技术。

当 CPU 处理数据时，它会先到 Cache 中去寻找，如果数据因之前的操作已经读取而被暂存其中，就不需要再从 RAM 中读取数据——由于 CPU 的运行速度一般比内存的读取速度快，而内存周期（访问内存所需要的时间）为数个时钟周期，因此若要访问内存，就必须等待数个 CPU 周期从而造成浪费。

提供"高速缓存"的目的是为了让数据访问的速度适应 CPU 的处理速度，其基本原理是内存中"程序执行与数据访问的局域性行为"，即一定程序执行时间和空间内，被访问的代码集中于一部分。为了充分发挥高速缓存的作用，不仅依靠"暂存刚刚访问过的数据"，还要使用硬件实现的指令预测与数据预取技术——尽可能把将要使用的数据预先从内存中取到高速缓存中。

现今计算机上使用的微处理器都在芯片内部集成了大小不等的数据高速缓存和指令高速缓存，通称为 L1 高速缓存（L1 Cache）。例如，Intel 80486 微处理器有一个 8K 的 L1 Cache，Pentium 则有一个 16K 的 L1 Cache；而比 L1 更大容量的 L2 高速缓存曾经被放在 CPU 外部（主板或者 CPU 接口卡上），但是现在已经成为 CPU 内部的标准组件；更昂贵的顶级家用、工作站、服务器的 CPU 甚至会配备比 L2 高速缓存还要大的 L3 高速缓存。内存 Cache 位于计算机 CPU 和 DRAM 之间。

（1）内存 Cache 有助于提高计算机的处理速度，因为其存储的是 CPU 经常使用的指令和数据。

（2）有三类内存 Cache：L1、L2 和 L3。

（3）当处理器需要指令和数据时，搜索顺序如下：
- L1；
- L2；
- L3；
- RAM；
- Hard disk/CD/DVD。

2.2.4.2 磁盘高速缓存

磁盘高速缓存是一种用来加快磁盘访问的高速 RAM，可以是磁盘驱动器本身的一部分（有时称为硬盘缓存或缓冲），也可以是计算机中为磁盘预留的 RAM（有时称为软盘高速缓存）。硬盘缓存更加高效，但也昂贵得多，因此容量较小。所有当今的磁盘几乎都包括内部缓存。

现在高速缓存的概念已被扩充，不仅在 CPU 和主内存之间有 Cache（磁盘高速缓存），而且在内存和硬盘之间也有 Cache，乃至在硬盘与网络之间也有某种意义上的 Cache——Internet 临时文件夹——凡是位于速度相差较大的两种硬件之间的、用于协调两者数据传输速度差异的结构，均可称为 Cache。

2.2.5 只读存储器

只读存储器（ROM）是一种半导体存储器，其特性是一旦存储数据就无法再将之改变或删除，即数据只能读，不能写入和修改，且其内容不会因为电源关闭而消失，通常用于存储不需经常变更的程序或数据，如 BIOS 和启动程序。

2.2.5.1 固件

固件（Firmware）是一种嵌入在硬件设备（写入 ROM）中的软件（程序或数据）。固件位于软件和硬件之间，是软件和硬件的合成品。通常它是位于特殊应用集成电路（ASIC）或可编程逻辑器件（PLD）之中的闪存或 EEPROM 或 PROM 里，有的可以让用户更新。固件广泛应用在各种电子产品中，从计算器，到计算机中的键盘、硬盘、内存卡，甚至科学计算和工业机器人。日常生活中的移动电话、数码相机、电视遥控器等，均包含固件以执行设备的基本操作和高级功能。

2.2.5.2 可编程只读存储器

可编程只读存储器（PROM）是可以存储程序的内存芯片。PROM 刚出厂时是空白的，可以使用 PROM 编程器写入数据。但是，只要数据写入 PROM，就不可擦除和更改。和 ROM 一样，PROM 也是非易失性存储器。

2.2.5.3 电子抹除式可复写只读存储器

电子抹除式可复写只读存储器（EEPROM，或称 E^2PROM）使用特定的电压来擦除芯片上的信息，以便写入新的数据。类似于 PROM，EEPROM 在断电时数据仍保留。EEPROM

和 ROM 类似，速度均比 RAM 慢。EEPROM 类似于闪存，因此又被称为 flash EEPROM。两者区别在于，EEPROM 一次只写入或擦除 1 个字节，而闪存则以块为单位写入或擦除数据，所以闪存比 EEPROM 快。

由于 EEPROM 的优秀性能，以及联机操作的便利性，它被广泛用于需要经常擦除的 BIOS 芯片以及闪存芯片，并逐步替代部分有断电保留需要的 RAM 芯片，甚至取代部分硬盘功能。它与高速 RAM 成为当前最常用且发展最快的两种存储技术。

2.2.5.4 互补式金属氧化物半导体（CMOS）

互补式金属氧化物半导体（CMOS）是一种集成电路半导体，可在硅晶圆上制作出 PMOS（正极）和 NMOS（负极）元件，由于 PMOS 与 NMOS 在特性上为互补性，因此称为 CMOS。此半导体可用来制作微处理器（Microprocessor）、微控制器（Microcontroller）、静态随机存取内存（SRAM）以及其他数位逻辑电路。

CMOS 具有只有在晶体管需要切换启闭时才需耗能的优点，非常省电且发热少，因此特别适合电池供电的设备，如便携式计算机。个人计算机中有一个电池供电的 CMOS，存储系统日期、时间、启动参数等。即 CMOS：

- 存储计算机的启动信息；
- 存储计算机的配置信息（如磁盘驱动器类型、键盘、显示器等）。

2.2.6 虚拟内存

虚拟内存是为多任务内核提供的内存管理技术。它使得应用程序认为它拥有连续的可用的内存（一个连续完整的地址空间），而实际上，它通常是被分隔成多个物理内存碎片，还有部分暂时存储在外部磁盘存储器上，在需要时进行数据交换。与没有使用虚拟内存技术的系统相比，使用这种技术的系统使得大型程序的编写变得更容易，对真正的物理内存（RAM）的使用也更有效率。

对虚拟内存的定义是基于对地址空间的重定义的，即把地址空间定义为"连续的虚拟内存地址"，借此"欺骗"程序，使它们以为自己正在使用一大块的"连续"地址。

- 虚拟内存使用硬盘来模拟内存。
- 虚拟内存的目的是扩大寻址空间，增加程序可用的地址集合。
- 为方便实现虚拟内存，操作系统将虚拟内存划分成若干页面，每个页面大小固定，存储在硬盘上。当需要页面时，操作系统将其从硬盘复制到内存，将虚拟地址转换为真实地址。
- 将虚拟内存转换为真实地址的过程称为映射；将虚拟页面从硬盘复制到内存的过程称为页面调度或交换。

2.3 端口和连接器

外部设备通常通过端口连接到计算机。端口有不同类型的连接器。连接器通过电缆与外围设备连接。

端口是计算机与其他计算机或设备之间的物理接口。计算机有各种不同类型的端口，用于连接磁盘驱动器、显示屏、键盘、调制解调器、打印机、鼠标和其他外围设备。

连接器是插入端口或接口以连接设备的媒介。有棒状或铜板状突出的是公接头，有插槽或凹洞的是母接头。

2.3.1 串行口

串行接口又称"串口"，主要用于串行式逐位（一次一个 bit）数据传输。常见的有一般计算机应用的 RS-232（使用 25 针或 9 针连接器）和工业计算机应用的半双工 RS-485 与全双工 RS-422。

串行接口按电气标准及协议来分，包括 RS-232-C、RS-422、RS-485、USB 等。
- COM 端口类型就是一种串口；
- 串口通常连接鼠标、键盘和调制解调器。

消费性电子产品已经由 USB 取代串口；但在非消费性用途，如网络设备等，串口仍是主要的传输控制方式。

2.3.2 并行口

并行端口，又称并口，是计算机上数据以并行方式（即一次多于一个 bit）传递的端口，也就是说至少应该有两条连接线用于传递数据。IEEE 1284 定义了端口双向传输的标准，允许端口同时既可以传输数据，又可以接收数据。与只使用一根线传递数据（不包括用于接地、控制等的连接线）的串行端口相比，并口在相同的数据传送速率下，可以更快地传输数据。所以在 21 世纪之前，在需要较大传输速度的地方，例如打印机，并口得到广泛使用。但是随着速度迅速提高，并口上导线之间数据同步成为一个很难处理的难题，导致并口在速度竞赛中逐渐被淘汰。目前 USB 等改进的串口逐渐代替了并口。
- LPT 端口类型就是一种并口；
- 并口通常连接打印机。

2.3.3 小型计算机系统接口

小型计算机系统接口（SCSI）是一种特殊的高速并行端口。SCSI 通常应用于硬盘、磁带和打印机，还可以连接诸如扫描仪和光盘驱动器等其他设备。

2.3.4 USB 接口

通用串行总线（USB）是连接计算机系统与外设的一个串口总线标准，也是一种输入输出接口技术规范，被广泛应用于个人计算机和移动设备等信息通信产品，并扩展至摄影器材、数字电视（机顶盒）、游戏机等其他相关领域。USB 速度比并行端口（例如 LPT）

与串行接口（例如 RS-232）等传统计算机用标准总线快许多。

USB 已有效地取代了很多串行接口和并行接口。一个单独的 USB 接口可用于连接多达 127 个外设，如鼠标、调制解调器、键盘、数码相机、打印机、媒体播放器、闪存、网络适配器和硬盘等。USB 支持热插拔和即插即用。

2.3.5 PC 卡接口

PC 卡（旧称 PCMCIA）插槽也被称为 PCMCIA 接口。PC 卡是可插入笔记本电脑并提供额外功能的电子产品。PC 卡是一种通用外设接口，不同类型和规模的 PC 卡能让外存、内存、传真、调制解调器、网络设备、声卡等外设与笔记本电脑连接。

最初是由"个人电脑储存卡国际联盟"（Personal Computer Memory Card International Association）设定了这种卡的标准，故命名为"PCMCIA"卡，根据它的首字母缩略。之后，此种接口卡的新版本称为"CardBus"。

第一种 PC 卡是"一型接口卡"（Type I），可支持记忆卡（例如 ATA 一型快闪存储卡）；第二种 PC 卡（"二型接口卡"，Type II）则增加了输入输出支持；第三种 PC 卡（"三型接口卡"，Type III）在此基础上进一步扩展。输入输出功能已经不止能支持存储卡，这一功能衍生出了一种快闪存储卡（依照 Type I 的标准）——CF 卡、Mini 卡和 SSFDC 卡（SM 卡）。

PC 卡已经发展出新一代 ExpressCard（曾命名为"新卡"）来代替现有的 CardBus 标准。ExpressCard 比 CardBus 更快更简单。通过 ExpressCard 插槽，主机可提供 PCI Express 和 USB 2.0 速度的连接。它支持热插拔。

课文 3 输入输出

输入/输出，或 I/O，通常指信号或数据在内存和外存或其他外设之间的输入和输出。输入是指输入设备将信号或数据发送到系统中，而输出则是系统将信号和数据发送到输出设备。

3.1 输入

3.1.1 输入设备

输入设备是将外界信息送入计算机内部的硬件。通常，用户通过输入设备输入数据到计算机；计算机处理输入的数据；通过输出设备将处理后的数据（信息）反馈给用户。输入设备和输出设备是计算机和用户之间的硬件接口。

（1）输入设备提供了如下功能：
- 输入数据；
- 将数据转换成计算机内的格式。

（2）常用的输入设备：键盘、鼠标、手写笔、数码笔、话筒、数码相机和扫描仪等。

3.1.2 键盘

计算机键盘是打字机式的按键设备，用户可以通过键盘将数据输入计算机中。计算机键盘上通常有如下几种。
- 字母数字键：字母和数字。
- 标点符号键：逗号、句号、分号等。
- 特殊键：功能键、控制键、方向键、大小写锁定键等。

3.1.3 定位设备

定位设备是一种输入设备，用户通过控制指针移动来选择显示屏上的操作对象。鼠标、轨迹球、触摸板、控制杆、触摸屏和光笔等都属于定位设备。

3.1.3.1 鼠标

鼠标是一种很常用的计算机输入设备,它可以对屏幕上的光标或指针进行定位,并通过按键和滚轮装置对光标或者指针所经过位置的屏幕元素进行操作。

鼠标有以下三种基本类型。
- 机械鼠标:在其底部有个橡胶或金属球,可以沿各个方向滚动。机械传感器检测鼠标球滚动的方向并相应地移动屏幕上的指针。
- 光电机械鼠标:与机械鼠标类似,但它使用光学传感器来检测鼠标球的运动。需要使用鼠标垫。
- 光电鼠标:光电鼠标是通过红外线或激光检测鼠标器的位移,将位移信号转换为电脉冲信号,再通过程序的处理和转换来控制屏幕上光标的移动。光电鼠标的光电传感器取代了传统的滚球。这类鼠标的反应比机械及光学鼠标更迅捷,更准确。

3.1.3.2 轨迹球

轨迹球,又叫跟踪球。和机械鼠标一样,轨迹球通过读取可滚动球的滚动方向和速度来定位。不同的是,机械鼠标是基座和球一起动,而轨迹球只是球在基座上滚动,基座相对桌面不动。这就减小了轨迹球所用的空间,而且用户可以将轨迹球置于任何表面,包括膝上。因此,轨迹球很受便携式计算机等设备的青睐。

3.1.3.3 触摸板

触摸板,也称作触控板,是一种广泛应用于笔记本电脑上的输入设备。利用用户手指的移动来控制指针的动作。触摸板可以视作是一种鼠标的替代物。在其他一些便携式设备上,如个人数码助理(PDA)与一些便携影音设备上也能找到触摸板。苹果公司设计和销售的便携式数码多媒体播放器 iPod 就是由中央滚轮(又称为点拨轮)操作的。

3.1.3.4 控制杆和方向盘

控制杆是一种输入设备,由基座和固定在上面作为枢轴的主控制杆组成,作用是向其控制的设备传递角度或方向信号。控制杆主要用来操纵电子游戏,通常有一个或多个按钮,按钮的状态也可被计算机识别。

方向盘是一个基于轴旋转的圆形装置,是现代汽车的一个重要组成部件。通过控制方向盘,可以控制设备的行进方向。

控制杆和方向盘常用于计算机游戏,也可用于 CAD/CAM 系统以及其他应用。

3.1.3.5 触摸屏

触摸屏,又称为触控屏幕、触控面板、轻触式屏幕,是可接收触头(无论是手指或胶笔尖等)等输入讯号的感应式液晶显示装置。当接触了屏幕上的图形按钮时,屏幕上的触觉反馈系统可根据预先编程的程序来驱动各种连接装置,可用于取代机械式的按钮面板,并借由液晶显示画面制造出生动的影音效果。触控屏幕的用途非常广泛,从常见的 PDA、提款机,到工业用的触控计算机。

3.1.3.6 光笔

光笔是一种输入设备，利用光敏感的探测器来选择屏幕上的对象。光笔类似于鼠标，但用户可以用光笔直接指向屏幕上的对象来选择操作对象。

3.1.3.7 手写笔或数码笔

手写笔或数码笔是电子书写器具，属于定位设备。看起来像一支圆珠笔，但使用压力而不是墨水来书写文字和画线。对于不喜欢使用键盘或者不习惯使用中文输入法的人来说非常有用。

3.1.3.8 指点杆

指点杆（又称触控点）是一种主要应用于笔记本电脑的定位设备，通常位于QWERTY键盘上的按键G、H、B之间，装有一个可替换的橡胶帽以提高其可操控性。ThinkPad用户群间有时戏称指点杆为"小红点"或"中原一点红"。大多数指点杆是压力敏感的，压力越大，指针移动得越快。

3.1.4 语音输入

语音输入通过麦克风将数据输入计算机中进行处理。

3.1.4.1 语音识别

语音识别技术，也被称为自动语音识别（Automatic Speech Recognition，ASR），其目标是将人类语音中的词汇内容转换为计算机可识别的输入，例如按键、二进制编码或者字符序列。注意语音识别只是听写功能而已，不能理解所识别的内容。

语音识别技术的应用包括语音拨号、语音导航、室内设备控制、语音文档检索、简单的听写数据录入等。语音识别技术与其他自然语言处理技术如机器翻译及语音合成技术相结合，可以构建出更加复杂的应用，例如语音到语音的翻译。

3.1.4.2 音频输入

音频输入是将语音、音乐和声效等任何声音输入计算机中的过程。

3.1.5 数码相机

数码相机也叫数字式相机，是集光学、机械、电子一体化的产品。它集成了影像信息的转换、存储和传输等部件，具有数字化存取模式、与计算机交互处理和实时拍摄等特点。传统照相机光线通过镜头、在底片上靠溴化银的化学变化来记录图像；数码相机是一种利用电子传感器把光学影像转换成电子数据的照相机，其传感器是一种光感应式的CCD或CMOS，用来取代底片的化学感光功能。在图像传输到计算机以前，通常会先储存在相机内部数码存储设备中（通常是存储卡）。

3.1.6 视频输入

视频输入将捕获到的全动态图像输入计算机中,并存储在硬盘或 DVD 等存储介质上。

3.1.6.1 数码摄像机

数码摄像机(又称数字摄像机)是指摄像机的图像处理及信号的记录全部使用数字信号(而非模拟信号)完成的摄像机。

3.1.6.2 网络摄像头

网络摄像头是传统摄像机与网络视频技术相结合的新一代产品,除了具备一般传统摄像机所有的图像捕捉功能外,还内置了数字化压缩控制器和基于 Web 的操作系统,使得视频数据经压缩加密后,通过万维网、即时消息或视频呼叫应用送至终端用户。而远端用户可在 PC 上使用标准的网络浏览器(如 Microsoft IE、Firefox),根据网络摄像头的 IP 地址,对网络摄像头进行访问,实时监控目标现场的情况,并可对图像资料实时编辑和存储。

3.1.6.3 视频会议

视频会议(也被称为视频远程会议)是一种交互通信技术,允许位于两个或多个地点的多个用户之间通过同步双向视频和音频传输进行交互。也被称为可视化协作,是一个组件系统。视频会议不同于电视电话,是为整个会议而非个人提供服务。

3.1.7 扫描和识别设备

3.1.7.1 光学扫描仪

光学扫描仪,通常称为扫描仪,是一种感光的输入设备,将图像、文本或实物扫描并转换为数字图像后输入计算机中,并形成文件保存起来。光学扫描仪将扫描结果表示为位图图像。因此,用户不能直接编辑扫描的文本,而需要光学字符识别(OCR)系统将其转换为文本字符。现今的大多光学扫描仪都附带 OCR 功能。

3.1.7.2 光学阅读器

光学阅读器是一种使用光源读取字符、标记和代码,然后将其转换成计算机可以处理的数字数据的设备。

两种使用光学阅读器技术的分别是光学字符识别和光学标记识别。

1)光学字符识别(OCR)

光学字符识别是一种将图像或手写的或印刷文本(通常由扫描仪获取)转化为计算机可编辑的本文或字符编码(ASCII、Unicode)格式的技术。光学字符识别主要应用于模式识别、人工智能和计算机视觉等研究领域。

2）光学标记识别（OMR）

光学标记识别，又称为光标阅读机、光学标记阅读机、光电阅读器、读卡机，是用光学扫描的方法对复选框、填空区域等按一定格式印刷或书写的标记进行电子读取和识别的技术。作为标准化数据专用计算机输入设备，这项技术非常适合大量手工填写表格又需要迅速准确处理的应用，如考试、调查、统计、选举、测评等。除了光标阅读机，还需要答题卡（选择卡）和填涂用笔（一般是 2B 铅笔，欧洲使用 HB）。学生在选择题考试的标准答题卡上用 2B 铅笔将答案或个人信息等涂黑，然后答题卡交由光标阅读机（阅卷机）自动阅卷。

光标阅读机是利用光电转换的原理，首先将信息卡上信息点的光信号转换为电信号，经模/数（A/D）转换，把电信号（模拟信号）变为数字信号，再利用数字滤波、格式预制、对比筛选等一系列技术，完成由涂点到符号的转化过程，同时也完成了计算机对数据录入的需求。

3.1.7.3 条码扫描仪

条码扫描仪，又称为条码扫描器、条码阅读器、条码扫描枪、条形码扫描器、条形码扫描枪或条形码阅读器，是一种光学阅读器，使用激光束来读取条码。条码是一种识别码，包含不同间隔的垂直线。条码中包含商品的名称、价格、制造商等相关信息。

3.1.7.4 磁墨水字符识别读写器

磁墨水字符识别读写器（MICR）是一种把 MICR 字符转换成计算机可以处理的格式的设备。MICR 是一种字符识别技术。银行业使用 MICR 处理银行支票。与条形码不同的是，MICR 码比较易懂。MICR 提供了一种安全的、高速的信息扫描和处理方法。

3.1.8 终端

终端是用户与计算机进行通信的设备。一般来说，终端是一个键盘和显示屏的组合。终端功能一般限于数据的输入和显示。具有可编程数据处理能力的终端称为"智能终端"或"胖客户端"。处理能力取决于主机的终端称为"瘦客户端"。个人计算机可以运行软件来模拟终端的功能，还允许本地程序访问远程终端主机系统。

3.1.8.1 POS 机

POS 机是一种广泛应用在零售业、餐饮业、旅馆等行业的电子系统，主要功能在于统计商品的销售、库存与顾客购买行为。借助 POS 系统，业者可以有效提升经营效率。POS 系统可以说是现代零售业界经营上不可或缺的必要工具。POS 系统除了计算机软件外，通常要具备下列的硬件设备：收银机、计算机主机、条码扫描仪、磁卡阅读器、打印机、客户显示器等，此外不同的零售业者为了管理的方便也会个别采用许多不同的装置，例如：

PDA 或是其他特殊规格的手持式装置，通常也具备网络以随时传输信息至企业总部。

3.1.8.2 自动取款机

自动取款机（ATM），又名自动提款机或自动柜员机，是银行设置的一种小型机器，利用一张信用卡大小的胶卡上的磁带或芯片卡上的芯片来记录卡号、截止期等基本账户信息，通过这些银行卡（或称金融卡或提款卡），客户可以提款、存款、转账等。为保障账户安全，客户必须输入个人识别码（自动提款机提款磁卡的个人密码）。

3.1.8.3 智能显示

智能显示是微软首创的、通过 Wi-Fi 连接的、作为瘦客户端的便携式触摸液晶显示屏。

3.1.9 生物识别输入

3.1.9.1 生物测定学

生物测定学，也称生物计量学、生物识别技术，原指用数理统计方法对生物进行分析，现在多指对生物体（一般特指人）本身的生物特征来区分生物体个体的计算机技术。研究领域主要包括脸、指纹、手掌纹、虹膜、视网膜、体形、个人习惯（例如敲击键盘的力度和频率、签字）、语音等，相应的识别技术就有人脸识别、指纹识别、掌纹识别、虹膜识别、视网膜识别、体形识别、键盘敲击识别、签字识别、说话人识别等。

3.1.9.2 生物识别设备

生物识别设备（又称生物特征辨识设备）是一种将生物体（一般特指人）本身的生物特征转换成数字代码并与存储在计算机中的数字代码进行比对的设备。

3.2 输出

3.2.1 输出设备

输出设备是人与计算机交互的一种硬件，用于数据的输出。它把各种计算结果数据或信息以数字、字符、图像、声音等形式表示出来。常见的有显示器、打印机、扬声器、耳麦/耳机等。

3.2.1.1 显示屏和显示器

显示屏，也称为视频显示单元、视频显示装置、终端、屏幕，是指计算机显示器用于显示图像及色彩的屏幕。而显示器，也称为监视器，是指整个显示器，包括底座、边框支架等。

3.2.1.2 显示器的分类

显示器有各种分类方法。

（1）根据色彩显示能力，显示器可分为以下三类。

- 单色：实际上显示两种颜色，一个是背景色，另一个是前景色。颜色可以是黑色和白色、绿色和黑色或者黄色和黑色。
- 灰度：是一种特殊的单色显示器，显示各种深浅不同的灰色。
- 彩色：可以显示多达 2^{32} 种不同的彩色。彩色显示器有时被称为 RGB 显示器，因为它们有三原色——红色、绿色和蓝色。

（2）显示器还可以根据屏幕大小来分类。屏幕尺寸依屏幕对角线计算，通常以英寸（inch）作单位，现时一般主流尺寸有 17"、19"、21"、22"、24"等。小型 VGA 显示器的尺寸是 14"。16"及以上的显示器通常被称为全页显示器。显示器根据屏幕大小还可以分为纵向（高度大于宽度）或横向（宽度大于高度）。大的横向显示器可以并排显示两个完整的网页。

（3）另一种常用的分类方式是根据显示器所接受的信号分类：模拟显示器或数字显示器。几乎所有现代显示器均接受模拟信号，这是 VGA、SVGA 和其他高分辨率彩色显示的标准要求。

3.2.1.3 显示屏的分类

1）阴极射线管显示器（CRT）

CRT 是桌面显示器，它类似于一个标准的电视，包含一个阴极射线管。它是利用阴极电子枪发射电子，在阳极高压的作用下，射向荧光屏，使荧光粉发光，同时电子束在偏转磁场的作用下，做上下左右的移动来达到扫描的目的。早期的 CRT 技术仅能显示光线的强弱，展现黑白画面。而彩色 CRT 具有红色、绿色和蓝色三支电子枪，三支电子枪同时发射电子打在屏幕玻璃的磷化物上来显示颜色。由于它笨重、耗电，所以在部分领域正在被轻巧、省电的液晶显示器取代。

2）液晶显示器（LCD）

液晶显示器是一种采用了液晶控制透光度技术来实现色彩的显示器。和 CRT 显示器相比，LCD 的优点是很明显的。由于通过控制是否透光来控制亮和暗，当色彩不变时，液晶也保持不变，这样就无须考虑刷新率的问题。画面稳定、清晰、无闪烁感，刷新率不高。LCD 显示器还通过液晶控制透光度的技术原理让底版整体发光，所以它做到了真正的完全平面。一些高档的数字 LCD 显示器采用了数字方式传输数据、显示图像，这样就不会产生由于显卡造成的色彩偏差或损失。完全没有辐射的优点，即使长时间观看 LCD 显示器屏幕也不会对眼睛造成很大伤害。体积小、能耗低也是 CRT 显示器无法比拟的，一般一台 15 英寸 LCD 显示器的耗电量也就相当于 17 英寸纯平 CRT 显示器的三分之一。

液晶的物理特性是：当通电时导通，排列变得有秩序，使光线容易通过；不通电时排列混乱，阻止光线通过。从技术上简单地说，液晶面板包含了两片相当精致的无钠玻璃素材，中间夹着一层液晶。当光束通过这层液晶时，液晶本身会排列站立或扭转呈不规则状，因而阻隔或使光束顺利通过。

液晶显示技术也存在弱点和技术瓶颈，与 CRT 显示器相比，亮度、画面均匀度、可视角度和反应时间上都存在明显的差距。其中反应时间和可视角度均取决于液晶面板的质量，画面均匀度和辅助光学模块有很大关系。

3）等离子显示器（Plasma）

等离子显示器（Plasma Display Panel，简称等离子）是一种平面显示屏幕，光线由两块玻璃之间的离子，射向磷质而发出。放出的气体并无水银成分，而是使用惰性气体氖及氙混合而成，这种气体是无害气体。

等离子显示器甚为光亮，可显示更多种颜色，也可制造出较大面积的显示屏，对角可达 381 厘米（150 英寸）。等离子显示屏的对比度也很高，可制造出全黑效果，观看电影尤为适合。

等离子的发光原理是在真空玻璃管中注入惰性气体或水银蒸气，加电压之后，使气体产生等离子效应，放出紫外线（Ultraviolet），激发荧光粉而产生可见光，利用激发时间的长短来产生不同的亮度。等离子显示屏中，每一个像素都是三个不同颜色（三原色）的等离子发光体所产生的。由于它是每个独立的发光体在同一时间一次点亮的，所以特别清晰鲜明。

3.2.2 打印机

打印机是一种输出设备，可在纸张或者透明胶片上输出文字或图形。

（1）根据所使用的技术，打印机可分为以下几种。

- 菊轮：打印的方式与打字机相同，即通过用铸造的字符撞击墨带（色带）来打印字符，但不能打印图形。噪音很大。
- 点阵：依靠一组像素或点的矩阵组合而形成更大的图像，点阵式打印机运用了击式打印机的原理，用一组小针来产生精确的点，它不但可以打印文本，还可以打印图形，相对而言便宜、速度快、比较通用。但是噪音也比较大、打印文本的质量通常要低于采用单独字模的击打式打印机。
- 喷墨：喷墨打印机可以把数量众多的微小墨滴精确地喷射在要打印的媒介上，对于彩色打印机包括照片打印机来说，喷墨方式是主流。由于喷墨打印机可以不仅局限于三种颜色的墨水，现在已有六色甚至七色墨盒的喷墨打印机，其颜色范围早已超出了传统 CMYK 的局限，也超过了四色印刷的效果，印出来的照片已经可以媲美传统冲洗的相片，甚至有防水特性的墨水上市。喷墨打印机比点阵打印机打印质量要好得多，而且噪音也小得多。喷墨打印机的价格一般比激光打印机便宜但打印速度慢。
- 激光：利用激光束扫描硒鼓（也称为感光鼓），通过控制激光束的开与关使传感光鼓吸与不吸墨粉，感光鼓再把吸附的墨粉转印到纸上而形成打印结果。打印速度快、打印质量高、工作噪声小。大多数激光打印机提供多种字体供选择，但价格比其他的打印机要贵得多。
- LCD 和 LED：类似于激光打印机，但使用液晶或发光二极管产生图像。

- 行式打印机：一次可以打印一整行的文字。这种打印机是所有击打式打印机中打印速度最快的，过去常用于大型计算机中心的海量打印。一个快速的行式打印机每分钟可打印多达 3000 行。缺点是无法打印图形、打印质量不高、噪声大。它从未用于个人计算机用户，现在已经被高速的激光打印机所取代。
- 热敏打印机（热感式打印机）：将印字头加热，再运用热度与停留时间来促使热敏纸显示出不同深浅的颜色。相对于针式打印机，热敏打印速度快、价格较低廉、噪音低、打印清晰、使用方便，但缺点是只能使用专用的热敏纸，热敏纸不耐光线照射，易造成纸上条码褪色，影响辨识率。热敏打印技术最早使用在传真机上。目前，热敏打印机已在 POS 终端系统、银行系统、医疗仪器等领域得到广泛应用。
- 照片打印机：照片打印机是一种彩色打印机，专门用于在相纸上打印高质量的数码照片。这些打印机通常有数量众多的喷嘴，可以达到 1 微微升/墨滴数的打印效果。

(2) 打印机也可以按以下特点分类。

- 打印质量：根据打印机产生的输出可以分为高品质（信函质量）、近高品质、草稿质量打印。菊花轮、喷墨和激光打印机可以产生高品质打印。
- 速度：以每秒打印的字符数（cps）或每分钟打印的页数（ppm）衡量。打印机的速度差别很大。菊花轮打印机通常最慢，约 30 cps。行式打印机较快（每分钟高达 3000 行）。点阵打印机速度可达 500cps，高档的激光打印机每分钟可以打印 30 页、40 页甚至更多的文档。
- 击打式或非击打式：击打式打印机通过打印头击打色带产生输出，菊花轮、点阵式、行式打印机是击打式打印机。非击打式打印机的印刷设备不接触纸张，热敏打印机、激光打印机和喷墨打印机属于非击打式打印机。击打式和非击打式打印机之间一个比较大的区别是击打式打印机噪声比较大。
- 图形：有些打印机（如菊轮和行式打印机）只能打印文本，有些打印机（如喷墨和激光打印机）同时可以打印文本和图形。
- 字体：某些打印机，特别是点阵打印机，仅提供一种或几种字体。而激光和喷墨打印机则提供丰富的字体。

3.2.3 扬声器和耳机

扬声器，俗称喇叭，是一种将电子信号转换成声音的换能器。作为将电能转变为声能的器材，扬声器的品质、特性，对整个音响系统的音质起着决定性作用。

耳机（又称耳筒、听筒），是一对转换单元，它接受媒体播放器或接收器所发出的电讯号，利用贴近耳朵的扬声器将其转化成可以听到的音波。耳机一般是与媒体播放器可分离的，利用一个插头连接。好处是在不影响旁人的情况下，可独自聆听音响；亦可隔开周围环境的声响，对在录音室、DJ 等在嘈杂环境下工作的人很有帮助。耳机原是给电话和无线电上使用的，但随着可携式电子装置的盛行，耳机多用于音频放大器、收音机、CD 播放器、手机、便携式媒体播放机等，也可用于计算机和 Hi-Fi 音响。

耳塞是把发音单元塞在耳孔之外，由于细小、轻巧、制作简单且造价廉宜，近年已成便携式音源的主流配备。耳麦则兼有耳机和麦克风的功能。

3.2.4 其他输出设备

3.2.4.1 传真机

传真机是一种使用电话网络传送文件复印本的电讯技术。传真机其实是影像扫描仪、调制解调器及计算机打印机的一种合体，扫描仪把文件的内容转化成数码影像，调制解调器则把影像资料透过电话线传送，在另一端的打印机则把影像变成原文件的复印本。现今在各企业中，传真功能多与扫描、打印、影印等功能整合至多功能打印机（或多功能事务机）中。家庭传真机的用途则由电子邮件（配合数码相机或扫描仪）取代。

3.2.4.2 绘图仪

绘图仪是一种优秀的输出设备。与打印机不同，打印机是用来打印文字和简单图形的。要想精确地绘图，如绘制工程中的各种图纸，最好用专业的绘图设备——绘图仪。彩色绘图仪使用不同的彩笔绘制不同的颜色。

3.2.4.3 数码投影仪

数码投影仪或数位投影仪，是一种可以通过不同的接口同计算机、VCD、DVD、游戏机、DV 相连接，在屏幕上播放相应的文本、图像以及视频的设备。广泛用于会议演示、课堂培训、家庭影院和现场活动。

3.2.4.4 高清电视

高清晰度电视（高清电视）是指传送的电视信号所达到的分辨率高于传统电视信号（NTSC、PAL）所允许的范围。其水平和垂直清晰度至少是常规电视的两倍，并且配有多路环绕声。高清电视显示屏幕纵横比为 16∶9。完整的高清数字电视体系包括高清电视节目源、高清机顶盒、高清电视机和必要的传输网络。

3.2.4.5 多功能外设

多功能外设（又称多功能辅助设备、多功能事务机、多功能设备、多功能一体机）是一个看起来像复印机，但同时提供了打印机、扫描仪、复印机，甚至传真机功能的设备。多功能外设日益成为 SOHO（小型办公室/家庭办公室）用户的青睐。

3.2.5 重要概念

3.2.5.1 像素

图像的每个点称为像素，又称画素，为图像显示的基本单位，译自英文 Pixel，Pix 是英语单词 Picture 的常用简写，加上英语单词"元素"Element，就得到 Pixel，故"像素"表示"图像元素"之意，有时亦被称为 Pel（Picture Element）。每个这样的信息元素不是一个点或者一个方块，而是一个抽象的采样。每个像素可有各自的颜色值，可用三原色显示，因而又分成红、绿、蓝三种子像素（RGB 色域），或者青、品红、黄和黑（在 CMYK 色域，印刷行业以及打印机中常见）。照片是一个个采样点的集合，故而单位面积内的像素越多代表分辨率越高，所显示的图像就接近于真实物体。

一个像素所能表达的不同颜色数取决于比特每像素（BPP，Bit Per Pixel）。这个最大数可以通过取 2 的色彩深度次幂来得到。例如，常见的取值如下。

- 1 bpp，$2^1 = 2$ colors （单色）
- 2 bpp，$2^2 = 4$ colors
- 3 bpp，$2^3 = 8$ colors
  ⋮
- 8 bpp，$2^8 = 256$ colors （称为"8 位色"）
- 16 bpp，$2^{16} = 65\,536$ colors （称为高彩色，亦称为"16 位色"）
- 24 bpp，$2^{24} \approx 16.8$ million colors （称为真彩色，亦称为"24 位色"）
- 32 bpp，$2^{32} \approx 4.3$ billion colors

3.2.5.2　分辨率

分辨率又称解像度、解析度，泛指量测或显示系统对细节的分辨能力。分辨率常用于描述显示器、打印机和位图图像的清晰度。

（1）对于点阵打印机和激光打印机，分辨率表示每英寸点数。例如，一个 300dpi 打印机每平方英寸可以打印 90 000 点。

（2）对于图形显示器，屏幕分辨率是指整个屏幕可以显示的像素点的数目。例如，
- CGA：640×200 像素；
- EGA：640×350 像素；
- VGA：640×480 像素；
- Super VGA：800×600 分辨率、256 色的显示器需要 512KB 大小的图形显示卡，1024×768 分辨率、256 色的显示器需要 1MB 大小的图形显示卡。

课文 4 辅助存储器

辅助存储器（辅存），也称为存储介质、二级存储或外部存储器（外存），是计算机主存（或称内存）之外的所有可访问数据存储器。与主存不同，外存不能直接被 CPU 访问。计算机通常使用输入/输出设备访问外存，然后将其数据传送到主存。当断电时外存数据不会丢失，因此，外存是非易失性的。外存每单位的售价通常比主存便宜。

1) 外存的特性
- 非易失（与 RAM 相比）；
- 每 MB 的单价便宜（与 RAM/ROM 相比）；
- 使用方便（与文件柜相比）；
- 相对快（与文件柜相比）；
- 速度比 RAM 慢。

2) 常见的外存
- 软盘；
- 硬盘；
- 光盘（例如，CD-ROM、CD-R、CD-RW、DVD-ROM、DVD-R、DVD-RW、蓝光盘）；
- 闪存（例如，U 盘，也叫闪盘）；
- 磁带；
- MO（磁光盘）；
- PC 卡；
- 智能卡。

4.1 软盘

软盘（Floppy Disk，FD）是个人计算机设备中最早使用的可移动存储介质。1981 年，日本索尼公司首次推出 3.5 英寸软盘，随着硬件加工技术的发展，软盘尺寸渐渐减小，容量渐渐增加。但是由于软盘读取方式的局限，磁头在读写磁片数据时必须接触盘片，而不是像硬盘那样悬空读写，因此软盘已经难以满足大量、高速的数据存储，而且软盘的存储稳定性也较差，容易受到外界环境影响，如受热、受潮、多次读写，均使之寿命减少。随着光盘、U 盘、移动硬盘等移动存储接口的应用，5.25 英寸及 8 英寸的软盘已极为罕见，

3.5 英寸的软盘使用也逐渐被淘汰。软盘的主要特性如下。
- 圆形的塑料磁性介质。
- 由磁道和扇区构成。
- 每个扇区存放 512 字节的数据。
- 软盘容量的计算：

例如，对于一张 3½" Macintosh HD 1.44MB 的软盘（双密度），其容量
= 2 面×80 磁道/面×18 扇区/磁道×512 字节/扇区
= 1 474 560 bytes ≈1.44MB

4.2 硬盘

硬盘是计算机上使用坚硬的旋转盘片为基础的非易失性的存储设备。它在平整的磁性表面存储和检索数据、指令和信息。通过离磁性表面很近的读写头、由电磁流来改变极性方式将数据写到磁盘上。硬盘容量比软盘大，读写速度也比软盘快。

一个硬盘通常由若干张盘片组成。每张盘片有两个读/写磁头，每面一个。所有的读/写磁头都连在一个磁头控制器上，由磁头控制器负责各个磁头的运动，磁头可沿盘片的半径方向运动。每张盘片有相同数量的磁道，各张盘片中相同的磁道构成了柱面。

4.2.1 格式化

格式化是指对磁盘或磁盘中的分区（Partition）进行初始化的一种操作，创建空的文件系统、清除磁盘或分区中所有的文件：
- 定义磁盘上的磁道和扇区；
- 擦除磁盘上的所有数据；
- 创建文件分配表 FAT（FAT 中包含每个文件所占用的扇区信息以及空扇区信息）和根目录表结构。

4.2.2 硬盘容量

硬盘的容量＝磁头数*柱面数*扇区数*每扇区字节数（通常为 512）。例如，一个由 8 张盘片、16 382 个柱面、63 个扇区构成的硬盘，其容量为 16*16 383*63*512 ≈ 8G。

当前硬盘容量一般有 36GB、40GB、45GB、60GB、75GB、80GB、120GB、150GB、160GB、200GB、250GB、300GB、320GB、400GB、500GB、640GB、750GB、808GB、1TB、1.5TB、2TB、2.5TB、3TB 以及 1PB 等多种规格。

4.2.3 硬盘转速

转速（Rotational Speed）是硬盘内电机主轴的旋转速度，也就是硬盘盘片在一分钟内

所能完成的最大转数。转速的快慢是标示硬盘档次的重要参数之一，它是决定硬盘内部传输率的关键因素之一，在很大程度上直接影响到硬盘的速度。硬盘的转速越快，硬盘寻找文件的速度也就越快，相对地，硬盘的传输速度也就得到了提高。硬盘转速以每分钟多少转来表示，单位表示为 RPM（Revolutions Per Minute），表示"转/分钟"。RPM 值越大，内部传输率就越快，访问时间就越短，硬盘的整体性能也就越好。

硬盘的主轴马达带动盘片高速旋转，产生浮力使磁头飘浮在盘片上方。要将所要存取数据的扇区带到磁头下方，转速越快，则等待时间也就越短。

一般家用台式计算机硬盘的转速为 5400rpm、7200rpm 等，笔记本硬盘转速则以 4200rpm、5400rpm 为主，而服务器中使用的 SCSI 硬盘转速基本都采用 10 000rpm，甚至还有 15 000rpm 的，性能要超出家用产品很多。

4.2.4 访问时间

访问时间（又称为读取时间），是磁头从起始位置到达目标磁道位置，并且从目标磁道上找到要读写的数据扇区所需时间，用于衡量存储设备和内存的读写速度。

（1）访问时间度量
- 在存储介质上定位到要读写的数据所花费的时间；
- 将所需数据从内存传送到 CPU 的时间；
- 通常为 5～80 毫秒，数值越小越好。

（2）访问时间=寻道时间+旋转延迟+数据传输时间
- 寻道时间：读/写磁头找到正确磁道所需时间。
- 旋转延迟：是指读/写头已处于要访问的磁道，等待所要访问的扇区旋转到读/写头下方的时间。平均旋转延迟=硬盘转 1 圈所需时间的一半。
- 数据传输时间：数据从磁盘传输到内存的时间。

例如，计算一个寻道时间为 25 毫秒、数据传输时间为 2 毫秒、转速为 4500rpm 的硬盘的平均访问时间。

4500 rev takes 1 min

1 rev takes $\frac{1}{4500}$ min = $\frac{1}{4500} \times 60 \times 1000\, m\sec s \approx 13.33\, m\sec s$

0.5 rev takes 6.67ms

访问时间 = 寻道时间 + 旋转延迟 + 数据传输时间
$\approx$ 25 ms + 6.67ms + 2ms
$\approx$ 33.67ms

4.2.5 硬盘的特性

- 由盘片、柱面、扇区构成。
- 转速主要有 4200rpm、5400rpm、7200rpm、10 000rpm、15 000rpm 等。
- 硬盘工作时，磁头悬浮在盘片上方。

- 可能会有坏扇区以及磁头划碰（又称磁头划道、磁头碰撞、磁头撞毁，指读/写磁头与盘面的碰撞，Head crash 一般不能修复，整个硬盘必须更换）。
- SCANDISK（或 CHKDSK）命令检测和修复磁盘。
- 容量一般为 36GB 至 1PB，或更多。

4.2.6　维护磁盘上的数据

4.2.6.1　备份

备份指将文件系统或数据库系统中的数据加以复制；一旦发生灾难或错误操作时，得以方便及时地恢复系统的有效数据和正常运作。最好将重要数据制作三个或三个以上的备份，并且放置在不同的场所，以利日后恢复之用。

4.2.6.2　碎片

碎片是一个文件的内容分散存放在磁盘的两个或多个不连续的扇区。这样，读写文件就需要到不同的地方去读取，降低了磁盘的访问速度。

4.2.6.3　碎片整理

磁盘碎片整理是使文件存放在连续的扇区，增加文件的连续性，优化文件的读取和写入速度，提高系统性能。

4.2.6.4　数据压缩

数据压缩技术是通过各种机制来降低数据的大小，以占用更少的存储空间。

4.2.7　软盘和硬盘的特性

（1）软盘是一种磁性介质，允许用户读取和写入数据、指令和信息。

（2）硬盘的基本参数：盘片、读/写磁头、柱面、扇区、容量、每分钟转速、访问时间和硬盘缓存（参见 2.2.4.2 节）。

（3）软盘和硬盘的主要区别主要有以下几点。
- 软盘：磁头接触磁盘表面。
- 硬盘：磁头悬浮在盘片上方。如果磁头接触盘面，磁盘和磁头都将受损，即所谓的"磁头撞毁"。

4.3　闪存

闪存是一种非易失性的计算机存储芯片，可电擦除和重新编程。闪存卡，或称快闪存储卡，是一种用于存储数字信息的电子闪存存储设备，多为卡片或者方块状。闪存主要用于内存卡、U 盘、数码相机、手机、笔记本电脑、MP3 播放器和游戏机。它们体积小、可重复读写、无须外部电源。

4.3.1 USB 闪存盘

USB 闪存盘（又称优盘、U 盘、电子盘、记忆棒）是一种用闪存来进行数据存储的介质，通常使用 USB 插头。U 盘体积极小、重量轻、可热插拔、可重复写入。面世后迅速普及并取代传统的软盘及软盘驱动器。有时读卡器也会被归类为闪存盘。2010 年的 U 盘存储容量可达 256GB。

相比于其他便携式存储设备（尤其是软盘），U 盘有许多优点：较不占空间，通常操作速度较快，能存储较多数据，读写时断开一般也不会损坏硬件（软盘在读写时断开马上损坏）、只会丢失数据。

U 盘通常使用塑胶或金属外壳，内部含有一张小的印刷电路板，U 盘尺寸可以小到像钥匙圈饰物一样能够放到口袋中，或是串在颈绳上。突出于保护壳外的 USB 连接头通常用一个小盖子盖住。

4.3.2 智能卡

智能卡，又称智慧卡、聪明卡、集成电路卡及 IC 卡，外形与信用卡一样，卡上含有一个集成电路芯片。智能卡包含了微处理器、I/O 接口及存储器，提供了数据的运算、访问控制及存储功能。常见的有：
- 电话 IC 卡。
- 员工考勤卡。
- 病人医疗卡。
- 数字现金（通过网络为产品或服务付款的电子化（无形）货币）。

4.4 光存储技术

光存储指以光电工程方法，将数据储存于光学可读的介质上，以进行数据的储存。计算机所使用的只读存储光盘以及蓝光光盘等光学盘片就是光存储的应用。光盘一般有三种格式：只读（CD、CD-ROM、DVD-ROM）、可写（一次性写入，如 CD-R、DVD-R）、可擦写（可重写，CD-RW、DVD-RW）。一张标准的蓝光光盘可以容纳约 25GB 的数据，一张 DVD 光盘约容纳 4.7GB 的数据，一张 CD 光盘约容纳 700MB 的数据。

4.4.1 CD 光盘

CD 光盘是一种用于储存数字数据的光学盘片，最初开发用作储存数码音乐。CD 于 1982 年面世，至今仍然是商业录音的标准储存格式。

标准 CD 直径为 120 毫米，可以容纳 80 分钟的无压缩音频（约 700MB 的数据）。迷你 CD 的直径为 60~80 毫米，可以存储约 24 分钟的音频。

4.4.2 DVD 光盘

DVD 光盘，也称数字视频光盘、数字多功能光盘、数字多用途光盘或数字通用光盘，是一种光盘存储器，通常用来播放标准电视机清晰度的电影、高质量的音乐等大容量存储数据用途。DVD 与 CD 的外观极为相似，它们的直径都是 120 毫米左右，但 DVD 的容量几乎是 CD 的 7 倍。DVD 的容量为 4.7GB~17GB。标准 DVD 使用 650 纳米（nm）波长的激光，此光为红色。

有以下几种 DVD：DVD-ROM——只能读、不能写的光盘；DVD-R 和 DVD+R——限写一次的 DVD；DVD–RW、DVD+RW 和 DVD–RAM——可多次读写擦除数据的光盘。

4.4.3 蓝光光盘

蓝光光盘（蓝光盘，也称为 BD）是 DVD 之后的下一代光盘格式之一，用于存储高品质的影音以及高容量的数据存储。蓝光光盘的命名是由于其采用波长 405 纳米（nm）的蓝色激光光束来进行读写操作(DVD 采用 650 纳米波长的红光，CD 则是采用 780 纳米波长)。可擦写蓝光光盘的数据传输率为 36Mbps。一个单层的蓝光光盘的容量为 25GB 或 27GB，足够录制一个长达 12 小时的标准视频或者 2 小时以上的高清晰影片。而双层的蓝光光盘容量可达到 46GB 或 54GB，可以刻录一个长达 8 小时的高清晰影片。4 层和 8 层的蓝光光盘容量分别为 100GB 和 200GB。

4.4.4 磁光盘

MO（磁光盘）是一种结合了磁盘技术和光学技术的存储技术。MO 像磁盘一样可读可写，像软盘一样可便携。存储容量可达 200MB。MO 的数据存取速度比软盘快，但比硬盘慢。

4.5 磁带

磁带是一种非易失性的存储媒体，由带有可磁化的塑料带状物组成（通常是卷起）。磁带有多种类型，可储存视频、音频等。磁带按用途可大致分成录音带、录像带、计算机磁带和仪表磁带四种。用于计算机的磁带在 20 世纪 80 年代曾被广泛应用，但现在已经不常用。

（1）磁带主要用于硬盘的备份和归档以及分散存储大程序。

（2）磁带是顺序存取的。

（3）QIC（1/4 英寸磁带盒）和 DAT（数字录音带、数码音频磁带）是当前最常用的格式。

磁带速度比较慢，一般用于长期的数据存储和备份。要经常使用的数据一般存储在磁盘上。

4.6 RAID 存储系统

RAID，称为独立磁盘冗余数组，简称磁盘数组，其基本思想就是把多个相对便宜的硬盘组合成一个磁盘数组，使性能达到甚至超过一个价格昂贵、容量巨大的硬盘。根据选择的版本不同，RAID 比单个硬盘有如下优势：数据集成度高、容错功能强、处理量或容量大。另外，RAID 对于计算机来说，看起来就像一个单独的硬盘或逻辑存储单元，操作系统也只会把它当作一个硬盘，使用方便。RAID 常被用在服务器上。RAID 存储系统设计的两个主要目的是提高数据的可靠性和增强系统的输入输出性能。

RAID 可以分成不同的等级，每种等级都有其理论上的优缺点。
- RAID 0——无容错设计的条带磁盘阵列，亦称为带区集。由两个或两个以上的硬盘组成，其容量是每个硬盘容量的总和，好处是磁盘空间利用率高、并行 I/O、读写性能高，缺点是不具有冗余、没有容错能力、可靠性差，如果一个磁盘（物理）损坏，则所有的数据都会丢失。
- RAID 1——镜像与双工。在主硬盘上存放数据的同时也在镜像硬盘上写一样的数据。当主硬盘（物理）损坏时，镜像硬盘则代替主硬盘的工作。RAID 1 数据可靠性和安全性高。但无论用多少磁盘做 RAID 1，仅算一个磁盘的容量，因此磁盘空间利用率低，存储成本高。
- RAID 5——块交叉分布式奇偶校验盘阵列。RAID 5 是一种存储性能、数据安全和存储成本兼顾的存储解决方案。不对存储的数据进行备份，而是把数据和相对应的奇偶校验信息存储到组成 RAID 5 的各个磁盘上，并且奇偶校验信息和相对应的数据分别存储于不同的磁盘上。当 RAID 5 的一个磁盘数据发生损坏后，利用剩下的数据和相应的奇偶校验信息去恢复被损坏的数据。RAID 5 可以理解为 RAID 0 和 RAID 1 的折中方案。RAID 5 可以为系统提供数据安全保障，但保障程度要比镜像低而磁盘空间利用率要比镜像高。RAID 5 具有和 RAID 0 相近似的数据读取速度，只是多了一个奇偶校验信息，写入数据的速度比对单个磁盘的写入操作稍慢。同时由于多个数据对应一个奇偶校验信息，RAID 5 的磁盘空间利用率要比 RAID 1 高，存储成本相对较低。

课文 5

软件

软件由使计算机完成特定任务的程序构成,是一系列按照特定顺序组织的数据和指令的集合。软件并不只是包括可以在计算机上运行的计算机程序,与这些计算机程序相关的文档,一般也被认为是软件的一部分。简单而言,软件就是程序加上有关的说明文档。

5.1 软件的分类

软件一般分为两大类:系统软件和应用软件。系统软件为计算机的使用提供最基本的功能,但是并不针对某一特定应用领域。而应用软件则恰好相反,不同的应用软件根据用户和所服务的领域提供不同的功能。

其他类型的软件还包括几种。

- 编程语言:又称程序设计语言(Program Design Language,PDL),定义计算机编程语言的语法和语义等规则。例如,Pascal、C、C++、VB/VB.NET、C#、Java 等。
- 中间件:是一种独立的系统软件或服务程序,分布式应用软件借助这种软件在不同的技术之间共享资源。中间件在操作系统、网络和数据库之上、应用软件之下,总的作用是管理计算资源和网络通信,为处于自己上层的应用软件提供运行与开发的环境,帮助用户灵活、高效地开发和集成复杂的应用软件。在基于 XML、SOAP、Web 服务、面向服务的体系结构等现代信息技术中应用比较广泛。数据库、Apache 的 Tomcat、IBM 公司的 WebSphere、BEA 公司的 WebLogic 应用服务器以及 Kingdee 公司的 Apusic 等都属于中间件。
- 测试件:也称测试元件、测试工具,是用于测试硬件和软件包的软件。测试件是软件的子集,主要用于软件测试,特别是自动化软件测试。测试件通过检验和确认方法生成测试结果。
- 固件:参见 2.2.5.1 节。
- 设备驱动程序:简称驱动程序,是一种可以使计算机和设备进行通信的程序,相当于硬件的接口,操作系统通过这个接口,才能控制硬件设备的工作。一般当操作系统安装完毕后,首先要安装的就是硬件设备的驱动程序。不同版本的操作系统对硬件设备的支持也是不同的,一般情况下版本越高所支持的硬件设备也越多。即插即用设备的驱动程序通常已包括在 Windows 系统中;其他新硬件设备通常附带驱动程序光盘或软盘。

系统软件主要指面向硬件或者开发者所设立的软件，如操作系统、解释系统、编译系统、数据库管理系统、中间件等。系统软件直接与硬件交互，管理、控制和协调硬件，并确保应用软件的正常工作。系统软件是用户、应用软件和计算机硬件的接口。

系统软件通常分为以下三类。

- BIOS 和设备固件：提供操纵和控制硬件的基本功能。
- 操作系统：如 Microsoft Windows、Mac OS X 和 Linux，使得计算机各部件协调工作，执行诸如在内存和磁盘间的数据传输、将数据输出到显示设备等任务，操作系统还提供了运行系统软件和应用软件的平台。
- 实用程序：分析、配置、优化和维护计算机的程序。

应用软件，也称为终端用户程序，是指针对用户的某种特定的应用目的而开发的软件。它可以是一个特定的程序，例如图像浏览器。也可以是一组功能联系紧密，可以互相协作的程序的集合，例如微软的 Office 软件。也可以是一个由众多独立程序组成的庞大的软件系统，例如数据库管理系统。

例如，数据库管理程序 Access、文字处理软件 Word、电子表格软件 Excel 等。

5.2 系统软件

5.2.1 BIOS

BIOS（基本输入输出系统）是一组被固化到计算机中的内置软件，为计算机提供最低级最直接的硬件控制程序。当打开计算机电源，BIOS 就会由主板上的闪存运行，并将芯片组和存储器子系统初始化。BIOS 会将自己从闪存中解压缩到系统的主存，然后开始运行。PC 的 BIOS 代码也包含诊断功能，以保证某些重要硬件组件，如键盘、磁盘、输出输入端口等，可以正常运作且正确地初始化。几乎所有的 BIOS 都可以选择性地运行 CMOS 存储器的设置程序；CMOS 存储器保存 BIOS 会访问的用户自定义设置数据（时间、日期、硬盘细节等）。BIOS 一般置于计算机的 ROM 芯片（通常称为 ROM BIOS），这确保了 BIOS 将总是有效，不会因磁盘故障而损坏，并保证计算机可以正常启动。BIOS 必须：

- 存储在 ROM 芯片中；
- 包含计算机启动的指令。

BIOS 和 CMOS 的区别与联系：BIOS 是主板上的一块 EPROM 或 EEPROM 芯片，里面装有系统的重要信息和设置系统参数的设置程序（BIOS Setup 程序）；CMOS 是主板上的一块可读写的 RAM 芯片，用于存储关于系统配置的具体参数，其内容可通过设置程序进行读写。CMOS RAM 芯片靠后备电池供电，即使系统掉电后信息也不会丢失。BIOS 中的系统设置程序是完成 CMOS 参数设置的手段；CMOS RAM 既是 BIOS 设定系统参数的存放场所，又是 BIOS 设定系统参数的结果。简而言之，通过 BIOS 设置程序对 CMOS 参数进行设置。

5.2.2 操作系统

操作系统（OS）是管理计算机硬件与软件资源的程序，同时也是计算机系统的核心与基石。操作系统的基本功能包括管理与配置内存、决定系统资源供需的优先次序、控制输入与输出设备、操作网络与管理文件系统等。操作系统也提供一个让用户与系统交互的操作接口。

操作系统位于底层硬件与用户之间，是两者沟通的桥梁。用户可以通过操作系统的用户界面，输入命令。操作系统则对命令进行解释，驱动硬件设备，实现用户要求。一个标准个人计算机的 OS 应该提供以下功能：

- 进程管理（Processing Management）；
- 内存管理（Memory Management）；
- 文件系统（File System）；
- 网络通信（Networking）；
- 安全机制（Security）；
- 用户界面（User Interface）；
- 驱动程序（Device Drivers）。

5.2.2.1 开机启动

启动过程是个自举过程，当用户开机时，即启动操作系统的运行，进行一系列的初始化操作。引导程序（开机管理程序，又称自举引导程序）将操作系统载入计算机。开机的主要步骤如下：

（1）计算机接通电源，电源供应器发送电子信号到计算机的各个部件。
（2）处理器芯片查找 ROM 芯片中的 BIOS。
（3）BIOS 执行 POST（开机自检）程序，检查各个部件，例如键盘、驱动器等。
（4）POST 自检结果与 CMOS 芯片中的数据相比较。
（5）BIOS 按 CMOS 设置的启动顺序（一般为 A 盘、硬盘）搜索启动磁盘。
（6）加载并执行启动盘的第一个物理扇区的主引导记录（Master Boot Record，MBR，又称为主引导扇区）。
（7）MBR 将活动分区的第一个扇区中的引导扇区（引导程序）载入内存（RAM）中。
（8）最后，操作系统被加载到内存中。

5.2.2.2 OS 的分类

（1）根据用户和任务分类。

- 单用户单任务操作系统：
在同一时间只允许一个用户运行一个程序，如 DOS。
- 单用户多任务操作系统：

允许一个用户同时运行两个或更多的程序。如 Windows 3.X、Windows 95/98 等。
- 多用户操作系统：

允许多个用户在同一时间内使用计算机。所有的大型机和小型机都是多用户操作系统，但大多个人计算机和工作站都不是。多用户的同义词是时间共享。多用户操作系统允许多个用户在同一时间内使用计算机。最明显的一个例子是远程登录服务器，多个远程用户有权在同一时刻进入 UNIX Shell 系统。如 UNIX、Linux 等。
- 并行操作系统：

并行操作系统支持在多 CPU 的计算机系统中同时运行程序。并行处理（multiprocessing）还包含多个 CPU 协调处理程序的能力。术语"并行程序（multiprogramming）"可以更恰当地表达这个概念。并行程序往往通过软件实现，而并行处理则侧重于多硬件 CPU 的实现方式。一个系统可以同时并行程序处理和多处理器处理。如 Windows 2000/2003、Windows XP/Vista、Windows 7/8/8.1/10、Microsoft 服务器操作系统、UNIX、Linux 等。

（2）独立/网络/嵌入式操作系统。
- 独立操作系统：

独立操作系统是台式机、笔记本电脑或移动计算设备上的完整的操作系统，而且还可以与网络操作系统协同工作。独立操作系统有时也被称为客户端操作系统。有些独立操作系统还包括网络功能，允许家庭和小型商务用户建立小型网络。如 DOS、Windows XP、Windows 10、Mac OS X、OS/2 Warp Client、UNIX 和 Linux 等。
- 网络操作系统：

网络操作系统（NOS）是在计算机操作系统下工作，使计算机操作系统增加了网络操作所需要的能力，主要为控制网络及其信息（如数据包）发送、控制多个用户访问网络资源、提供特定的管理功能，如安全性等。其目标是相互通信及资源共享。如 NetWare、Windows 2003/XP、Vista、Windows 7/8/10、OS/2 Warp Server for e-business、UNIX、Linux 和 Solaris 等。
- 嵌入式操作系统：

嵌入式操作系统是嵌入式计算机系统的操作系统。这些操作系统设计得非常紧凑，高效，除了具备一般操作系统最基本的功能，如任务调度、同步机制、中断处理、文件功能等外，还有以下特点：开放性、可伸缩性的体系结构、强实时性、统一的接口、强大的网络功能、固化代码、良好的移植性等。嵌入式操作系统一般属于实时操作系统。大多数 PDA 和小型设备的操作系统是嵌入式操作系统。如流行的嵌入式操作系统包括 Pocket PC(P/PC、PPC)、Windows Mobile 2003/2003 SE、Windows Mobile 5/6/6.1/6.5、Windows Phone CE/7/8/8.1、Windows 10 Mobile、Palm OS、Symbian OS、Windows IoT(原来的 Windows Embedded)、VxWorks、Android 和 iOS(原来的 iPhone OS)。

5.2.2.3　命令行 OS 和图形界面 OS

1）基于命令行界面的 OS

命令行界面（Command Line Interface，CLI）是在图形用户界面得到普及之前使用最为广泛的用户界面，它通常不支持鼠标，用户通过键盘输入指令，计算机接收到指令后，予以执行。也有人称为字符用户界面（CUI）。

通常认为，命令行界面（CLI）没有图形用户界面（GUI）那么方便用户操作。因为，命令行界面通常需要用户记忆操作的命令，但是，由于其本身的特点，命令行界面要较图形用户界面节约计算机系统的资源。在熟记命令的前提下，使用命令行界面往往要较使用图形用户界面的操作速度要快。所以，在现在的图形用户界面的操作系统中，通常都保留着可选的命令行界面。

- 提供有限的命令集合；
- 如果不知道语法，就无法运行命令。

2）图形界面的 OS

图形用户界面（Graphical User Interface，简称 GUI，又称图形用户接口）是指采用图形方式显示的计算机操作用户接口。与早期计算机使用的命令行界面相比，图形界面对于用户来说在视觉上更易于接受。

在图形用户界面中，计算机画面上显示窗口、图标、按钮等图形，表示不同目的之动作，用户通过鼠标等指针设备进行选择。

- 通过菜单、按钮、快捷菜单等执行命令；
- 操作方便快捷。

5.2.2.4 节和 5.2.2.5 节译文略。

5.2.3 实用程序

实用程序是用来分析、配置、优化和维护计算机系统的软件。

5.2.3.1 Windows 实用程序

通过 Start→Windows Administrative Tools，可以找到并执行 Windows 实用程序。

1）磁盘清理

磁盘清理是 Microsoft Windows 内置的应用程序，可清除硬盘内不必要的文件以提供更多的剩余空间，从而令 Windows 的运作更畅顺。磁盘清理会分析硬盘，然后确定哪些文件可能不再需要，并由用户决定是否删除这些文件。它进行了初步的磁盘分析，并分成多项不同类型的清理对象。Windows 磁盘清理会清除的项目包括：

- 长期没有访问过的压缩文件；
- 临时 Internet 文件；
- 临时 Windows 文件；
- 下载的程序文件；
- 回收站内的文件；
- 用户不用的可选 Windows 组件；
- 用户不用的安装程序文件；
- 安装日志文件；
- 脱机文件。

2）磁盘碎片整理

磁盘碎片整理是 Microsoft Windows 内置的程序，重新组织硬盘上的文件和剩余空间，使文件尽量存放在连续的扇区，以提高文件的读取和写入速度以及程序的运行速度。

5.2.3.2 实用程序软件包

1）计算机维护软件

计算机维护软件是确认和修复操作系统问题、检测并修复磁盘问题、提高计算机性能的实用程序。

例如，Norton Utilities。

2）杀毒程序

杀毒程序是用来搜索并删除硬盘中病毒的实用程序。大多数杀毒程序包括自动升级功能，可下载新的病毒配置文件，以检测和清除新病毒。

例如，Norton Antivirus、McAfee VirusScan。

3）文件压缩

文件压缩工具是缩小文件大小，使其占用尽量少的存储空间的实用程序。

例如，WinRar、WinZip、PKZIP。

5.3 编程语言

编程语言，又称程序设计语言，是一组用来定义计算机程序的语法规则。每一种语言都有一套独特的关键字和程序指令语法。编程语言随硬件的发展而发展。

编程语言分两类：低级语言和高级语言。

- 低级语言与特定的机器有关。每种低级语言都运行在特定的计算机上，与 CPU 的机器语言或指令直接对应，很难移植到其他类型的计算机上。低级语言由于无须大量的编译即可被 CPU 运行，因此以该类编程语言编写的源代码一般比高级语言的源代码编译和运行效率高。
- 高级语言独立于机器，一种高级语言可以在多种计算机和操作系统上运行。高级语言是以人类的日常语言为基础的一种编程语言，使用一般人易于接受的文字来表示，使程序编写员编写更容易，亦有较高的可读性。

机器语言和汇编语言属于低级语言，过程语言、非过程语言、面向对象的程序设计语言和可视化程序设计语言均属于高级语言。

5.3.1 机器语言

- 第一代程序设计语言。
- 使用二进制代码编写程序，可读性差，但能够直接被计算机识别和执行。
- 其他编程语言必须转换为机器语言才能被执行。

- 不同类型的 CPU 都有自己独特的机器语言。

5.3.2 汇编语言

- 第二代程序设计语言。
- 使用简单的缩写或符号来表示指令。
- 与特定的物理（或虚拟）计算机体系结构相关，而大多数高级编程语言则易于移植。
- 由称为汇编器的特殊程序将源代码转换为机器语言。

5.3.3 过程程序设计语言

- 高级程序设计语言。
- 第三代程序设计语言（3GL）。
- 类似自然语言的形式描述对问题的处理过程，更像英语表达方式。
- 关注于所解决问题的逻辑和求解过程，而不关心计算机的内部结构。
- 编译程序将指令转换成机器语言，并以 EXE（可执行文件）形式存储。

5.3.4 非过程程序设计语言

- 高级程序设计语言。
- 第四代程序设计语言（4GL）。
- 程序员只要告诉计算机做什么但不是如何去做。
- 程序员的编程效率可以比编制第三代程序语言提高 10 倍。

5.3.5 面向对象程序设计语言

- 使用"对象"数据结构来设计应用和计算机程序，将数据和方法封装其中，以提高软件的重用性、灵活性和扩展性。
- 使用事件驱动的程序来检查和响应事件。
- 事件的例子：按键、单击按钮、双击按钮等。
- 编程技术：数据抽象、封装、模块化、多态和继承。

5.3.6 可视化程序设计语言

- 第五代程序设计语言（5GL）。
- 提供可视或图形界面来创建源代码。
- 程序设计人员通过拖放对象（控件等）构建程序。

5.3.7 程序设计语言的执行

不管使用什么程序设计语言，最终都需要转换成机器语言，计算机才能理解和执行。有以下两种转换方法。

- 编译：编译器（Compiler）将源代码翻译成目标语言。源代码一般为高级程序设计语言，而目标语言则是汇编语言或目标机器的目标代码（机器代码）。编译器一般执行语法分析、预处理、语义分析、代码生成、代码优化等操作。编译器的主要工作流程如下：

源代码（source code） —编译器（compiler）→ 目标代码（object code） —链接器（linker）→ 可执行程序（executables）

- 解释：解释器（Interpreter）直接解释执行高级编程语言。相对地，编译器并不运行程序或源代码，而是首先将其翻译成另一种语言（机器码），以供多次运行而无须再编译。运行编译生成的目标代码（机器码）比使用解释器运行程序要快。解释器不会一次把整个程序翻译出来。它每翻译一行程序语句就立刻执行，然后翻译下一行并执行，直至完成。

许多程序设计语言同时使用编译器和解释器来实现，例如，Java 和 C#。

课文 6 和课文 7 译文略。

课文 8 电子表格介绍

8.1 Excel 介绍

8.1.1 电子表格

电子表格（spreadsheet）又称电子数据表，是一类模拟纸上计算表格的计算机程序。单元格是工作表的基本组成元素，是由工作表中行与列交叉所围成的网格。单元格内可以存放文本、数值或公式等内容。公式为"="引导的表达式，公式表达式中可以使用各种运算符和 Excel 函数。电子表格通常用于财务报表，因为它能够频繁地重新计算整个表格。

Microsoft Excel 由 Microsoft 为使用 Windows 和 Apple Macintosh 操作系统的计算机而编写和运行的一款电子表格软件。直观的界面、出色的计算功能和图表工具，再加上成功的市场营销，使 Excel 成为最流行的微机数据处理软件。在 1993 年，作为 Microsoft Office 的组件发布 5.0 版之后，Excel 就开始成为所适用操作平台上的电子制表软件的霸主。

一个 Excel 文件可以包含多个工作簿（workbooks）；一个工作簿包含多个工作表（worksheets）；一个工作表包含 1 048 576×16 384 个单元格（在 Excel 2002/2003 等版本中，一个工作表包含 65 536×256 个单元格）。

利用 Excel，用户可以跟踪和审核数据，构建模型以分析数据，使用公式计算数据，创建数据图表，通过排序、筛选、分类汇总和数据透视表等透视数据等。

8.1.2 单元格的内容

单元格是存储信息的基本单元。在单元格中可以输入数值、文本、公式等。

8.1.2.1 数值

数值常量由数字 0~9 及一些符号组成。数值默认的对齐方式为右对齐。构成数值的常用符号主要有以下几个。

（1）+：数字前加"+"号，表示正数。例如，"+10"表示正 10，显示为"10"。
（2）-：数字前加"-"号，表示负数。例如，"-10"表示负 10，显示为"-10"。
（3）()：数字包含在括号中也表示负数。例如，"(10)"表示负 10，显示为"-10"。
（4）.：小数点表示法。如"10.12"。

（5）$：数字前加"$"号，表示货币。例如，"$10"表示10美元，显示为"$10"。

（6）%：百分比表示法。例如，"32%"表示百分之三十二，显示为"32%"。

（7）/：分数表示法。例如，"1 1/3"表示一又三分之一，显示为"1 1/3"。注意：输入真分数时，前面需要加0和空格，否则，Excel解释为日期。例如，"0 1/2"表示二分之一；而"1/2"表示1月2日。

（8）E 或 e：科学记数表示法。例如，"1E3"表示 1000。如果单元格中输入的数字位数过大，则 Excel 将使用科学记数法表示。例如，"1 200 000 000 000"将自动显示为"1.2E＋12"。

8.1.2.2 公式

Excel 中公式为"="引导的表达式。公式表达式中可以使用各种运算符和 Excel 函数。例如：在单元格 A11 中输入公式"=sum(A1:A10)"，用于计算单元格 A1 到 A10 范围的所有数字的和。再如：如果 A1 单元格的内容为"Hello"，A2 单元格的内容为"World"，则在单元格 A3 中输入公式"=A1&A2"，用于连接文本，结果为"HelloWorld"。

8.1.2.3 文本

文本常量是由字母、数字和符号组成的字符串。文本默认的对齐方式为左对齐。注意，纯数字组成的字符串为数字常量，通过在其前面加西文半角的"'"号，可以转换为文本常量。例如"'12345"表示文本字符串 12345。

8.1.2.4 节、8.1.3 节、8.2 节和 8.3 节译文略。

课文 9

数据库介绍

9.1 数据库

数据库，通常缩写为 DB，是存储数据的仓库，即存储在计算机系统中结构化的、可共享的相关数据的集合，计算机程序能够对其进行查询和维护。

传统数据库由字段、记录和文件组成。字段由同类的信息组成。记录指包括了若干列（字段）信息项的一行数据。一个数据库文件由若干条记录组成。

例如，一本电话簿就类似于一个数据库文件。它包含一个记录列表，每条记录由三个字段组成：姓名、地址和电话号码。

用户需要数据库管理系统（DBMS）来访问数据库中的信息。DBMS 是用于输入、组织和选取数据库中的数据的计算机软件。

9.2 数据库管理系统

数据库管理系统（DBMS）是为管理数据库而设计的计算机软件，使用户可以存储、修改和抽取数据库中的信息。其主要功能包括以下几点。

（1）数据定义功能。使用数据定义语言（Data Definition Language，DDL），可以生成和维护各种数据对象的定义。

（2）数据操作功能。使用数据操作语言（Data Manipulation Language，DML），可以对数据库进行查询、插入、删除和修改等基本操作。

（3）数据库的管理和维护。包括数据库的安全性、完整性、并发性、备份和恢复等功能。

常用的数据库模型包括关系模型（Relational Model）、层次模型（Hierarchical Model）、网状模型（Network Model）、多维模型（Multidimensional Model）和面向对象的数据模型（Object Oriented Model）。

关系模型具有完备的数学基础、简单灵活、易学易用等特点，已经成为数据库的标准。目前流行的 DBMS 都是基于关系模型的关系数据库管理系统。

有许多不同类型的数据库管理系统，从运行在个人计算机上的小型系统到运行在大型机上的大型系统，一应俱全。DBMS 典型的例子包括 Oracle、IBM DB2、Microsoft Access、Microsoft SQL Server 和 MySQL 等。通常由数据库管理员（DBA）创建和维护数据库系统。

借助 DBMS，用户可以创建数据库；增加、修改、删除数据库中的数据；排序和检索数据库中的数据；对数据库中的数据创建表单和报表。

9.2.1 关系数据库

关系数据库是符合关系模型的数据库。关系数据库理论是由美国 IBM 公司的研究员 E.F. Codd 提出和定义的。

- 一个数据库包含一系列相关文件；
- 每个文件包含若干条记录；
- 每条记录由若干个字段构成；
- 每个字段包含字符等信息。

例如，大学数据库包括：

- file1 是学生文件；
- file2 是学生考试成绩文件；
- file3 是教学人员文件；
- file4 是教学人员薪酬文件。

数据库文件中的每条记录必须是相同的字段结构。

1）数据文件（数据表）

每个数据库包含一系列相关的数据文件。一个数据文件是存储在磁盘（硬盘、CD-ROM 或 DVD-ROM）上的相关记录的集合。数据文件即数据表。

2）记录

字段值的集合称为记录。记录是包含若干字段信息的一行数据。一组记录组成一个数据库文件。数据库文件中的每条记录包含不同的数据。在关系数据库管理系统中，记录又称为元组（Tuple）。

例如，人事档案文件包含若干条记录，每条记录包含四个字段：工号、姓名、部门以及电子邮箱。

3）字段

字段是指特定的信息。字段的定义包括名称、类型和长度。

4）主键

在关系数据库设计中，主键（或称主码）、唯一键（或称唯一码）或关键字段是一个候选键，用来唯一标识数据表的每一行（即每一条记录）。主键可以是一个字段或多个字段的组合。数据表中没有两行记录的值完全相同。一个数据表可以有多个唯一键，但最多只有一个主键。主键不能为空。

5）外键

外键（或称外码）是两个表之间的参照约束。外键是另一个表的主键。

9.2.2 数据库文件实例

学生档案文件——学籍卡的集合。

记录=一个学生的详细信息。每条记录都有一个唯一的标识。Student ID 是主键。
1）对学籍卡的操作
（1）维护：添加记录、修改记录、删除记录。
（2）信息输出：查询、报表。
2）学籍卡的排序
按学生学号（Student ID）或学生姓名（Student Name）对学籍卡进行排序，以方便检索。
3）索引
数据库索引是 DBMS 中按指定的字段进行排序的数据结构，以协助快速查询、更新数据表中的数据。例如，数据表按 Student ID 字段做索引，就是按 Student ID 排序。通过 Student ID 进行查询，可以快速找到学生记录。

9.2.3　数据完整性

数据完整性确保数据是"整体的"或完整的，即确保数据在进行任何操作（如传输、存储和检索）时保持一致。简言之，数据完整性是指数据的完整性、完全性、一致性和正确性。

数据完整性包括三种：实体完整性、参照完整性和自定义完整性约束。

实体完整性要求关系的主键属性值不能为空（NULL）且不能有相同值。

参照完整性是数据表之间关联关系的约束，也就是对外键的约束。它不允许关系引用不存在的元组：即外键的取值必须是其所关联关系中主键的值，或者是空值。

自定义完整性（又称域完整性）是对数据表中字段属性的约束，它包括字段的值域、字段的类型及字段的有效规则等约束，它是由确定关系结构时所定义的字段的属性决定的。例如，Gender 字段的取值要么是空，要么只能是"男"或"女"。

9.2.4　数据冗余

数据冗余是指数据之间的重复，也可以说是同一数据存储在不同数据文件中的现象。这将增加系统维护的成本，降低系统的运行速度。数据库规范化可以解决数据冗余的问题。

9.3 节、9.4 节和 9.5 节译文略。

课文 10

数据通信和网络

10.1 数据通信

数据通信是通信技术和计算机技术相结合而产生的一种新的通信方式。要在两地间传输信息必须有传输信道,根据传输媒体的不同,有有线数据通信与无线数据通信之分。数据通信通过传输信道将数据终端与计算机联结起来,使不同地点的数据终端实现软、硬件和信息资源的共享。通信设备用于数据传输。常用的通信设备有调制解调器、集线器、路由器等。通信软件是指可以传输数据的程序。

10.1.1 数据通信的组成

发送部件	→	传输信道	→	接收部件
计算机终端		电话线		计算机终端
+		无线电波		+
软件		微波		软件
		同轴电缆		
		光纤		

10.1.2 调制解调器

调制解调器对于使用电话线进行数据通信是必不可少的设备,用于计算机数据和电话信号之间的转换。它是一个将数字信号调制到模拟载波信号上进行传输,并解调收到的模拟信号以得到数字信息的电子设备。其目标是产生能够方便传输的模拟信号并且能够通过解码还原原来的数字数据。根据不同的应用场合,调制解调器可以使用不同的手段来传送模拟信号,例如使用电话线、无线电波或光纤等。

使用普通电话线音频波段进行数据通信的电话调制解调器是人们最常接触到的调制解调器。调制解调器俗称"猫"。其他常见的调制解调器还包括用于宽带数据接入的有线电视电缆调制解调器、ADSL 调制解调器。数字式移动电话实际上也是一种无线方式的调制解调器。现代电信传输设备是为了在不同的介质上远距离传输大量信息,因此也都以调制解调器的功能为核心。其中,微波调制解调器速率可以达上百万比特每秒;而使用光纤作为传输介质的光调制解调器可以达到几十吉比特每秒（Gbps）以上，是现在电信传输手段的骨干。

（1）调制解调器的特性：
- 自动应答；
- 自动拨号；
- 自动断开连接；
- 错误纠正；
- 数据压缩。

（2）调制解调器的数据传输率：

单位是 bps，即每秒传输的二进制位。

10.1.3　通信软件

通信软件为不同计算机或用户提供对系统的远程访问、文件和信息的交换等功能。通信软件包含如下功能程序：
- 帮助用户建立与另一台计算机或网络的连接；
- 管理数据、指令和信息的传输；
- 为用户提供通信接口。

前两个是系统软件，第三个是应用软件。用于通信的应用软件：电子邮件、FTP、Web 浏览器、新闻组/留言板、聊天室、即时消息、视频会议和视频电话等。

10.1.4　数据传输

数据通常以数据包的方式传输，数据包由一组二进制数据组成。为了成功传输数据，发送设备和接收设备必须同步。

有两种数据传输方法：异步传输和同步传输。

1）异步传输
- 有时又称为启停同步传送，数据传输不是连续流的方式，而是间歇式发送；
- 数据传输时每发送一个字节，都需要起始位和终止位；
- 通常用于最简单的数据链路，即终端到计算机的数据传输。

2）同步传输
- 定期传输数据；
- 没有起始位和终止位；
- 发送字符时没有停顿；
- 适合发送大批数据；
- 每批数据前有特殊的同步字符和控制信息，并以错误校验信息结束。

10.1.5　协议

网络传输协议或简称为传送协议或协议，是指计算机通信的共同语言。现在最普及的计算机通信为网络通信，所以"传送协议"一般都指计算机通信的传送协议，如 TCP/IP、NetBEUI 等，是信息交换的计算机共同遵守的格式、规则和标准。协议包括发送信号、认证、错误检测和纠正等功能。协议描述了计算机通信的语法、语义和同步等规则，可以采用硬件实现，也可以采用软件实现，或者同时采用软硬件实现。协议定义了如下事项：

- 所采用的错误检测方法；
- 数据压缩方法；
- 如何确定发送设备是否完成信息发送；
- 如何确定接收设备是否接收到信息。

TCP/IP 协议族，包含了一系列构成互联网基础的网络协议。这些协议最早发源于美国国防部的 ARPA 网项目。TCP/IP 协议族中的主要协议有以下几种。

1）传输控制协议（Transmission Control Protocol，TCP）

TCP 是一种面向连接的、可靠的、基于字节流的运输层通信协议。需要可靠的数据流服务的应用，如万维网、电子邮件和文件传输等 Internet 应用均依赖于 TCP。而不需要可靠的数据流服务的其他应用，则使用用户数据报协议（UDP）。

2）用户数据报协议（User Datagram Protocol，UDP）

UDP 是一个简单的面向数据报的传输层协议。UDP 只提供数据的不可靠传递，它将数据发送出去后不保留数据备份。所以 UDP 有时也被认为是不可靠的数据报协议。UDP 是一种主要用于网络信息广播的无连接协议。流媒体、实时多媒体游戏和 IP 电话（VoIP）就是典型的 UDP 应用。

3）网际协议（Internet Protocol，IP）

它又称为互联网协议，是用于报文交换网络的一种面向数据的协议。IP 是在 TCP/IP 协议中网络层的主要协议，它根据源主机和目的主机的地址传送数据。第一个架构的主要版本，现在称为 IPv4，仍然是最主要的互联网协议，尽管世界各地正在积极部署 IPv6。IP 是建立 Internet 的主要协议。

4）互联网控制消息协议（Internet Control Message Protocol，ICMP）

ICMP 是 IP 协议的扩展，其目的是用于在 TCP/IP 网络中发送控制消息，提供可能发生在通信环境中的各种问题反馈，通过这些信息，令管理者可以对所发生的问题作出诊断，然后采取适当的措施去解决问题。例如，PING 命令使用 ICMP 来测试 Internet 的连接。

5）超文本传输协议（HyperText Transfer Protocol，HTTP）

HTTP 是互联网上应用最为广泛的一种网络协议。所有的 WWW 文件都必须遵守这个标准。设计 HTTP 最初的目的是为了提供一种发布和接收 HTML 页面的方法。例如，当用户在浏览器中输入一个 URL，即发送一个 HTTP 命令给 Web 服务器并引导服务器获取和传输所请求的页面。

6）邮局协议版本 3（Post Office Protocol-Version 3，POP3）

POP3 协议主要用于支持使用客户端远程管理在服务器上的电子邮件。大多 Web 邮件服务，例如，Gmail 和 Yahoo! Mail 均使用 POP3 协议。

7）因特网信息访问协议（Internet Message Access Protocol，IMAP）

它又称互联网消息访问协议，是一个应用层协议，用来从本地邮件客户端（如 Microsoft Outlook、Outlook Express、Foxmail）访问远程服务器上的邮件。IMAP 和 POP3 是邮件访问最为普遍的 Internet 标准协议。所有现代的邮件客户端和服务器都对两者给予支持。

8）文件传输协议（File Transfer Protocol，FTP）

FTP 是用于在网络上进行文件传输的一套标准协议。它属于网络传输协议的应用层。FTP 的工作方式与 HTTP（从服务器传输网页到用户的浏览器）和 SMTP（在 Internet 上传输电子邮件）类似。FTP 通常用于从服务器上下载文件，或将文件上传到服务器上。

10.1.6 数据通信的方向

1）单工——只有一个传输方向

数据信息在通信线上始终向一个方向传输，只能由发送器向接收器传输数据。例如，广播和电视就是单工传输方式，收音机、电视机只能分别接收来自无线电台、电视台的信号，不能进行相反方向的信息传输。

2）半双工——双向传输，但同一时刻只能向一个方向传输

数据信息可以双向传输，但必须交替进行，即同一时刻只能向一个方向传送数据。例如，对讲机就是半双工设备，因为在同一个时刻只能有一方说话。

3）全双工——同时双向传输

发送器和接收器可以同时进行双向数据传输。例如，电话、手机就是全双工设备，因为双方可同时说话。

10.2 网络

计算机网络，简称网络，是利用通信设备和线路将地理位置不同的、功能独立的多个计算机系统连接起来，以功能完善的网络软件实现网络的硬件、软件及资源共享和信息传递的系统。简单地说即连接两台或多台计算机进行通信的系统。

- 若干计算机或终端互连构成了网络。
- 每台计算机称为节点。
- 为网络分配资源、提供服务的计算机和设备称为服务器。
- 网络的几何连接形状，即网络的物理结构，称为拓扑结构。

10.2.1 计算机网络的分类

按照其覆盖范围，计算机网络分成以下几类。

- 局域网（LAN）：指覆盖有限的地理区域（如家庭、学校、计算机实验室、办公室或楼层等）的计算机网络。

- 广域网（WAN）：通常跨接很大的物理范围，使用电话线、电缆和空气波等通信通道连接多个城市或国家，或横跨几个洲并能提供远距离通信，形成国际性的远程网络。
- 城域网（MAN）：介于 LAN 和 WAN 之间的计算机网络，通常涵盖一个大学校园、一座城镇或城市。

10.2.1.1 局域网

局域网（LAN）指覆盖局部区域（如家庭、办公室或楼层）的计算机网络。网络中称为节点的计算机或设备共享诸如打印机、大容量硬盘和程序等资源。节点通常通过电缆互连。

局域网包括：

- 硬件如 PC+通信板；
- 软件包，如 Windows 7；
- 连接计算机和外设的电缆。

局域网的主要特性：

- 允许软件、硬件（服务器、打印机等）和信息的共享；
- 服务器工作站有大容量硬盘，提供可访问的应用程序和数据；
- 局域网通常位于同一建筑物内；
- 传输电缆通常是双绞线、同轴电缆或光纤。

10.2.1.2 广域网

广域网（WAN）是一个覆盖较大地理区域的计算机网络，甚至覆盖整个世界。WANs 通常连接多种小的网络，例如局域网（LANs）或城域网（MANs）。广域网通常通过公用网络（如电话线）或者租用线（专线）或卫星连接。

- 使用通信通道连接电话线、电缆和无线电波等各类通信介质。
- Internet（因特网）是世界上最大的广域网。

10.2.1.3 因特网

因特网（又称互联网），是世界上最大的全球性互联网络。可以是任何分离的实体网络之集合，使用标准的互联网协议套（TCP/IP 协议），形成逻辑上的单一网络。Internet 包含成百上千通过电子或者光学网络技术相连的私有网、公共网、学术网、商务网和政府网。Internet 的主要服务项目有万维网（World Wide Web，缩写为 WWW，又称全球网）、电子邮件（E-mail）、远程登录（Telnet）、查询服务（Finger）、文件传输（FTP）、文档服务器（Archive）、新闻论坛（Usenet）、电子公告牌（BBS）、新闻群组（News Group）等。

10.2.1.4 内特网

内特网（Intranet），或称企业内部网、内部网、内网，是一个基于因特网的 TCP/IP 协议的计算机网络，它通常建立在一个企业或组织的内部并为其成员或有权限的用户提供信息的共享和交流等服务，如万维网、文件传输、电子邮件等。Intranet 网站就像其他网站一样，只是 Intranet 的防火墙会拦截未授权的非法访问。

与 Internet 类似，Intranet 也用于信息共享。Intranet 的安全性使其迅速发展，其构建和管理费用比基于专有协议的私有网络低。

10.2.1.5 外特网

外特网（Extranet），或称企业外部网、外部网、外网，是一个使用 Internet 协议、网络互连、公共电信系统的私人网络，为供应商、零售商、合作伙伴或客户提供安全的企业或组织共享信息。用户只要有有效的用户名和密码，即可以访问 Extranet。Extranet 正成为商业合作伙伴信息共享和沟通交流的流行方式。

10.2.2 网络拓扑结构

网络拓扑是计算机网络中相互连接的各种元素（链接、节点等）的布局模式。网络节点相互之间的连接方式和通信方式均取决于网络的拓扑结构。网络拓扑结构可以是物理或逻辑的。物理拓扑结构是指网络设备、位置和电缆等的物理设计方式；而逻辑拓扑指数据在物理设计模式下的传输方式。常见的网络拓扑结构包括环型、星型、总线型、线型、树型、全连接型等。

10.2.2.1 环型拓扑

通过通信线路将设备"手拉手"连接形成一个闭合的环。环型拓扑安装相对比较昂贵但带宽比较高，可以跨越比较大的距离。

- 传入的数据取自令牌。
- 传出的数据附加到空的令牌上。
- 通信通常是单向的环，即没有冲突。
- 一个节点发生故障，整个网络即瘫痪。
- 传输介质一般采用双绞线或同轴电缆。
- 需要令牌环接口卡。

10.2.2.2 星形拓扑

星形拓扑是指网络中的各节点设备通过一个网络集中设备（如集线器 Hub 或者交换机 Switch）连接在一起，各节点呈星状分布的网络连接方式。星形拓扑中如果一个节点出现故障，不会影响其他节点的正常运作。但中央节点的负担较重，易形成瓶颈。中央节点一旦发生故障，则整个网络不能工作。

10.2.2.3 总线型拓扑

所有设备都连接到一个被称为总线或骨干的共享通信电缆上。总线网相对便宜，易于布线。以太网采用总线拓扑结构。

10.2.3 网络通信技术

10.2.3.1 以太网

以太网（Ethernet）是施乐公司于 1976 年与 DEC 和 Intel 公司合作开发的计算机局域网组网技术。以太网采用总线型或星型拓扑结构，并支持 10Mbps 的数据传输速率。IEEE 制定的 IEEE 802.3 标准给出了以太网的技术标准，它规定了包括物理层的连线、电信号和介质访问层协议的内容。以太网是当前应用最普遍的局域网技术。

以太网的标准拓扑结构为总线型拓扑，但目前的快速以太网（100BASE-T、1 000BASE-T 标准）为了最大程度地减少冲突、最大程度地提高网络速度和使用效率，使用交换机/集线器（Switch/Hub）来进行网络连接和组织，这样，以太网的拓扑结构就成了星型，但在逻辑上，以太网仍然使用总线型拓扑和 CSMA/CD（Carrier Sense Multiple Access/Collision Detect，即带冲突检测的载波监听多路访问）的总线争用技术。

100Base-T（快速以太网）支持 100Mbps 的数据传输率，而吉比特以太网或称千兆以太网（GbE，Gigabit Ethernet 或 1GigE）支持 1Gbps（1000Mbps）的数据传输率。

10.2.3.2 令牌环

令牌环（Token-Ring）网将各节点连成一个环形拓扑结构。为了解决数据发送竞争问题，令牌环网使用一个称为令牌（Token）的特殊数据帧，使其沿着环路循环。规定只有获得令牌的站点才有权发送数据帧，完成数据发送后立即释放令牌以供其他站点使用。由于环路中只有一个令牌，因此任何时刻至多只有一个站点发送数据，不会产生冲突，而且，令牌环上各站点均有相同的机会公平地获取令牌。

10.2.3.3 IEEE 802.11

IEEE 802.11 和 IEEE 802.11x 是当今无线局域网通用的标准，它是由 IEEE（电气与电子工程师协会）所定义的无线网络通信的标准。

10.2.3.4 蓝牙

蓝牙是一个开放的无线通信技术标准，为移动电话、笔记本电脑、个人计算机、打印机、数码相机、游戏机等创建高安全性的无线个人局域网（PAN），以提供近距离数据交换功能（使用短波无线电传输）。蓝牙应用于：
- 蓝牙耳机移动电话和免提设备之间的无线通信，这也是最初流行的应用。
- 特定距离内计算机间的无线网络。
- 计算机与外设的无线连接，如鼠标、耳机、打印机等。
- 蓝牙设备之间的文件传输。
- 传统有线设备的无线化，如医用器材、GPS、条形码扫描仪、交管设备等。
- 数个以太网之间的无线桥架。
- 家用游戏机的手柄。

- 依靠蓝牙支持，使 PC 或 PDA 能通过手机的调制解调器实现拨号上网。
- 实时定位系统（RTLS）应用："节点"或"标签"嵌入被跟踪物品中，"读卡器"从标签接收并处理无线信号以确定物品位置。

10.2.3.5 红外数据协会

红外数据协会（Infrared Data Association，IrDA）是一个为使用红外线传输的硬件和软件提供标准的工业组织。IrDA 标准已经在各种各样的计算机平台上使用，越来越多的嵌入型应用上也得到了实现。通过红外线，设备之间可以无须线缆即可通信。IrDA 端口的数据传输率与传统并口的传输率相当。唯一的限制是使用 IrDA 的两个设备必须相距比较近（几英尺之内）而且相互之间清晰可见。

IrDA 接口广泛应用于医疗设备、测试和测量设备、掌上电脑、移动电话和笔记本电脑（目前大多笔记本电脑和手机均提供蓝牙无线通信技术）等设备中。

10.2.3.6 无线应用协议

无线应用协议（WAP）是使移动用户使用无线设备（例如移动电话、寻呼机、收发两用无线电通信设备、智能电话和发报机等）随时使用互联网信息和服务的安全规范。WAP 的主要意图是使得无线终端设备能够获得类似网页浏览器的功能。WAP 支持大多数无线网络，而且所有的操作系统均支持 WAP。

10.2.3.7 无线网络

无线网络指无线的计算机网络，一般和电信网络结合在一起，无须电缆节点之间即可相互连接。无线电信网络一般被应用于使用无线电波、微波等电磁波的遥控信息传输系统。

10.2.4 网络互连

10.2.4.1 中继器和集线器

中继器（Repeater）又叫转发器，是对电缆上传输的数据信号进行再生放大，然后转发到其他电缆上，从而避免长距离的信号传输衰减而失真，起到延长信号的传输距离、扩展网络覆盖范围的功能。

中继器主要用于扩展局域网的连接距离。用中继器连接起来的网段所形成的网络，逻辑上等同于单一网段的网络。使用中继器扩充网络距离，其优点是简单、廉价；缺点是当负载增加时，网络性能急剧下降。所以只适合网络负载很轻和网络时延要求不高的场合。中继器不具备网桥和路由器所具备的智能路由功能。

集线器（Hub）又称集中器，是一个共享设备，其实质是一个多端口的中继器，可以对接收到的信号进行再生放大，以扩大网络的传输距离。现代网络技术中，通常使用集线器代替中继器进行网络互连。集线器不具备自动寻址能力，所有的数据均被广播到与之相连的各个端口，容易形成数据堵塞。所以，当网络较大时，应该考虑采用交换机来代替集线器。

10.2.4.2 网桥和交换机

网桥（Bridge）也称桥接器，主要用于连接两个寻址方案兼容的局域网，或采用相同或兼容的网络协议的网络，如两个以太网、两个令牌环网或以太网和令牌环网。

使用网桥进行网络互连，可以扩充网络的规模，其主要优点是提高网络性能以及增强网络的健壮性。当网络负载重而导致性能下降时，如果用网桥将其分为多个网络段，可以大幅度缓解网络通信繁忙的程度，提高通信效率；网桥还可以起到隔离作用，一个网络段上的故障不会影响另一个网络段，从而提高网络的可靠性。其缺点是由于网桥在执行转发前先接收帧并进行缓冲，与中继器相比会带来一定时延。

交换机（Switch）也叫交换式集线器，是局域网中的一种重要设备。它可将用户收到的数据包根据目的地址转发到相应的端口。它与一般集线器的不同之处是：集线器是将数据转发到所有的集线器端口，即同一网段的计算机共享固有的带宽，传输通过碰撞检测进行，同一网段计算机越多，传输碰撞也越多，传输速率会变慢；交换机则具备自动寻址能力，只需将数据转发到目的端口，所以每个端口为固定带宽，传输速率不受计算机台数的增加而受影响，具有更好的性能。

交换机工作原理等同于多端口网桥，现代网络技术中，通常使用交换机代替网桥进行网络互连。而在高性能的网络中，交换机慢慢地取代了集线器。

10.2.4.3 路由器

路由器（Router，又称路径器或宽带分享器）是一种多端口的网络设备，它能够连接多个不同网络或网段，并能将不同网络或网段之间的数据信息进行传输，从而构成一个更大的网络。路由器主要用于异种网络互联或多个子网互联。为了能路由数据包，路由器之间会通过路由协议进行通信并创建和维护各自的路由表。路由表存储了去往某一网络的最佳路径、该路径的"路由度量值"以及下一跳路由器。

路由器是互联网络中使用最广泛的网络互连设备。其优点主要是负载共享和最优路径，能更好地处理多媒体、安全性高、可隔离不需要的通信量、节省局域网的带宽、减少主机负担等，适用于大规模的复杂拓扑结构的网络；其缺点是不支持非路由协议、处理速度比网桥慢、安装复杂、价格高等。

路由器的目的就是"路由"，即把数据包从一个网络路由到另外一个网络。路由器的主要功能：数据包转发、路由选择、拆分和包装数据包、ICMP 消息的处理、防火墙功能、管理功能等。

10.2.4.4 网关

网关（Gateway）亦称网间协议转换器，不仅具有路由器的全部功能，还能实现不同体系结构网络之间的互联。网关支持不同协议之间的转换，实现不同协议网络之间的通信和信息共享。网关可用于 LAN-LAN、LAN 与大型机以及 LAN 与 WAN 的互连。网关提供的服务是全方位的。

网关经常在家庭中或者小型企业网络中使用，用于连接局域网和 Internet。网关也可以是一个地球上的计算机系统，用于在卫星和地面网络之间交换数据信号和语音信号。

10.2.5 网络架构

10.2.5.1 客户端-服务器架构

客户端-服务器结构或主从式架构，简称 C/S 结构，把客户端与服务器区分开来。每一个客户端软件的实例都可以向一个服务器或应用程序服务器发出请求。有很多不同类型的服务器，例如文件服务器、邮件服务器、打印服务器和网络服务器等。虽然它们的目的不同，但基本构架是一样的。

客户端-服务器结构通过不同的途径应用于很多不同类型的应用程序，最常见就是目前在因特网上用的网页。例如，当用户在百度（www.baidu.com）查阅资料时，用户的计算机和网页浏览器就被当作一个客户端，同时，组成百度的计算机、数据库和应用程序就被当作服务器。当用户的网页浏览器向百度请求一个指定的文章时，百度服务器从百度的数据库中找出所有该文章需要的信息，结合成一个网页，再发送回用户的浏览器。

客户端-服务器结构有时又称为两层架构。

10.2.5.2 对等网架构

端对端（peer-to-peer，P2P）又称对等互联网络、对等网，网络上的每个使用端或实体都拥有相同的等级，同时扮演客户端与服务器的角色。这种网络设计架构不同于客户端-服务器架构，客户端-服务器中的资源通常来源于服务器。

10.2.6 信道

通信通道（简称信道）是数据、指令和信息等模拟或数字信号在通信系统中传输的通道，是信号从发射端传输到接收端所经过的传输媒质。广义的信道定义除了包括传输媒质，还包括传输信号的相关设备。有线电视线路和电话线就是两种信道。

10.2.6.1 基本概念

1）带宽

带宽通常指信号所占据的频带宽度；在被用来描述信道时，带宽是指能够有效通过该信道的信号的最大频带宽度。对于模拟信号而言，带宽又称为频宽，以赫兹（Hz）为单位。对于数字信号而言，带宽是指单位时间内链路能够通过的数据量。由于数字信号的传输是通过模拟信号的调制完成的，为了与模拟带宽进行区分，数字信道的带宽一般直接用波特率或符号率来描述。

2）延迟

延迟是指数据包从发送源抵达接收目的地所需要的时间。延迟和带宽分别用于衡量网络的速度和容量。

3）基带

基带，也称为基本频带，是信源（信息源）发出的没有经过调制（频谱移动和变换）

的原始电信号所固有的频带（频率带宽）。以太网属于基带网络。

4）宽带

宽带是能够支持宽范围通信频率的传输媒介。它将媒介的总传输能力划分为多个相互独立的带宽信道，每个信道只工作于特定的频率范围，因而宽带可传播多路信号。有线电视服务就是基于宽带通信线路（主要是光纤）工作的。

5）传输介质

传输介质是网络中传输数据、连接各网络节点的实体（可以是固态、液态或气态等）。传输介质有物理（有线）传输介质和无线传输介质两大类。

10.2.6.2 物理传输介质

物理传输介质使用电线、电缆和其他有形材料发送通信信号。在局域网中常见的物理传输介质有双绞线、同轴电缆、光缆。其中，双绞线是常用的传输介质，它一般用于星型网络中，同轴电缆一般用于总线型网络，光缆一般用于主干网的连接。

1）双绞线

双绞线（Twisted Pair）是目前局域网最常用的一种布线材料。所谓双绞线，就是由两根相互绝缘的铜导线按照一定的规格互相缠绕在一起而成的网络传输介质。使用双绞线可以降低信号的干扰程度：如果外界电磁信号在两条导线上产生的干扰大小相等而相位相反，那么这个干扰信号会相互抵消。

2）同轴电缆

同轴电缆（Coaxial Cable）由四个部分构成。第一部分是最里层的传导物，为铜质或铝质的电缆芯线，用于传输信号；第二部分为包裹着线的绝缘材料；第三部分为一层紧密缠绕的网状编织导线，起着屏蔽层的作用，保护电缆免受电磁干扰；第四部分为同轴电缆的最外一层，是起保护作用的塑料外皮。

同轴电缆传输容量大，抗干扰性好，但价格比较高。

3）光纤

光导纤维（Fiber-optic Cable，简称"光纤"）是一种能利用光的全反射作用来传导光线的透光度极高的光学玻璃纤维。光纤线由纤芯、包层和套层组成。纤芯一般由纯净的玻璃制成；包层包围着核心部分，一般由玻璃或塑料制成，其光密度要比核心部分低；最外层是起保护作用的套层。纤芯的光折射率比包层的光折射率略大些。据光的全反射原理：光从折射率大的介质（纤芯）射向折射率小的介质（包层）的界面时，光在界面处全部被反射回原介质（纤芯）中。故光波束从光导纤维一端进入芯线后，在芯线与包层的界面上作多次全反射而曲折前进。

因为光波在光纤里传输数据的速率很快，且传输衰减极小，所以通过光纤能够实现远距离传输。此外，光波不受电磁干扰的影响，对湿气等环境因素有很强的抵抗能力，因此是理想的通信传输媒体。其缺点是实现的代价较高。光缆一般用于高速的骨干网络中。

10.2.6.3 无线传输介质

1）红外线

红外线，俗称红外光，是波长介乎微波与可见光之间的电磁波，波长在 770 纳米至

1毫米之间，在光谱上位于红色光外侧，是不可见光。具有很强热效应，并易于被物体吸收，通常被作为热源。透过云雾能力比可见光强。在通信、探测、医疗、军事等方面有广泛的用途。

红外传输使用红外线为载体传输数据，因为红外线无法穿过物体，因而要求发射方和接收方彼此处在视线范围内。红外线传输数据的速度很快，可达到100Mbps（100兆比特/秒）。红外传输常用在近距离的通信传输系统中，例如在具有红外传输功能的笔记本电脑和台式电脑之间、手机或掌上电脑和PC之间，都可以通过红外传输来进行数据交换。

红外传输的优点是使用方便、速度较快、传输安全；其缺点是传输距离近，通常在几十到数米的范围内。

2）广播无线电

广播无线电波或射频波是指在自由空间（包括空气和真空）传播的电磁波，是单向无线传输介质，其频率在300GHz以下。无线电技术是通过无线电波传播信号的技术。

无线电技术的原理在于，导体中电流强弱的改变会产生无线电波。利用这一现象，通过调制可将信息加载于无线电波之上。当电波通过空间传播到达收信端，电波引起的电磁场变化又会在导体中产生电流。通过解调将信息从电流变化中提取出来，就达到了信息传递的目的。

3）蜂窝无线电

蜂窝无线电波是一种广泛用于移动通信特别是无线调制解调器和移动电话的双向无线通信方式。

- 蜂窝电话：蜂窝电话或移动电话允许用户通过手持或车载电话进行通信。移动电话其实是一个能够和单元站点进行通信的全双工双向无线电设备。蜂窝电话覆盖区通常分为多个小区。每个小区由一个基站发射机覆盖。理论上，小区的形状为蜂窝状六边形，这也是蜂窝电话名称的来源。当前广泛使用的移动电话系统标准包括GSM、CDMAOne和TDMA。运营商已经开始提供下一代的3G移动通信服务，其主导标准为CDMA2000和UMTS。移动电话除了可以进行语音通信以外，还提供收发短信（短消息SMS）、Web浏览、电子邮件、Internet访问、游戏、蓝牙、红外线、摄影、MMS（彩信、多媒体短信）、MP3播放器、广播以及GPS等功能。高档移动电话（智能手机）提供很多高级计算机的功能。

- GSM：全球移动通信系统，俗称"全球通"，是当前应用最为广泛的移动电话标准。GSM标准的广泛使用使得在移动电话运营商之间签署"漫游协定"后用户的国际漫游变得很平常。GSM较之它以前的标准最大的不同是其信令和语音信道都是数字的，因此GSM被看作是第二代（2G）移动电话系统。从用户观点出发，GSM的主要优势在于提供更高的数字语音质量和替代呼叫的低成本的新选择（例如短信）。从网络运营商角度来看，其优势是能够部署来自不同厂商的设备，因为GSM作为开放标准提供了更容易的互操作性，而且，标准就允许网络运营商提供漫游服务，用户就可以在全球使用他们的移动电话了。GSM使用窄带TDMA，允许在同一无线电频率上同时处理八个通话。

- CDMA：码分多址或分码多重进接、码分复存，被称为第2.5代移动通信技术，是使用扩频技术的数码蜂窝技术，即将需传送的具有一定信号带宽信息数据，用一个带宽远大于信号带宽的高速伪随机码进行调制，使原数据信号的带宽被扩展，再经载波调制并发送出去。接收端使用完全相同的伪随机码，与接收的带宽信号作相关处理，把宽带信号换成原信息数据的窄带信号即解扩，以实现信息通信。与 GSM 等使用 TDMA 技术不同的是，CDMA 并不为每一个用户指定一个特定的频率，而是每个信道使用全部可用频谱。CDMA 手机具有话音清晰、不易掉话、发射功率低和保密性强等特点，最大的优点就是相同的带宽下可以容纳更多的呼叫，而且它还可以随语音传送数据信息。更为重要的是，基于宽带技术的 CDMA 使得移动通信中视频应用成为可能，而且是构建 3G 通信技术的通用平台。

- 3G：第三代移动通信技术，也就是 IMT-2000（国际移动通信-2000），是下一代移动通信系统的通称，指支持高速数据传输的蜂窝移动通信技术。3G 服务能够同时传送声音（通话）及数据信息（电子邮件、即时通信等）。3G 的代表特征是提供高速数据业务，速率至少在 200kbps 以上。第一代手机是指模拟信号手机；第二代手机是指数字信号手机，如 GSM，提供低速率数据业务；2.5G 是指在第二代手机上提供中等速率的数据服务，传输率一般在几十至一百多"kbps"，如 CDMA1x。由于采用了更高的频带和更先进的无线（空中接口）接入技术，3G 标准的移动通信网络通信质量较 2G、2.5G 网络有了很大提高。更高的频带范围和用户分级规则使得单位区域内的网络容量大大提高，同时通话允许量大大增加。3.5G 和 3.75G 甚至为笔记本电脑和智能手机提供"Mbps"的移动宽带访问能力。3G 系统致力于为用户提供更好的语音、文本和数据服务。与现有的技术相比较而言，3G 技术的主要优点是能极大地增加系统容量、提高通信质量和数据传输速率。此外利用在不同网络间的无缝漫游技术，可将无线通信系统和 Internet 连接起来，从而可对移动终端用户提供更多更高级的服务。3G 将在 GSM、TDMA 和 CDMA 之类的无线空中界面上工作。
- 4G：移动电话系统的第四代，是继 3G 之后的无线通信系统。4G 并不是一个明确的定义，在人们的构想中，4G 是集 3G 与 WLAN 于一体的，并能够传输高质量视频图像，它的图像传输质量与高清晰度电视不相上下。4G 系统应该能够以 100Mbps（对于火车、汽车等高速移动通信）、1Gbps（对于行人、静止状态的低速通信）的速度下载，上传的速度也能达到 20Mbps，并能够满足几乎所有用户对于无线服务的要求。4G 应该为智能手机、笔记本无线调制解调器和其他移动设备提供全面的安全的全 IP 移动宽带解决方案，并为用户提供超宽带 Internet 访问、IP 电话、游戏以及流媒体等服务。
- 5G：第五代移动电话行动通信标准，也称第五代移动通信技术。也是 4G 之后的延伸，正在研究中。由于物联网尤其是互联网汽车等产业的快速发展，其对网络速度有着更高的要求，这无疑成为了推动 5G 网络发展的重要因素。因此全球各地政府均在大力推进 5G 网络，以迎接下一波科技浪潮。

4）微波

微波是指波长介于红外线和特高频（UHF）之间的射频电磁波。微波的波长范围大约在 1m 至 1mm 之间，所对应的频率范围是 300 MHz（0.3 GHz）至 300GHz。

微波在雷达科技、微波炉、等离子发生器、无线网络系统（如手机网络、蓝牙、卫星电视及 WLAN 技术等)、传感器系统上均有广泛的应用。微波传输一般发生在两个地面站之间。微波传输有两个显著的特点：一是直线传播；二是受环境条件的影响较大，如大气层的条件、固体物阻挡（如建筑物）等，都会妨碍微波的传播，因而微波通信的使用范围受到一定的限制。

5）通信卫星

通信卫星是太空中用于通信的人造卫星。现代通信卫星使用地球同步轨道、莫尼亚（Molniya）轨道或低纬度轨道。卫星通信传输突破了地域的界限，有海量的带宽，因此具有广泛的应用市场，如长途电话、天气预报、广播电视节目以及军事用途等。卫星通信传输的中继站点设立在绕地球轨道运行的卫星上。

10.2.7 数据处理

计算机数据处理是指使用计算机程序对数据进行输入、汇总、分析并转换成有用的信息的过程。可以是自动化处理系统，包括对数据的记录、分析、排序、汇总、计算、传输和存储等处理。数据处理系统有时又称为信息系统。

10.2.7.1 集中式数据处理

- 将处理、硬件和软件等所有内容放置在一个中心位置。
- 处理效率比较低。
- 数据必须物理传输到计算机。
- 处理后的数据（信息）必须交付给用户。

10.2.7.2 分布式数据处理

- 中央大型机系统与位于不同地点的小型机和微型机互连。
- （位于不同地点的）用户可以控制本地数据的处理。
- 中央大型机处理整个公司（包括分布在不同地点的分公司）的业务。

课文 11

万维网简介

11.1 万维网

万维网（也称为 Web、WWW、W3、3W）是一个由许多互相链接的超文本文档组成的系统，通过 Internet 访问。借助 Web 浏览器，用户可以浏览包含文本、图像、音频、视频以及其他多媒体信息的 Web 页面，并通过超链接在信息之间导航。

在万维网中，每个有用的事物，称为"资源"；并且由一个全局"统一资源标识符"（URL）标识；这些资源通过超文本传输协议（Hypertext Transfer Protocol）传送给用户，而后者通过点击链接来获得资源。

万维网常被当成 Internet 的同义词，这是一种误解，万维网是依靠 Internet 运行的一项服务。

11.1.1 Web 浏览器

Web 浏览器，或简称浏览器，是用来查找并显示万维网上的网页的应用软件。浏览器在超文本传输协议（HTTP）的控制下显示网页。HTTP 定义了信息的格式和传输方式，以及为了响应各种命令 Web 服务器和浏览器所采取的行为。Web 客户端和 Web 服务器通过 HTTP 进行通信。当用户在浏览器中输入网址（URL），实际上就是发送一条 HTTP 命令给 Web 服务器，索取和传输请求的网页并在用户浏览器上显示信息。

两种最流行的浏览器是微软的 Internet Explorer 和 Mozilla 的 Firefox。这两者均为图形浏览器，可以同时显示文字和图形。目前大多数浏览器还支持包括声音和视频的多媒体信息，但有些格式需要插件。

虽然浏览器主要用来访问万维网，但也可以访问由私有网络的 Web 服务器提供的信息或文件系统所提供的文件。

11.1.2 统一资源定位器

统一资源定位器（URL，或称统一资源定位符）是万维网上的文件和其他资源的全球地址。统一资源定位器的标准格式如下：

协议类型://服务器地址（必要时需加上端口号）/路径/文件名

URL 地址的第一部分称为协议标识符，它表示所使用的协议类型名称；第二部分是资源名称，用于指定资源所在的服务器 IP 地址或域名、所在路径以及文件名。协议类型和资源名称是由一个冒号和两个正斜杠分隔。

例如，下面的两个 URL 分别指向 www.cc.ecnu.edu.cn 中的两个不同文件：第一个使用 HTTP 协议获取指定的网页（网站主页、首页）；第二个使用 FTP 协议获取一个可执行文件。

http://www.cc.ecnu.edu.cn/index.html

ftp://ftp.cc.ecnu.edu.cn/exercises.exe

11.1.3 域名

域名用于标识一个或多个 IP 地址，是由一串用点分隔的字符串组成。基于友好的域名，用户可以更方便地访问 Internet。例如，域名 www.microsoft.com 表示微软公司的 IP 地址。URL 可以使用域名来识别特定的网页。例如，http://www.microsoft.com/index.html 中的域名是 www.microsoft.com。

每个域名有一个后缀表示其所属的顶级域（TLD）。常用的顶级域名如下。

- gov——政府部门；
- edu——教育机构；
- org——非盈利组织；
- mil——军事部门；
- com——商业机构；
- net——网络组织；
- int——国际组织；
- cn——中国；
- ca——加拿大；
- jp——日本。

由于互联网是基于 IP 地址而不是域名的，所以每个 Web 服务器需要一个域名系统（DNS）服务器实现 IP 地址与域名的相互转换。例如，域名 www.baidu.com 可转化为 IP 地址：119.75.217.56。

11.1.4 Web 服务器

Web 服务器是一种在万维网上使用超文本传输协议（HTTP）提供给用户网页等信息的计算机程序。每个 Web 服务器都有一个 IP 地址，也可能会有域名。例如，用户在浏览器中输入如下 URL：http://www.cc.ecnu.edu.cn/index.html，则意味着发送一个请求给域名为 www.cc.ecnu.edu.cn 的服务器，服务器获取到名为 index.html 的网页后，将其发送给浏览器。

通过安装服务器软件并连接到 Internet，任何一台计算机均可以充当服务器。常用的 Web 服务器软件包括 NCSA 和 Apache 提供的公共域软件（开源软件包）以及 Microsoft 和 Netscape 等公司提供的商业软件包（如 Internet 信息服务）等。

11.2 HTML 简介

万维网是一种支持电子文档的 Internet 服务器系统。电子文档采用 HTML（超文本标记语言）格式，支持图像、音频、视频以及超链接。网站是存储在 Web 服务器上的相关网页、文档、图片等的集合。

11.2.1 超文本标记语言

超文本标记语言（HTML）是用来创建 Web 网页的一种标准，是一种编辑语言，用于创建包含 Internet Explorer 或 Firefox 等万维网浏览器所使用的样式和链接的文本文件。

通过使用标记（Tags）和属性（Attributes），HTML 定义了 Web 文档的结构和布局。HTML 文档的结构一般如下：

```
<HTML>
<HEAD>
    <TITLE>  页面标题：IT Fundamentals  </TITLE>
</HEAD>
<BODY>
     Everything you need to know about <B> IT Fundamentals </B> is on the Web Page
</BODY>
</HTML>
```

包含 HTML 内容的文件最常用的扩展名是.html 或.htm。

11.2.2 基本要求

11.2.2.1 编写工具

可以用文本编辑器或所见即所得的 HTML 编辑器来编辑 HTML 文件。文本编辑器记事本的启动方法如下：Start→All programs→Accessories→NOTEPAD。Dreamweaver 和 FrontPage 是常用的所见即所得的可视化网页编辑器。

11.2.2.2 浏览器

浏览器软件可以测试所编写的网页是否成功。Internet Explorer 和 Firefox 是常用的浏览器。

11.2.2.3 服务器

如果要使世界各地都能访问网页，需要连接到服务器并保证与其他网站的可链接性。

11.2.3 标记

超文本标记语言（HTML）是用来创建 Web 网页的一种标准。HTML 使用名为标记(tag,

也称标签）的特殊记号，以表明 Web 浏览器显示文本和图形等网页元素的方式。

标记用于指定网页元素的类型、格式和外观。标记通常成对出现，有开始标记和结束标记。每个标签用一对尖括号（< >）括起。

例 1：<HTML> 和 </HTML>，其中
<HTML> 表示 HTML 文件的开始；
</HTML> 表示 HTML 文件的结束。

例 2：<B> This text will be bold </B> 表示文本内容加粗显示。

符号"/"表示标记的结束。标记不区分大小写。

11.2.4 节～11.2.7 节译文略。

11.2.4　编制网页的步骤

（1）启动记事本；
（2）输入网页文件内容；
（3）保存为"文件名.html"或者"文件名.htm"；
（4）启动 Windows 资源管理器（Windows Explorer/File Explorer）；
（5）双击"文件名.html"或者"文件名.htm"，测试网页；
（6）返回记事本—修改—保存；
（7）在 Web 浏览器（例如 Microsoft Edge/Internet Explorer）中执行"刷新（Refresh）"命令，测试网页；
（8）重复第（6）和第（7）步，继续编制和完善网页，直至满意为止。

11.3 节和 11.4 节译文略。

课文 12 计算机和社会

12.1 电子商务

12.1.1 电子商务基本概念

电子商务（E-Business，或 eBusiness）通过计算机网络以电子交易方式进行交易活动和相关服务活动，不仅指在网上实施商品和服务的买卖行为，还包括通过 Internet 提供技术或客户支持，是传统商业活动各环节的电子化、网络化。电子商务包括电子货币交换、供应链管理、电子交易市场、网络营销、联机事务处理、电子数据交换（EDI）、库存管理和自动数据收集系统等。电子商务利用到的信息和通信技术（ICT）包括 Internet、Intranet、Extranet、电子邮件、数据库、电子目录和移动电话等。

电子商贸（E-commerce）是电子商务（E-Business）的子集，仅指通过 Internet 实施商品和服务的买卖行为，即商品或服务的购买、运输和交换等。

12.1.2 电子商务模型

12.1.2.1 B2B

B2B 指的是企业对企业（如生产商和批发商之间，或批发商和零售商）通过电子商务的方式进行商品或服务的交易。例如，一间卖免洗餐具的工厂，若其主要的销售对象为餐厅，这就是 B2B 的模式。B2B 着重于企业间网络的建立、供应链体系的稳固。不同于 B2C，B2B 不用靠规模，而是靠企业间网络的建立稳固其销售。

12.1.2.2 B2C

B2C 指的是企业对顾客（有时也称企业对消费者）透过电子商务的方式进行商品或服务的交易。仍以一间卖免洗餐具的工厂为例，若餐厅提供免洗餐具给用餐人，餐厅对用餐人这种模式为 B2C。某人从零售商处购买了一双鞋也是 B2C 模式。B2C 需要靠规模经济的方式，吸引购买、降低售价来增加利润。

12.1.2.3 C2B

C2B 指的是顾客对企业（有时也称消费者对企业）透过电子商务的方式进行商品或服

务的交易。消费者提供商品或服务，公司付钱。

12.1.2.4　C2C

C2C（消费者对消费者，或称顾客对顾客）是一种网络交易的方式，它是指通过第三方进行个人对个人的电子交易形式。较著名的例子有 EBay 和 Taobao（淘宝）等。

12.1.3　电子购物车

电子购物车是一种电子商务工具（软件或服务），是消费者在网上商店购物的界面。允许用户将商品放在购物车中，购物车会在预定的时间段内保存这些商品。从购物车中还可以超链接到所选商品的颜色、大小、订购数量等信息页面。一旦消费者输入其送货地址，税金和运费等信息会自动计算并显示。卖方还给购物车提供了编号等重要信息，方便消费者跟踪订单。

12.2　电子数据交换

电子数据交换（EDI）是一种利用计算机进行商务处理的方式。EDI 是将贸易、运输、保险、银行和海关等行业的商务信息（包括采购订单、发票、发货通知和目录清单等），用一种国际公认的通用标准格式，形成结构化的事务处理的报文数据格式，通过计算机通信网络（如增值网 VAN 等），使各有关部门、公司与企业之间进行数据交换与自动处理，并完成以贸易为中心的全部业务过程。EDI 包括买卖双方数据交换、企业内部数据交换等。

EDI 不是用户之间简单的数据交换，EDI 用户需要按照国际通用的消息格式发送信息，接收方也需要按国际统一规定的语法规则，对消息进行处理，并引起其他相关系统的 EDI 综合处理。整个过程都是自动完成，无须人工干预，减少了差错，提高了效率。

使用 EDI 的主要优点：① 降低了纸张文件的消费。② 减少了许多重复劳动，提高了工作效率。③ 使得贸易双方能够以更迅速、有效的方式进行贸易，大大简化了订货过程或存货过程，使双方能及时地充分利用各自的人力和物力资源。④ 可以改善贸易双方的关系，厂商可以准确地估计日后商品的需求量，货运代理商可以简化大量的出口文书工作，商业用户可以提高存货的效率，提高竞争能力。

由于 EDI 的使用可以完全代替传统的纸张文件的交换，因此，有人称其为"无纸贸易"或"电子贸易"。

12.3　电子邮件

电子邮件（E-mail，又称电子函件，有时会被简称为电邮或邮件）是指通过 Internet 上的电子通信系统进行信件的书写、发送、存储和接收，属于存储-转发机制。电子邮件是互联网上最受欢迎且最常用到的功能之一。

早期的电子邮件大多是文本格式，其他文件只能以附件的方式发送。随着技术的发展，电子邮件中已经可以包含各种照片、视频等多媒体文件。邮件的发送可以是个人之间的通信，也可以是个人到计算机或计算机到个人，甚至计算机程序之间也可以通信。

电子邮件地址的格式是用户名@域名。@是英文 at 的意思，所以电子邮件地址是表示在某部主机上的一个使用者账号（例 hjiang@gmail.com）。

电子邮件的信息包括两部分：信息头和信息体。信息头包括发件人地址、收件人地址、邮件主题等信息。信息体即电子邮件的内容。

电子邮件的优点：

- 便宜；
- 快捷；
- 方便；
- 一封电子邮件可以发送到许多收件地址；
- 可以添加附件。

12.4 即时消息

即时消息（Instant Messaging，IM），也称即时讯息、即时通信、即时通信，是一个实时通信系统，允许两人或多人使用网络实时地传递文字消息、文件、语音与视频交流。大部分的即时消息服务可提供状态信息——显示联络人名单、联络人是否在线等。在互联网上受欢迎的即时消息服务有 WeChat（微信）、AOL Instant Messenger、Windows Live Messenger、Yahoo! Messenger、QQ、飞信、飞鸽传书和 Skype 等。

即时消息属于在线聊天的一种方式，所不同的是，即时消息是朋友（事先建立的"好友列表"、"朋友列表"或"联系人"）之间的实时的基于文本的网络通信系统，而在线聊天还包括多用户环境下甚至匿名用户之间的基于 Web 的通信。

12.5 新技术和新模式

12.5.1 大数据

大数据（Big Data），或称巨量资料，指的是所涉及的资料非常复杂或者规模巨大到无法通过目前主流软件工具以及传统的数据处理应用，在合理的时间内达到获取、分析、管理、检索、共享、存储、传输、可视化、查询、更新以及信息隐私等，并整理成为帮助企业经营决策更积极目的的数据集合。大数据是需要新处理模式才能具有更强的决策力、洞察发现力和流程优化能力的海量、高增长率和多样化的信息资产。

狭义而言，大数据通常指使用预测分析、用户行为分析，或者特定的高级数据分析方法从海量的数据中抽取有价值的信息。

大数据具有如下特点：Volume（大量）、Variety（多样）、Velocity（高速）、Variability

(可变性)、Veracity（真实性）、Value（价值）等。

12.5.2 互联网+

"互联网+"（Internet Plus 或 Internet +）类似于高速公路和工业四代，是李克强总理在 2015 年 3 月 5 日的政府工作报告中提出，目的是紧跟信息发展的趋势：通过"制定'互联网+'行动计划，推动移动互联网、云计算、大数据、物联网等与现代制造业结合，促进电子商务、工业互联网和互联网金融健康发展，引导互联网企业拓展国际市场。

"互联网+"是创新 2.0 下的互联网发展的新业态，是知识社会创新 2.0 推动下的互联网形态演进及其催生的经济社会发展新形态。"互联网+"是互联网思维的进一步实践成果，推动经济形态不断地发生演变，从而带动社会经济实体的生命力，为改革、创新、发展提供广阔的网络平台。

通俗地说，"互联网+"就是"互联网+各个传统行业"，但这并不是简单的两者相加，而是利用信息通信技术以及互联网平台，让互联网与传统行业进行深度融合，创造新的发展生态。它代表一种新的社会形态，即充分发挥互联网在社会资源配置中的优化和集成作用，将互联网的创新成果深度融合于经济、社会各领域之中，提升全社会的创新力和生产力，形成更广泛的以互联网为基础设施和实现工具的经济发展新形态。

"互联网+"代表着一种新的经济形态，它指的是依托互联网信息技术实现互联网与传统产业的联合，以优化生产要素、更新业务体系、重构商业模式等途径来完成经济转型和升级。"互联网+"计划的目的在于充分发挥互联网的优势，将互联网与传统产业深入融合，以产业升级提升经济生产力，最后实现社会财富的增加。

12.5.3 云计算

云计算（Cloud Computing）是基于互联网的相关服务的增加、使用和交付模式，通常涉及通过互联网来提供动态易扩展且经常是虚拟化的共享资源。云是网络、互联网的一种比喻说法。

对云计算的定义有多种说法。现阶段广为接受的是美国国家标准与技术研究院（NIST）定义：云计算是一种按使用量付费的模式，这种模式提供可用的、便捷的、按需的网络访问，进入可配置的计算资源共享池（资源包括网络，服务器，存储，应用软件，服务），这些资源能够被快速提供，只需投入很少的管理工作，或与服务供应商进行很少的交互。

云计算和存储解决方案提供给用户和企业通过第三方数据中心（可能远离用户，城市之间甚至跨越全球的距离范围）存储和处理数据的各种不同能力。云计算依赖于资源的共享以达到一致性和规模经济，就如同电网应用效果。

12.5.4 物联网

物联网（The Internet of Things，IoT）是指通过各种信息传感设备，实时采集任何需要监控、连接、互动的物体或过程等各种需要的信息，与互联网结合形成的一个巨大网络。物联网将物理设备、传播媒介（也指互连设备或者智能设备）、楼宇等对象，通过电子器件、软件、传感器、调节器以及网络连接进行相互联网，从而实现这些对象之间的数据收集和交换功能。2013 年，"物联网全球标准化行动(简称 IoT-GSI)"将物联网定义为："信息社会的基础设施"。物联网允许对象之间跨当前的网络架构相互感知和控制，并提供物理世界与计算机系统直接一体化的机会，从而提高效率、正确性和经济效益。融合了传感器和调节器的物联网是一种赛博物理系统，其中还包含智能网格、智能家居、智能运输以及智能城市等技术。物联网中的每个对象都是通过其嵌入的计算系统来识别，这些对象可以与当前的互联网基本架构相融合。专家预测，到 2020 年，物联网将包含多达 500 亿个的对象。

一般而言，物联网可以为设备、系统和服务提供比机器对机器（M2M）通信还要先进的连接技术，可以覆盖更加广泛的协议、域名和应用。物联网所嵌入设备（包括智能对象）之间的互连，可以带给几乎所有领域的自动化技能，同时提供类似于智能网格的高级应用，甚至可以扩展到智能城市。

12.5.5　移动网络

移动网络（Mobile Web）是指通过移动网络或者无线网络，将基于浏览器的互联网服务用于手持式移动设备，例如智能手机、平板电脑等，从而实现互联网访问。移动网络主要指的是基于浏览器的 Web 服务。

传统上，通过笔记本和台式机上的固定线路网络服务来链接到万维网。但是，便携式和无线设备提供了更易于访问互联网的手段和方式。国际电信联盟（ITU）2010 年初的一份报告就表明，按照当前的发展趋势，在未来的 5 年内，使用笔记本电脑和智能手机设备访问 Web 的用户将远远超过使用台式机设备访问 Web 的用户。

移动 Web 应用和本地应用（native applications，一种基于智能手机本地操作系统如 iOS、Android、WP 并使用原生程序编写运行的第三方应用程序）的区别越来越小，因为移动浏览器可以直接访问移动设备（包括加速器和 GPS 芯片）的硬件，因此基于浏览器的应用的速度和能力得到大大提高。对复杂用户界面图像功能的永久存储和访问可以进一步降低对特定平台本地应用研发的需求。

12.5.6　人工智能

人工智能（Artificial Intelligence，AI）是指对人的意识、思维的信息过程的模拟，是机器所展示的智能，它是研究、开发用于模拟、延伸和扩展人的智能的理论、方法、技术及应用系统的一门新的技术科学。

在计算机科学中，一个"理想"的人工智能机器是一个灵活的理性的代理，可以感知环境，并且采取相应的行为，以最大程度地实现其目的。通俗而言，"人工智能"这个术语是指使用机器来模拟人脑的"认知"功能，例如，"学习"和"问题解决"能力。当机器变得越来越有才能，脑力行为就不再需要智能了。例如，作为一种常规技术的光学字符识别不再被视为"人工智能"。目前被视为人工智能的技术包括成功理解人类语言、高水平战略博弈系统（例如围棋和国际象棋）、自驾车、演绎复杂数据等。人工智能还被认为如果一直持续发展下去的话，很有可能给人类带来危害。

人工智能研究领域可以分为如下几个子类：特定问题的研究、特定方法的研究、特定工具的使用以及如何满足特定的应用。

人工智能研究的主要问题和目标包括：推理、认知、规划、学习、自然语言处理（沟通）、感知以及移动和操纵对象的能力。实现通用智能是一种长期目的，研究方法包括：统计方法、计算智能、软计算（例如：机器学习）以及符号人工智能。

许多方法和工具可以用于人工智能，包括：各种搜索和数学优化方法、逻辑方法、概率以及经济学方法。人工智能领域充分融合了计算机科学、数学、心理学、语言学、哲学、神经科学以及人工心理等理论。

12.5.7 商务智能

商务智能（Business Intelligence，BI），又称商业智慧或商业智能，是一套完整的解决方案，使用现代数据仓库技术、在线分析技术、数据挖掘和数据展现技术进行数据分析，将企业中现有的数据进行有效的整合，快速准确地提供报表并提出决策依据，帮助企业做出明智的业务经营决策，以实现商业价值。

商务智能是一种使用战略、过程、应用、数据、产品、技术以及技术架构来支持商务信息的收集、分析、展示和传播的解决方案。商务智能技术可以用于处理大量结构化以及非结构化的数据，从而确定、研发以及创建新的战略性的商机。商务智能的目的是便于演绎这些大量的商务数据。基于洞察力来确定新的时机以及实现有效的战略，可以提供具有竞争力的市场优势以及长期的稳定性。

商务智能技术为商务运作提供历史性的、当前的以及预测性的信息。商务智能技术的主要功能包括：报表功能、联机分析处理、数据分析、数据挖掘、过程挖掘、复杂事件处理、商务性能管理、基准化分析、文本挖掘、预测分析以及规范性分析等。

商务智能可用于支持广泛的商务决策，包括运行型决策和战略型决策。基本的运作决策包括产品定位或定价。战略型决策包括在最广泛的层次中确定优先性、目标和方向。总之，当外部数据（来源于市场）和内部数据（金融数据和运作数据）有机结合，最能体现商务智能的有效性，这种有机结合，可以提供更强的竞争力，从而创建任何单一数据集所不能体现的"智能"。商务智能可以用于很多的应用领域，借助商务智能，机构可以洞悉新市场、评估产品和服务在各类市场中的需求和适用性，以及判断营销行为的影响力。

12.5.8 深度学习

深度学习（Deep Learning，DL 也称为深度结构化学习、分层次式学习、深度机器学习），是机器学习研究的一个分支，它使用具有复杂处理层的深度图，基于算法集同时结合多重线性和非线性变换，对数据进行高级抽象。

深度学习基于对具有代表性的数据的学习实现机器学习。每个观察（例如，一幅图像）可以采用各种方式表示，例如每个像素的强度值向量，或者使用一种更抽象的方法例如边的集合、特定形状所在的区域等。某些表示方法可能比其他方法更好，更能简化学习任务（例如，脸部识别或者脸部表情识别）。对深度学习的其中一种期望是，希望此方法可以为无监督或半监督特性学习和分层特征提取提供更有效的算法。

对深度学习领域的研究试图为大规模的无标签数据提供更好的表示方法、创建更好的表示模型。其中一些表示的灵感来源于神经科学的研究，并且基于神经系统中信息处理和通信模式的演绎，例如用于定义大脑中不同刺激和相应神经元响应之间的关系的神经编码。

不同的深度学习架构，例如深度神经网络、深度卷积神经网络、深度信念网络、循环神经网络等，均已经应用于诸如计算机视觉、自动语音识别、自然语言处理、音频识别和生物信息学等领域，并在不同的任务中产生了目前技术发展水平下最佳的处理结果。

12.5.9　Docker

Docker 是一个开源的应用容器引擎，可以自动配置软件容器中的 Linux 应用，让开发者可以打包他们的应用以及依赖包到一个可移植的容器中，然后发布到任何流行的 Linux 机器上，也可以实现虚拟化。

Docker 容器采用完整的文件系统形式打包了一系列软件，以支持其运行所需要的代码、运行时、系统工具、系统库等等任何需要安装在服务器上的内容，目的是确保在任何环境中都能够正常运行。

容器为 Linux 上系统层虚拟化提供了额外的抽象层和自动化机制。Docker 使用 Linux 内核中的资源隔离特性，例如 Cgroups 和 kernel 命名空间，以及联合文件系统，例如 aufs 以及其他允许独立的"容器"在单一 Linux 实例中运行的机制，从而避免启动和维护虚拟机的开销。

容器完全使用沙箱机制，相互之间不会有任何接口。几乎没有性能开销,可以很容易地在机器和数据中心中运行。最重要的是,他们不依赖于任何语言、框架包括系统。

12.5.10　3D 打印

3D 打印（3D Printing，3DP）技术，又称为增材制造，是快速成形技术的一种，它是一种以数字模型文件为基础，将计算机设计出的三维数字模型分解成若干层平面切片，然后由 3D 打印机把粉末状、液状或丝状塑料、金属、陶瓷或砂等可黏合材料按切片图形逐层叠加，最终堆积成完整三维物体的技术。该技术综合了数字建模技术、信息技术、机电控制技术、材料科学与化学等诸多方面的前沿技术知识，是一种具有很高科技含量的综合性应用技术。

3D 打印通常是采用数字技术材料打印机来实现的，可以实现大规模的个性化生产，制造出传统生产技术无法制造出的外形，并且可以实现首件的净型成形，大大减小了后期的辅助加工量，避免了委外加工的数据泄密和时间跨度。另外，由于其制造准备和数据转换的时间大幅减少，使得单件试制、小批量生产的周期和成本降低，特别适合新产品的开发和单件小批量零件的生产。这些优势使 3D 打印成为一种潮流，目前已在建筑、工业设计、珠宝、鞋类、模型制造、汽车、航空航天、医疗、教育、地理信息系统等诸多领域都得到了广泛的应用。

12.6 社会问题

12.6.1 网络犯罪

计算机犯罪,又称网络犯罪、电子犯罪、高科技犯罪或电子犯罪,是将计算机或网络作为犯罪工具、犯罪目标或者犯罪地点而实施的犯罪活动,例如最近比较流行的"网络钓鱼"就是一种利用计算机与网络来实施的犯罪行为。犯罪分子使用了计算机或网络来进行欺诈、偷盗、勒索、恐吓、造假和非法侵占等犯罪行为。

12.6.1.1 计算机安全风险

计算机安全风险是指可能导致计算机硬件、软件、数据、信息或处理能力等的损坏或损失的任何事件或行为。计算机安全风险包括以下几种。

1)系统故障

系统故障是计算机故障。

2)信息失窃

信息失窃是指个人或机密信息被盗。

3)软件失窃

软件失窃是指软件被盗。例如,程序被删或非法拷贝等。

4)硬件失窃

硬件盗窃是指计算机设备被盗。

5)计算机病毒、蠕虫和特洛伊木马

计算机病毒、蠕虫和特洛伊木马是导致计算机和计算机上的信息损坏的恶意程序。它们可能使用户的网络和操作系统变慢,危害严重时甚至会完全破坏计算机系统,并且,还可能使用户计算机将病毒传播给朋友、家人、同事以及到互联网的其他地方,在更大范围内造成危害。

- 计算机病毒:计算机病毒是一种在用户不知情或未经批准下,能自我复制及运行的计算机程序或代码片段。计算机病毒往往会影响受感染计算机的正常运作。所有的计算机病毒都是人为的。一个最简单的病毒就是不断地自我复制,从而耗尽内存和硬盘空间,甚至导致系统崩溃。更危险的病毒是通过网络不断传播,给计算机系统带来安全隐患。计算机病毒在传播期间一般会隐蔽自己,由特定的条件触发,然后开始产生破坏。计算机病毒具有传播性、隐蔽性、感染性、潜伏性、可激发性、表现性或破坏性等特征。
- 蠕虫病毒:蠕虫病毒是一种通过计算机网络不断自我复制的程序或算法,通常会执行恶意行为,例如耗尽计算机的资源,导致系统崩溃或网络中断。

蠕虫病毒与普通病毒之间有着很大的区别。一般认为,蠕虫病毒是一种通过网络传播的恶性病毒,它具有病毒的一些共性,如传播性、隐蔽性、破坏性等等,同时具有自己的一些特征，如不利用文件寄生（有的只存在于内存中），对网络的访问造成拒绝服务，以及和黑客技术相结合，等等。

普通病毒需要传播受感染的驻留文件来进行复制，而蠕虫不使用驻留文件即可在系统之间进行自我复制，普通病毒的传染能力主要是针对计算机内的文件系统而言，而蠕虫病毒的传染目标是互联网内的所有计算机。它能控制计算机上可以传输文件或信息的功能，一旦用户系统感染蠕虫，蠕虫即可自行传播，将自己从一台计算机复制到另一台计算机，更危险的是，它还可大量复制。因而在产生的破坏性上，蠕虫病毒也不是普通病毒所能比拟的，网络的发展使得蠕虫可以在短短的时间内蔓延整个网络，造成网络瘫痪。

局域网条件下的共享文件夹、电子邮件 E-mail、网络中的恶意网页、大量存在着漏洞的服务器等，都成为蠕虫传播的良好途径，蠕虫病毒可以在几个小时内蔓延全球，而且蠕虫的主动攻击性和突然爆发性将使得人们措手不及。此外，蠕虫会消耗内存或网络带宽，从而导致计算机变慢或崩溃或网络中断，而且它的传播不必通过"宿主"程序或文件，因此可潜入用户系统并允许其他人远程控制用户计算机，这也使其危害远比普通病毒大。典型的蠕虫病毒有尼姆达、振荡波等。

- 特洛伊木马病毒：特洛伊木马指的是一种恶意的后门程序，黑客用其盗取用户的个人信息，甚至远程控制对方的计算机，然后通过各种手段传播或者骗取用户执行该程序，以盗取用户的账户、密码等各种数据信息。与病毒相似，木马程序有很强的隐秘性、伪装性，随操作系统启动而启动，看起来是合法程序，但一旦执行，会在不经意间对计算机系统产生破坏、删除文件或窃取数据。

特洛伊木马与病毒的重大区别是特洛伊木马不具传染性，它并不能像病毒那样能够自我复制，也并不"刻意"地去感染其他文件，它主要通过将自身伪装起来，吸引用户下载执行。特洛伊木马中包含能够在触发时导致数据丢失甚至被窃的恶意代码，要使特洛伊木马传播，必须在计算机上有效地启用这些程序，例如打开电子邮件附件或者将木马捆绑在软件中放到网络吸引用户下载执行等。现在的木马一般主要以窃取用户相关信息为主要目的，相对一般病毒而言，病毒破坏用户信息，而木马窃取用户信息。典型的特洛伊木马有灰鸽子、网银大盗等。

木马的植入通常是利用了操作系统的漏洞，绕过了对方的防御措施（如防火墙）。中了特洛伊木马病毒的计算机，因为资源被占用，速度会减慢，莫名死机，且用户信息可能会被窃取，导致数据外泄等情况发生。

6）黑客

黑客是为了谋利或者挑战而入侵到计算机和计算机网络中的人。黑客往往是狂热的电脑迷，即热衷于学习程序设计语言和研究计算机系统（特别是计算机网络）内部运作秘密，是水平高超的计算机专家。

黑客通常具有硬件和软件的高级知识，并有能力通过创新的方法剖析系统。"黑客"能使更多的网络趋于完善和安全，他们以保护网络为目的，而以不正当侵入为手段找出网络漏洞。

- 白帽黑客：白帽黑客是有能力破坏计算机的安全但并不具恶意目的的黑客。白帽黑客一般有清楚的道德规范并常常试图同企业合作去发现并改善安全漏洞。白帽黑客测试网络和系统的性能以判定它们是否能够承受入侵。通常，白帽黑客攻击自己的系统，或被聘请来攻击客户的系统以便进行安全审查。这类黑客常被称为道德黑客。学术研究人员和专职安全顾问就属于白帽黑客。
- 黑帽黑客：黑帽黑客是利用公共通信网路，如互联网和电话系统，在未经许可的情况下入侵其他人的计算机系统，破坏计算机安全，实施信用卡欺诈、身份盗窃、侵犯隐私等不合法行为的黑客。
- 黑客激进分子：或称激进黑客、黑客活动分子、倡权黑客、黑客行动主义者，是利用黑客技术发布社会、思想、宗教或政治信息。不像为了经济利益恶意破坏系统的黑客，黑客激进分子的破坏性行动（如拒绝服务攻击）是为了引起关注。对于黑客激进分子而言，这是一种利用互联网功能的叛逆和抗议行为。
- 脚本小子：脚本小子是指那些完全没有或仅有一点点黑客技巧，而只是按照指示或运行某种骇客程序来达到破解目的的人。脚本小子是利用他人所编辑的程序来发起网络攻击的网络闹事者，他们通常不懂得攻击对象的设计和攻击程序的原理，不能自己调试系统发现漏洞，实际职业知识远远不如他们通常冒充的黑帽黑客。不少青少年因为网络入侵、传播病毒木马偷窃获利、破坏报复攻击等原因而实施计算机犯罪。
- 网络恐怖分子：网络恐怖分子利用黑客技术实现其政治目的，但它又不像普通的黑客攻击那样，只是出于个人喜好或想成名的愿望，造成一般的拒绝服务和麻烦，而是出于政治目的，想引起物理侵害或造成巨大的经济损失，其可能的目标包括语音通话系统、金融行业、电力设施、供水系统、油气能源、机场指挥中心、铁路调度、军事装备等国家基础设施，因此它也是一种暴力行为。网络恐怖分子企图通过制造能引起社会足够关注的伤亡来实现其政治或宗教目的。借助网络，网络恐怖分子不仅将信息技术用作武器来进行破坏或扰乱，而且还利用信息技术在网上招兵买马，并且通过网络来实现管理、指挥和联络。与传统的恐怖活动相比，它并不会造成直接的人员伤亡，而且它使用的科技手段更为高明、隐蔽。借助散布病毒、猜测口令进而攻入系统、置入木马程序进行后门攻击、截取程序等黑客攻击方法扰乱和摧毁信息接收和传递机制。
- 骇客：骇客是"Cracker"的音译，就是"破解者"的意思，是恶意地（一般是非法地）企图破解或破坏数据、程序、系统及网络安全、窃取信息的黑客。黑客和骇客根本的区别：黑客建设，而骇客破坏。

12.6.1.2 Internet 安全风险

Internet 上没有中央管理员进行统一管理，因此安全风险更大。

1）拒绝服务攻击

拒绝服务攻击（DoS）或分布式拒绝服务攻击（DDoS），亦称作洪水攻击，是使用无意义的网络流量对网络进行密集式的如洪水般的攻击，造成网络瘫痪甚至主机死机，从而阻止合法用户对正常网络资源的访问。DoS 要么迫使目标计算机重启，要么消耗目标计算机的资源使其不能提供服务，要么破坏用户间的通信介质使其不能有效通信。

黑客使用 DoS 攻击以阻止合法用户使用计算机资源。DoS 攻击试图用超出被攻击目标处理能力的海量数据消耗磁盘空间、内存、进程甚至网络带宽等系统资源，破坏两台计算机之间的连接，阻止用户的合法访问，破坏主机对特定系统的服务。DoS 攻击会导致用户丢失有价值的资源，如 E-mail 服务、对 Internet 或 Web 服务器的访问等。

2）数字证书

数字证书（Digital Certificate）又称为身份证书、公共密钥证书或电子证书，是一个包含个人身份信息的特殊电子文件，是国内外普遍采用的一整套成熟的信息安全保护机制的体现。数字证书是由权威公正的第三方机构即 CA（Certificate Authority）中心签发的，以数字证书为核心的加密技术可以对网络上传输的信息进行加密和解密、数字签名和签名验证，确保网上传递信息的机密性、完整性，以及交易实体身份的真实性，签名信息的不可否认性，从而保障网络应用的安全性。

3）数字签名

数字签名（Digital Signature）是数字证书的重要应用功能之一，所谓数字签名是指证书用户（甲）用自己的签名私钥对原始信息进行密码变换后所得的数据。信息接收者（乙）使用信息发送者的签名证书对附在原始信息后的数字签名进行解密后获得消息摘要，并对收到的原始数据采用相同的密码变换算法计算其消息摘要，将二者进行对比，即可校验原始信息是否被篡改。数字签名可以完成对数据完整性和传输数据行为不可抵赖性的保护。

数字签名采用了规范化的程序和科学化的方法，用于鉴定签名人的身份以及对一项电子数据内容的认可。它还能验证出文件的原文在传输过程中有无变动，确保传输电子文件的完整性、真实性和不可抵赖性。

数字签名对于电子商务以及大多数认证机制很重要。为保证有效性，数字签名必须是不可伪造的，有许多加密技术可保证其安全级别。

12.6.2　信息安全

信息安全是指对于信息或信息系统的安全保障，以防止其在未经授权情况下的使用、泄露或破坏。计算机用户会选择和构建各种控制和保护措施以防止系统和数据被有意或无意地破坏。这些安全措施包括：

- 身份 ID（用户名）；
- 密码、语音识别、指纹识别、激光眼扫描等；
- 版权保护软件（防止软件复制）；
- 系统的审计检查；
- 通信数据的加密；
- 灾难恢复计划的研发。

12.6.2.1　用户 ID

用户 ID（UID，或用户名称）是英文字母和/或数字字符的组合，用于标识一个特定的用户，用户 ID 一般具有唯一性。

12.6.2.2　密码

密码是一个用于身份验证的保密的字符串，它被用来保护不想被别人看到的隐私以及防止未经授权的操作。

用户名（账号）和密码经常被用来登录受到保护的操作系统、手机、有线电视解码器、自动取款机等。通常，计算机用户需要密码来登录系统、收发电子邮件，以及控制程序、数据库、网络和网站，甚至在线浏览新闻等。

12.6.2.3 生物识别设备

生物识别设备是一种将生物体（一般特指人）本身的物理或行为特征转换成数字代码并与存储在计算机中的数字代码进行比对的设备。

生物识别设备和系统包括指纹扫描仪、掌型识别系统、面部识别系统、语音认验系统、签名认验系统、虹膜识别系统等。

12.6.2.4 加密

加密是将数据转换成秘密代码。加密是实现数据安全性最有效的方法之一。必须有密钥或密码解密后才能读取加密的文件。未加密的数据称为纯文本（或明文），加密的数据称为密文。

12.6.2.5 防火墙

防火墙是一种计算机系统或网络，旨在防止未经授权的访问，同时允许授权的通信。它是基于一系列规则和标准允许或拒绝网络传输的设备或设备的集合。

防火墙可以采用硬件或软件或两者组合来实现。防火墙通常用于阻止未经授权的 Internet 用户访问连接到 Internet（特别是 Intranet）的专用网。所有进入或离开 Intranet 的消息必须经由防火墙，由防火墙检测每条信息，并阻截那些不符合指定安全标准的信息通过。防火墙是内部网和外部网之间、专用网与公共网之间构造的保护屏障；是不可信的 Internet 与可信任的 Intranet 之间建立起的安全网关，从而保护 Intranet 免受非法用户的侵入。

防火墙技术有以下几种类型。

- 数据包过滤：数据包过滤检查网络上的每个数据包，并基于用户定义的规则接受或拒绝数据包。虽然难以配置，但相当有效，而且对用户基本是透明的。但很容易受到 IP 欺骗的攻击。
- 应用网关：将安全机制用于特定的应用中，如 FTP 和 Telnet 服务器。这种方法非常有效，但可能会导致系统性能下降。
- 电路级网关：当建立 TCP 或 UDP 连接时所应用的安全机制，此时主机间传输的数据包无须进一步检测。
- 代理服务器：拦截进出网络的所有消息。代理服务器可有效地隐藏真实的网络地址。

12.6.3 隐私权

信息隐私权是个人和公司有权拒绝或限制对其信息的收集和使用。隐私有时也指不想被公众注意或认出的愿望。隐私是"不受干扰的权利"（right to be let alone）。如果某事对某人而言是私有的，则意味着其中一定存在特殊性或敏感性。隐私信息的曝光度取决于公众对其的接受程度，可能因地点和时间的不同而异。隐私除了涵盖信息安全外，还包括对信息的合法使用和保护。

隐私保护是计算机伦理学最早的课题。传统的个人隐私包括姓名、出生日期、身份证号码、婚姻、家庭、教育、病历、职业、财务情况等数据，现代个人数据还包括电子邮件地址、个人域名、IP 地址、手机号码以及在各个网站登录所需用户名和密码等信息。随着计算机信息管理系统的普及，越来越多的计算机从业者能够接触到各种各样的保密数据。这些数据不仅仅局限为个人信息，更多的是企业或单位用户的业务数据，它们同样是需要保护的对象。

12.6.3.1 cookies

cookie，又称网络 cookie，浏览器 cookie 和 HTTP cookie，指某些网站为了辨别用户身份而储存在用户本地终端（Client Side）上的数据（通常经过加密）。

服务器可以利用 cookies 包含的信息筛选并维护这些信息，以判断在 HTTP 传输中的状态。cookies 的一个重要应用是判定注册用户是否已经登录网站，用户可能会得到提示，是否在下一次进入此网站时保留用户信息以便简化登录手续，这些都是 cookies 的功用。另一个重要应用场合是"购物车"之类的处理。用户可能会在一段时间内在同一家网站的不同页面中选择不同的商品，这些信息都会写入 cookies，以便在最后付款时提取信息。

cookies 可以保持登录信息到用户下次与服务器的会话，换句话说，下次访问同一网站时，用户会发现不必输入用户名和密码就已经登录了（当然，不排除用户手工删除 cookies）。而还有一些 cookies 在用户退出会话时就被删除了，这样可以有效保护个人隐私。

如果在一台计算机中安装多个浏览器，每个浏览器都会以独立的空间存放 cookies。因为，一方面，cookies 不但可以确认用户，还能包含计算机和浏览器的信息，所以一个用户用不同的浏览器登录或者用不同的计算机登录，都会得到不同的 cookies 信息，另一方面，对于在同一台计算机上使用同一浏览器的多用户群，cookies 不会区分他们的身份，除非他们使用不同的用户名登录。

cookies 会影响 Web 用户的隐私和匿名。因为 cookies 包含了一些敏感消息：用户名、电脑名、使用的浏览器和曾经访问的网站。用户不希望这些内容泄露出去，尤其是当其中还包含私人信息时。

除了隐私问题，cookies 也有一些技术缺陷。特别地，它们并不总是可以准确地识别用户，它们还可能被用于安全攻击。

12.6.3.2 间谍软件

间谍软件（Spyware）是在未经用户许可的情况下搜集用户个人信息的计算机程序。间谍软件本身就是一种恶意软件，用来侵入用户计算机，在用户不知情或未经用户准许的情况下有意或者无意对用户的计算机系统和隐私权进行控制和破坏。间谍软件所收集的数据范围可以很广泛，从该用户平日浏览的网站、电子邮箱地址、到诸如用户名、密码、信用卡号等个人数据。

间谍软件通常包装为免费软件或共享软件，诱使用户从 Internet 上下载。一旦安装，间谍软件就会监视用户的行为，并将信息传输给间谍软件的作者。

间谍软件采用一系列技术来记录用户的个人信息，如监控击键、扫描用户计算机上的文件、窥探诸如聊天程序或文字处理软件、安装其他间谍软件、解读 cookies、更改 Web 浏览器的默认主页、录制用户访问 Internet 的行为等，并将信息及时反馈给间谍软件的作者，间谍软件的作者再以宣传或市场目的使用这些信息，或者兜售以牟取暴利。间谍软件的用途也多种多样，从盗窃用户的网上账户（主要是银行信用卡账户）和密码到统计用户的网络行为以作为广告用途。一些间谍软件统计用户访问的网站并且不断在用户计算机上弹出广告窗口，但是更多间谍软件搜集用户的密码信息以侵占用户的财产。

cookies 不被视为间谍软件，因为人们知道它们的存在。

12.6.3.3　垃圾邮件

一般来说，凡是未经用户许可就强行发送到用户邮箱中的任何电子邮件都可称作是垃圾邮件。垃圾邮件一般具有批量发送的特征。内容包括赚钱信息、成人广告、商业或个人网站广告、电子杂志、连环信等。垃圾邮件可以分为良性和恶性的。良性垃圾邮件是各种宣传广告等对收件人影响不大的信息邮件，恶性垃圾邮件是指具有破坏性的电子邮件。

12.6.3.4　后门

后门指未授权的用户（或恶意软件）绕过软件的安全性控制而从比较隐秘的通道获取对程序或系统访问权的黑客方法。在软件开发时，设置后门可以方便修改和测试程序中的缺陷。但如果后门被其他人知道（可以是泄密或者被探测到后门），或是在发布软件之前没有去除后门，那么它就对计算机系统安全造成了威胁。

12.6.3.5　内容过滤

内容过滤是限制访问 Web 上某些内容的过程，常用于过滤电子邮件和 Web 访问等。

1）互联网内容分级协会（ICRA）

互联网内容分级协会（ICRA）是为了让儿童安全地使用互联网，推进网络上内容过滤的国际性非营利组织，于 1999 年由英国的 IWF（Internet Watch Foundation）、美国的 RASC（Recreational Software Advisory Council）和德国的 ECO（Electronic Commerce Forum）的 3 个非营利组织共同设立，本部设在英国。ICRA 是在保证网络上内容自由表现的同时，以监护人能够掌握孩子连接到哪个网站为目标建立的业界内自主性限制团体。ICRA 主要提供了描述性的词汇，也就是所谓"ICRA 问卷"，问卷上简单定义有没有特定的元素。内容提供者检查他们的网站包含或不包含那些问卷中的元素,据此生成一个包含标签的小文件。关键的一点是，由内容提供者通过 ICRA 的标签系统对网络内容分级，而 ICRA 自身不对网站进行价值评判。主观上判断是否打开由用户决定。

2）Web 过滤软件

网页过滤软件，或称内容控制软件，是一个限制对 Web 上某些内容进行访问的程序。有些还使用特定的单词来过滤网站，还可以过滤电子邮件、聊天室和程序等。

12.6.4　计算机伦理

计算机伦理（Computer Ethics），又称信息伦理、网络伦理学，属于实践哲学的一个分支，是计算机行业人员开发和使用计算机相关技术和产品以及 IT 系统时的行为规范和道德指引。随着互联网的发展，隐私问题、间谍软件以及 Web 浏览器的 cookies 等日益成为技术方向的伦理道德问题。

12.6.4.1 知识产权

知识产权，指权利人对其原创的智力劳动成果所享有的专有权利。常见的知识产权包括著作权、商标、专利、工业设计权、商业秘密等。

知识产权事实上并非真正意义上的产权，它更像是一种垄断特权——在一段时间内对于智慧活动成果的垄断。实证经验表明，知识产权的保护能够确保智慧活动创造者的利益受到保护，并鼓励更多智慧活动的产生，从而对社会经济的发展起到推进作用，其他公众也能从知识产权保护中受益。知识产权所有者的利益可以通过对使用者收取费用，或在一段时间内禁止他人抄袭、竞争来获得保护。

知识产权一般分为三类：

（1）著作权。它包括文学和艺术作品，诸如小说、诗歌、戏剧、电影、音乐作品、绘图、绘画、摄影、雕塑以及建筑设计等。与著作权相关的权利包括表演艺术家对其表演的权利、录音制品制作者对其录音制品的权利以及广播电视组织对其广播和电视节目的权利。著作权持续到作者逝世后至少 50 年。

（2）工业产权。它包括发明（专利）、商标、工业品外观设计以及原产地地理标志等。专利保护期一般 20 年，工业设计保护至少 10 年，而商标则可无限期保护。

（3）商业秘密。企业可以认定任何信息为"商业秘密"，禁止能够接触这些机密的人将秘密透露出去，一般是通过合约的形式来达到这种目的。只要接触到这些秘密的人在获取这些机密前签署合约或同意保密，他们就必须守约。商业秘密的好处是没有时限，而且任何东西都可被认定为商业秘密。例如可口可乐的配方就属商业秘密，100 多年来外界都无法获知可口可乐的全部成分。

12.6.4.2 著作权

著作权，也称版权，是法律赋予作家、作曲家、剧作家、发行商或经销商独家出版、制作、销售文学、音乐、戏剧或艺术作品的权利。

著作权分为著作人格权与著作财产权。其中著作人格权的内涵包括了公开发表权、姓名表示权及禁止他人以扭曲、变更方式，利用著作损害著作人名誉的权利。著作财产权是无形的财产权，是基于人类智识所产生的权利，包括重制权、公开口述权、公开播送权、公开上映权、公开演出权、公开传输权、公开展示权、改作权、散布权、出租权等。

著作权是有期限的权利，在一定期限经过后，著作财产权即消失，而属公共领域，任何人皆可自由利用。在著作权的保护期间内，即使未获作者同意，只要符合"合理使用"的规定，亦可利用。

12.6.4.3 剽窃

剽窃，或称抄袭，是非法挪用、模仿、盗取或出版其他作者的语言、思想、观点或表述，并作为自己的原创作品，是一种侵犯著作权的行为。

剽窃被认为是一种学术欺骗行为，违反了新闻职业道德，会受到开除公职等制裁。

剽窃不是犯罪，而是违反了道德规范。

12.6.4.4 盗版

软件盗版就是在没有得到授权的情况下对有著作权的软件进行使用、复制、分发或销售。大多数零售程序只允许在一台计算机上使用或只由一个用户在任何时间使用。通过购买软件，购买者不仅是软件的所有者，更是软件的授权用户。用户可以复制该程序作为备份，但复制给朋友和同事则是违法的。

常见的软件盗版包括仿冒软件（伪正版软件）、软偷（随意复制，即只购买了一个正版软件，然后违反认证条款安装到多台计算机上。与朋友或同伴分享程序就是软偷）、OEM 分拆（OEM 即 Original Equipment Manufacture，原始设备制造商。OEM 分拆指本应随特定计算机一同销售的软件却被分开单独销售）、硬盘装载（安装未授权的软件到计算机硬盘上）和互联网软件盗版（软件被非法下载；通过点对点的对等网络的文件共享网络；假冒盗版软件通过拍卖网站和分类广告出售等）等。

Part III　Exercises

（第三部分　习题与思考）

Tutorial 1
Introduction to Information Technology

1. What is computer literacy?
2. Describe all the parts of an information system.
3. Distinguish between the term "data" and the term "information" in IT.
4. List and describe the components of a computer.
5. List and describe the commonly used input devices.
6. List and describe the commonly used output devices.
7. List and describe the categories of computers.

Tutorial 2
System Unit

1. Explain what a system unit is.
2. A computer company claims to have the fastest microcomputer at 6 GHz. What does this claim mean?
3. There are many different processors on the market today. Common examples include Pentium 4, Celeron and Athlon. Compare the 3 processors (speed, cost, characteristics etc.).
4. What is a heat sink and what is its purpose?
5. What is
   (a) ROM
   (b) RAM
   (c) cache memory?
6. Distinguish between static RAM and dynamic RAM. What are the different versions of each?
7. How much, many or long is a ...
   bit, byte, kilobyte, megabyte, gigabyte, terabyte, petabyte, word, millisecond, microsecond, nanosecond, picosecond.
8. The ASCII coding scheme uses the decimal number 65 to represent the letter "A", 90 as "Z", 97 as "a" and 122 as "z". What is the word processor operator typing if the following codes are received from the keyboard by the microprocessor. (See Appendix for the ASCII code)
   84 114 121 32 105 116 33 10 13 78 111 116 104 105 110 103 32 105 115 32 105 109 112 111 115 115 105 98 108 101 46 32 77 101 108 105 115 115 97 32 74 105 97 110 103 46
9. An expansion card is a circuit board that fits into an expansion slot in the motherboard. What is the purpose of the following types of expansion cards: NIC, modem card, graphics card, accelerator, sound card, PC to TV, memory card.
10. What are the three main factors that influence the speed of a processor?
11. Describe the differences between RISC and CISC technologies including the advantages and disadvantages.
12. Give a description of
    (a) flash memory
    (b) pipelining.
13. Describe the following four different types of ports: Serial, Parallel, USB and SCSI. Give an example of a device which can be connected to each particular type of port.

Tutorial 3
Input and Output Devices

1. Describe the basic types of mouse designs.
2. Describe any pointing device besides a mouse.
3. Data input and control input are distinctly different types of input mechanisms.
   (a) Describe the difference between the two, with some examples.
   (b) Devices such as the mouse, light pen and touch screen are suitable for control input, but not for data input. Why do you think this is so?
   (c) Two other input devices are the trackpad and the trackball. What are these, and what are they useful for?
4. Voice recognition is becoming very popular and is being used more widely, however the best software packages are still 90%~95% accurate. Name some voice recognition packages and distinguish between speaker dependent and speaker independent systems. When would voice recognition be an advantage to use? Might it ever be a disadvantage? Why?
5. When a color graphics terminal has colored pixels, more than one bit of memory is needed for each pixel. How much memory is required for the following displays:
   (a) Monochrome Apple Mac SE—that is, two colors, with 512 × 342 pixels.
   (b) IBM-PC EGA+, having 16 colors and 640 × 480 pixels.
   (c) IBM-PC SVGA having 256 colors and 800 × 600 pixels.
6. Why are data projectors, fax machines and multifunction devices used?
7. Describe the concept of virtual reality and how it is created. Be prepared to discuss your virtual reality experiences in class.
8. Describe the terms multimedia and interactive multimedia. Explain how multimedia can be used in the business world, the classroom and for entertainment.
9. Briefly describe the following types of monitors: CRT, LCD and Gas Plasma.
10. Name and describe different types of printers.
11. What are some of the alternative input and output devices for physically challenged users?

Tutorial 4
Secondary Storage

1. What is secondary storage? What is the difference between hard disks and floppy disks?

2. If you wish to expand the ability of your computer to store data on a long-term basis should you buy some more RAM and expand your primary storage or a larger disk drive for your secondary storage? Distinguish between the two types of storage. How does each type store data?

3. Define the terms: track, sector, cylinder, seek time, rotational delay, access time, WORM.

4. A company wishes to store records on its 2 000 customers and 600 stock items. Each customer record requires 10KB, each stock item 20KB and each sales order 25KB. Each customer has on average 50 sale orders per year. Estimate the size of the files and recommend the type and minimum size of secondary storage device the company should purchase.

5. Describe the basic types of CDs.

6. What are the advantages/disadvantages of CD-ROM/DVD? What is the storage capacity of each? What storage techniques enable the DVD to have more storage capacity? Do you believe they will eventually replace most printed matter?

7. Discuss methods used to backup data, especially with the modern very large disk sizes. Describe the concept of RAID storage.

8. Most hard disks rotate at something in the range of 3 600 to 7200 rpm, whilst flexible media drives rotate the disk at either 300 or 360 rpm.
   (a) Why do these different types of drives use different rotational speeds?
   (b) What is meant by rotational delay in a disk device? How can it be minimized?
   (c) Calculate the average access time of a hard drive which has a seek time of 15 milliseconds and rotates at 7200 rpm (ignoring data transfer time).

9. Suppose a 3 ½ double-sided floppy disk contained 80 tracks, each of which is divided into sectors capable of holding 512 characters. How many characters will the disk hold if each track is divided into 9 sectors? What if each track contains 18 sectors? How do these capacities compare to the size of a 600-page novel in which each page contains 3000 characters?

10. What are some uses for tape, microfilm and microfiche?

11. How is a smart card useful?

Tutorial 5

Software

1. What is the difference between system software and application software?
2. What is an operating system? Give some examples.
3. What are the basic functions of an operating system?
4. Describe the main categories of operating systems.
5. What are utility programs? Give some examples.
6. What is application software? Describe the common categories of the application software.
7. What steps are involved in the process of booting? Distinguish between a cold boot and a warm boot.
8. Discuss how the operating system provides an interface between the hardware and application programs.
9. Two features of the Microsoft Windows operating system are plug_and_play, and object linking and embedding (OLE). Explain these terms.
10. You have your PC set up so that immediately after booting a window loads which displays a list of your programs. You are currently editing a document on your PC using Word. You mark a block of text which you require to move so you also press the cut icon. At this point in time outline to the best of your knowledge the contents of RAM.
11. Describe the process of a disk defragmenter. You should use diagrams.
12. Describe the terms multiprocessing, multiprogramming and timesharing.
13. What are device drivers and why are they needed. If you do not have the correct device driver what is a commonly used way of finding the correct one?
14. Describe the concept of virtual memory, how it works and why it is used.
15. List and describe the different types of programming languages.
16. Compare and contrast a compiler and an interpreter.

Tutorial 6
Data Communications

1. Describe the use of modem in data communications.

2. What are the benefits of a computer network? Do you believe the rapid expansion of computer network usage is beneficial or detrimental to our society? How do you envisage society changing if this expansion continues?

3. A certain company wants to transfer a file from one computer to another using a data link. The file is 5.5 Mbytes in length, and must be transferred in 12 minutes or less. What bit rate should they use?

4. Describe the terms:
   (a) bandwidth
   (b) baseband
   (c) broadband

5. What sort of data transmission (simplex, half-duplex or full-duplex) is required for the following situations?
   (a) MSN Chat & WeChat are two software packages that allow users to talk to each other. Users can chat with more than one person and can have multiple windows open at any one time.
   (b) An auto teller machine.
   (c) The flight information display screen at an airport.
   (d) Fax machines.

6. Suppose you want to send a block of 2000 characters (8 bit bytes) using asynchronous transmission at a speed of 4800 bit/sec. The asynchronous format uses one start bit, one parity bit and two stop bits per byte, with no intercharacter delay. How long will it take to send the data?

7. The same block of data is to be sent using synchronous transmission at 4800 bit/sec. The synchronous data block begins with 2 SYN characters, 10 bytes of control information, and then the 2000 bytes of data and finishes with 2 bytes for error checking. How long will it take to send the data?

8. Computers talk to each other using a protocol. What is a protocol? And what is TCP/IP?

9. Describe the concept of a Global Positioning System (GPS) and how it works?

10. Briefly explain how instant messaging works.

Tutorial 7

Networks

1. As with most computer related topics, data communications and applications developed from data communications involve many initials. Explain the purpose of the following.
   EFT, POS, AFD, PIN, ATM, ECR and LAN.

2. The University network has a bus topology. Draw a diagram to represent the configuration of the network including the computers, the printers and the file server.

3. Distinguish between a wide area network (WAN) and a local area network (LAN).

4. Distinguish between a peer-to-peer network and a client/server network.

5. There is a growing market for people downloading music MP3 files across the internet using internet peer-to-peer (P2P). Describe how P2P works.

6. Provide a description of the following data communications terms:
   (a) Firewall
   (b) Hub
   (c) Intranet

7. Describe the concept of:
   (a) Telecommuting
   (b) Video conferencing
   (c) Telephony

8. Explain how intranets and extranets are useful in supporting communication in an organization?

9. If you were to network up a group of 5 computers in a doctor's surgery, what would you need to do to the existing 5 standalone computers?

10. Compare and contrast Internet, Intranet, and Extranet.

11. Describe both the advantages, and the disadvantages associated with electronic commerce.

Part IV Labs

（第四部分 实践指导）

Warm-up Exercise 1
Windows Practice Exercises

Section I   File Management

1. Create the directory structure and files in D:\.

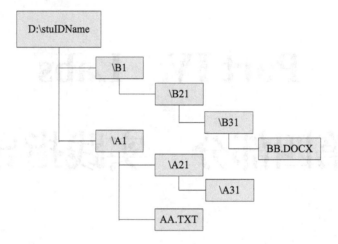

where

(1) **AA.TXT** is a text file which is created by Notebook, with the contents:

   My name is Xxx Xxxx (your real name in PinYin, or the given name for the foreign students), and my student ID is YYYYYYY (your real stuID).

(2) **BB.docx** is a word document, with the contents:

   Practical Tasks for Windows

   Today is 2017-09-XX (the real date of today)

2. Let the author of **BB.docx** be your name.

3. Copy the file **AA.TXT** into **\B1**, and rename it as **newAA.txt**.

4. Set the read-only attribute for **BB.DOCX** and let its author be **QWERT**.

5. Try to copy the file **win.ini** in drive C into **\A1**, then set the read-only and hidden attributes for it.

6. Create the folder D:\stuIDName\pics. Retrieve all the files with the extension **BMP** in drive C, and copy the three smallest retrieved files into D:\stuIDName\pics.

7. Retrieve all the text files in drive C whose size is MUDIUM. Copy the first two retrieved files into \B1.

8. Display the detailed information and the file extensions for the files in windows explorer.

Print screenshots (named separately as 8_1.jpg and 8_2.jpg in the directory of D:\stuIDName) for the two option setting dialogs.

9. Obtain and save the size (in GB) of available space on each hard drive into \A1\SIZE.TXT.

Section II   Printer Management (All the results should be saved in D:\stuIDName\printers)

1. Install the printer of **Dell 1130 Laser Printer** and print the test paper to the file D:\stuIDName\printers\Dell.prn.

2. Install the printer of **Canon Inkjet iP90** and print the test paper to the file D:\stuIDName\printers\Canon.prn.

3. Install the printer of **HP LaserJet 1022 Class Driver** and print the file C:\Windows\win.ini to the file D:\stuIDName\printers\win.prn.

Section III   Data security measures (WINRAR) (All the results should be saved in D:\stuIDName\winrars)

1. Compress the files view11.jpg and view22.jpg into D:\stuIDName\winrars\pic12.rar.

2. Compress the files view11.jpg and view22.jpg into D:\stuIDName\winrars\pic1122.rar with the password 1122.

3. Compress the files view33.jpg and view44.jpg into D:\stuIDName\winrars\pic34.exe.

4. Add the two files view11.jpg and view22.jpg into D:\stuIDName\winrars\pic4.rar.

5. Extract all the files included in pic1.rar into D:\stuIDName\winrars\.

6. Extract all the files included in pic2.exe into D:\stuIDName\winrars\.

7. Extract the file view88.jpg included in pic4.rar into D:\stuIDName\winrars\.

8. Extract the file view9.jpg included in pic3.exe into D:\stuIDName\winrars\.

Section IV   Data security measures (Virus Scan)

Carry out a virus scan of your U disk and hard disk drive C. Take screenshots (named as VirusScan1.jpg and VirusScan2.jpg in D:\stuIDName\viruses) of the virus scan set up screen and virus scanning screen.

Section V   Hardware and Software Problems Resolving

Create a file named Solution.docx in the D:\stuIDName\solutions to explain what you would do to resolve the following problems:

(1) Your computer freezes (where the mouse and keyboard become unresponsive and applications hang).

(2) Your document does not print.

(3) You send the wrong files to be printed.

Warm-up Exercise 2
Word Practice Exercises

1. Open and edit word1.docx with the final result as the provided sample.

(1) Title: WordArt (Fill-Red, Accent 2, Outline-Accent 2).

(2) Bold the first paragraph. Apply a 0.5" first-line indent to it; Change **Spacing** to 8pt **Before** and **Spacing** to 6pt **After**.

(3) Edit the **Planet** Picture: (a) Picture Styles: Metal Frame; (b) Color: Washout Recolor; (c) Wrap Text: Behind Text; (d) Resize it as the sample.

(4) Arrange the last two paragraphs into two columns with a line between them.

(5) Change the Bulleted List into the Numbered List as the sample.

(6) Create an **Italic** two-line **dropped** capital letter for the 4th paragraph.

【Word Sample 1】

The knowledge of planet

A planet is an astronomical object orbiting a star or stellar remnant that

(1) is massive enough to be rounded by its own gravity,
(2) is not massive enough to cause thermonuclear fusion, and
(3) has cleared its neighbouring region of planetesimals.

The term planet is ancient, with ties to history, science, mythology, and religion. The planets were originally seen by many early cultures as divine, or as emissaries of deities. As scientific knowledge advanced, human perception of the planets changed, incorporating a number of disparate objects. In 2006, the International Astronomical Union (IAU) officially adopted a resolution defining planets within the Solar System. This definition is controversial because it excludes many objects of planetary mass based on where or what they orbit. Although eight of the planetary bodies discovered before 1950 remain "planets" under the modern definition, some celestial bodies, such as Ceres, Pallas, Juno, Vesta (each an object in the Solar asteroid belt), and Pluto (the first-discovered trans-Neptunian object), that were once considered planets by the scientific community are no longer viewed as such.

*T*he planets were thought by Ptolemy to orbit Earth in deferent and epicycle motions. Although the idea that the planets orbited the Sun had been suggested many times, it was not until the 17th century that this view was supported by evidence from the first telescopic astronomical observations, performed by Galileo Galilei. By careful analysis of the observation data, Johannes Kepler found the planets' orbits were not circular but elliptical. As observational tools improved, astronomers saw that, like Earth, the planets rotated around tilted axes, and some shared such features as ice caps and seasons. Since the dawn of the Space Age, close observation by space probes has found that Earth and the other planets share characteristics such as volcanism, hurricanes, tectonics, and even hydrology.

Planets are generally divided into two main types: large, low-density gas giants and smaller, rocky terrestrials. Under IAU definitions, there are eight planets in the Solar System. In order of increasing distance from the Sun, they are the four terrestrials, Mercury, Venus, Earth, and Mars, then the four gas giants, Jupiter, Saturn, Uranus, and Neptune. Six of the planets are orbited by one or more natural satellites.

2. Open and edit word2.docx with the final result as the provided sample.

(1) Title: Centered, Bold, Heading 1 Style.

(2) Edit the **Planet** Picture: (a) Picture Styles: Soft Edge Oval; (b) Color: 66% Saturation; (c) Position in Middle Center with Square Text Wrapping; (d) Resize and relocate it as the sample.

(3) Apply red 1½ pt double-line shadow border and yellow shading to the last paragraph.

(4) Change the Bulleted List into the newly-defined (Bold and Red) Bulleted Character List (a symbol of an opened book).

(5) Replace all **Planet** in the body text with the format of Green, Bold, Italic, and Blue wave double underline.

(6) Add a Bold, Underline, 20pt Header **Protect our planet**, and page numbers to the bottom center of each page with the format of Page XXX.

【Word Sample 2】

3. Open and edit word3.docx with the final result as the provided sample.

(1) Title: Bold, Centered, 28pt, and with Text Effects: Gradient Fill—Purple, Accent 4, Outline—Accent 4; Inner Shadow—Inside Diagonal Bottom Left.

(2) Edit the **Planet** Picture: (a) Picture Styles: Snip Diagonal Corner, white; (b) Color: Blue, Accent color 1 Light; (c) Position in Middle Right with Square Text Wrapping; (d) Resize and relocate it as the sample.

(3) Apply and edit 3pt red 3D border and orange shading to the 4th paragraph.

(4) Change the Bulleted List into the newly-defined (Bold, Italic and Red) Bulleted Character List (a symbol of the red heart suit in poker).

(5) Replace all the **Planet** in the 3rd paragraph with the format of Red, Bold, Small caps, and Highlight.

(6) Create a blue five-line **in margin** dropped capital letter for the last paragraph.

(7) Add page numbers to the top center of each page started from 10.

【Word Sample 3】

4. Open and edit word4.docx with the final result as the provided sample.

(1) Title: WordArt (Fill—Olive Green, Accent 3, Sharp Bevel).

(2) Edit the **Planet** Picture: (a) Picture Styles: Metal Rounded Rectangle; (b) Color: Washout Recolor; (c) Wrap Text: Behind Text; (d) Resize and relocate it as the sample.

(3) Change the Bulleted List into the newly-defined (Bold and Green) Bulleted Character List (an arrow symbol).

(4) Delete the last two paragraphs, then copy and paste the excel spreadsheet in **Daily.xlsx**. Merge cells and delete the extra empty lines in the first row.

(5) Add a Horizontal Scroll shape with Shape Styles: Shape Fill—Orange, Accent 6, and a Bold Red 26pt text **Protect Our Planet** in it.

(6) Add a Bold Header **Better Planet, Better Life**, and page numbers to the bottom center of each page with the format of Page—XXX, started from 5.

【**Word Sample 4**】

Better Planet, Better Life

The knowledge of planet

A planet is an astronomical object orbiting a star or stellar remnant that
- is massive enough to be rounded by its own gravity,
- is not massive enough to cause thermonuclear fusion, and
- has cleared its neighbouring region of planetesimals.

The term planet is ancient, with ties to history, science, mythology, and religion. The planets were originally seen by many early cultures as divine, or as emissaries of deities. As scientific knowledge advanced, human perception of the planets changed, incorporating a number of disparate objects. In 2006, the International Astronomical Union (IAU) officially adopted a resolution defining planets within the Solar System. This definition is controversial because it excludes many objects of planetary mass based on where or what they orbit. Although eight of the planetary bodies discovered before 1950 remain "planets" under the modern definition, some celestial bodies, such as Ceres, Pallas, Juno, Vesta (each an object in the Solar asteroid belt), and Pluto (the first-discovered trans-Neptunian object), that were once considered planets by the scientific community are no longer viewed as such.

Klassy Kow Ice Cream Daily Customer Count							
	Mon	Tue	Wed	Thu	Fri	Sat	Sun
Week 1	9000	11000	12000	10000	12000	13000	12500
Week 2	10000	12300	13000	9000	12500	13500	12500
Week 3	19000	14500	14000	8000	11500	15000	11000
Week 4	95000	16000	15000	11000	12500	13000	10000
Week 5	9000	11000	12000	10000	12000	13000	12500
Week 6	10000	12300	13000	9000	12500	13500	12500
Week 7	19000	14500	14000	8000	11500	15500	11000
Week 8	95000	16000	15000	11000	12500	13000	10000

Protect Our Planet

Warm-up Exercise 3
PowerPoint Practice Exercises

1. Open and edit Power1.pptx with the final result as the provided sample.

(1) Apply the background style of Preset gradients—Bottom Spotlight—Accent 6 to all slides. Change the layout of the second slide to **Two Content**.

(2) Insert the picture **Logo1.jpg** into the right placeholder of the second slide. For this picture, apply the **Grow & Turn Entrance** Animation effect. For the body text in Slide 2, apply the **From Right, Fly In Entrance** Animation effect.

(3) Insert Date (updated automatically) at the bottom left of each slide, and slide number at the bottom right of each slide, with font size 24, red, bold.

(4) Add **your name** as the subtitle for the first slide. For the left picture in Slide 1, create a hyperlink linking to the third slide. Add a **Return** action button at the bottom center of the third slide, which links to the first slide.

(5) Apply the **Fly Through** transition with **Breeze** sound to all slides, and move automatically to the next slide after two seconds. Set up slide show to loop continuously until "Esc".

(6) Copy Salesperson and Total Monthly Sales from **Sales1.xlsx** and display as a graph.

【PPT Sample 1】

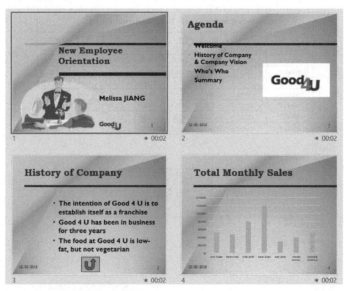

2. Open and edit Power2.pptx with the final result as the provided sample.

(1) Change the layout of the 2nd slide to **Two Content**. Insert the picture **cake2.jpg** into the right placeholder. Create a hyperlink for the picture linking to the 5th slide.

(2) Apply the **Facet** theme for the 3rd slide. Apply the **Ferris Wheel** transition to all slides, and move automatically to the next slide after two seconds.

(3) For the 5th slide, apply the **Teeter** emphasis animation to the left picture, with duration time of two seconds. Apply the **Bounce Entrance** animation to the right text, **As One Object** and **With Previous**.

(4) Change the title of the 4th slide into a **WordArt** object with the style of Gradient Fill—Brown, Accent 4, Outline—Accent 4; text effect of Transform—Warp—Chevron Down; Height of 1.8", and Width of 9.6".

(5) Display the slide number for each slide except the 1st one.

【PPT Sample 2】

3. Open and edit Power3.pptx with the final result as the provided sample.

(1) Change the layout of the 4th slide to **Two Content**. In the right placeholder, copy Salesperson and Bonus from **Bonus3.xlsx** and display as a bar graph. Move this slide before the 2nd one.

(2) Change the theme of all slides except the 1st one to **Main Event**.

(3) For the 6th slide, apply the **Grow/Shrink emphasis** animation for the **Computer** picture, with duration time of 1.5 seconds. Apply the **Grow & Turn Entrance** animation for the body text, **All at Once** and **After Previous**.

(4) Create a hyperlink for the text **Meeting** in Slide 1 linking to the 3rd slide. Insert a customized rectangle action button at the bottom right of the 3rd slide with text **Back** in it, and when clicked it can return to slide 1.

(5) Display Date & Time (update automatically) for each slide.

【PPT Sample 3】

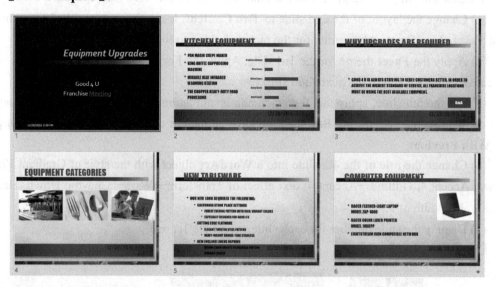

4. Open and edit Power4.pptx with the final result as the provided sample.

(1) Display the slide footer (**Healthy Food**) and automatically updated Date for each slide except the title slide.

(2) Apply the **Retrospect** theme for all slides, Change the theme color to **Marquee**.

(3) Apply the **All at Once**, **Float Down**, **Float in Entrance** animation to the body text in Slide 2. Insert a **Forward or Next** action button at the bottom left of the 3rd slide, which links to the last slide.

(4) Apply the **Horizontal Random Bars** transition to all slides, and move automatically to the next slide after two seconds.

(5) Create a hyperlink for the **Drink** picture in Slide 4 linking to the website **http://www.foods.org**.

【PPT Sample 4】

Lab 1
Introduction to Spreadsheets—Excel

1. Start **Microsoft Excel**.
2. Home Budget example:

Enter the following Home Budget model (ref. Figure IV-1-1):

Notes   Left column is all in column A

             Jan starts in column B

             Enter months up to Dec (column M)

	A	B	C	D	E	F	G	H	I	J	K	L	M
1	Home Budget by *Your Name*												
2													
3	TAX RATE=												
4													
5		Jan	Feb	Mar	Apr	May	Jun	Jul	Aug	Sep	Oct	Nov	Dec
6	GROSS INCOME												
7	TAX PAYABLE												
8	**NET PAY AMOUNT**												
9													
10	**EXPENSES**												
11													
12	RENT												
13	UTILITIES												
14	TELEPHONE												
15	LIFE INSURANCE												
16	FOOD												
17	LEISURE												
18	**TOTAL EXPENSES**												
19													
20	**SAVINGS DEPOSITS**												
21													
22	SAVINGS ACCOUNT												

Figure IV-1-1   Home Budget Spreadsheet

3. CHANGING THE WIDTH OF COLUMN A

Drag the line between the column A and B heading to make column A wide enough for the text. **SAVE** your spreadsheet as outlined in step 15.

4. LINES and BORDERS

(1) To underline the month headings first highlight them (use the mouse or **Shift**-arrow).

(2) Select **Borders** in the **Font** group on the **Home** tab.

(3) Choose the underline border.

(4) Similarly underline rows 8, 18 and 20.

5. COPYING A FORMULA (The Gross Pay Amounts)

(1) Enter **1855** in cell B6 for January.

(2) In cell C6 (February) enter: **=B6**

(This formula copies the contents of B6 into C6, i.e. Jan into Feb)

(3) The following steps copy C6 across to D6:M6   (i.e. Mar→Apr…Dec):

Highlight C6

drag the fill handle (the "+" symbol at the bottom right hand corner of cell C6) across to cell M6 and let the left mouse button go.

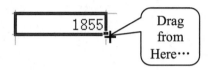

Look at the contents of these cells. (Move over them)

(4) Change MAY's figure to **2055**. (A pay rise)

(Now you see the magic of formulae)

6. COPYING WITH ABSOLUTE REFERENCES

(1) The TAX RATE is 23% of Gross Pay, so enter **23%** into cell B3.

(2) Move to cell B7.

(3) The taxes are 23% of Gross Pay Amount, which we can calculate by a formula:

Into cell B7 enter

=B3 * B6

If the formatting is incorrect you may have to choose one of the following buttons $ ▾ %, in the **Number** group on the **Home** tab to display it correctly.

(4) Now copy this across into cells C7:M7

Didn't give the required result, did it?

(5) Fix it up by changing one of the above **RELATIVE** references to an **ABSOLUTE**:

Move to cell B7 again and enter the formula

=$B$3 * B6

Try copying this across into C7:M7

(You MUST learn what the difference between **RELATIVE** reference and **ABSOLUTE** reference is!)

7. COPYING WITH RELATIVE REFERENCES:

(1) January's NET PAY AMOUNT is income-tax, so in cell B8 enter

= B6–B7.

(2) Copy this across to C8:M8.

(Move across row 8 to see that the references are automatically altered column by column as the formula is copied across)

8. REJUSTIFYING THE MONTH HEADINGS

(1) Block the month headings by using the mouse.

(2) Use one of the following horizontal alignment buttons " ≡, ≡, ≡ " in the **Alignment** group on the **Home** tab.

9. EXPENSES DATA

Enter the following data in column B then copy across (be sure not to type the $ signs!):

RENT is $375 per month all year.

UTILITIES are $100 per month all year.

TELEPHONE is $18 per month all year.

LIFE INSURANCE is $28.65 per month all year.

FOOD is 30% of NET PAY AMOUNT each month, so enter **FOOD** into cell D3; enter **30%** into cell E3; enter = **$E$3*B8** into cell B16.

LEISURE is 15% of NET PAY AMOUNT each month, so enter **LEISURE** into cell D4; enter **15%** into cell E4; enter =**$E$4*B8** into cell B17.

10. MONTHLY EXPENSES TOTALS

(1) In cell B18 type the formula (or use the **Sum** button Σ in the **Editing** group on the **Home** tab) =**SUM(B12:B17).**

(2) Copy this across into C18:M18.

The numbers 1 164.407 5 and 1 233.707 5 should appear.

11. FORMATTING THE CELLS OF YOUR MODEL

A range of cells can have their display format changed by

(1) Block the range of B6:M23.

(2) On the **Home** tab, in the **Number** group, click ▫ to launch the **Format Cells** dialogue box (ref. Figure Ⅳ-1-2): on the **Number** tab, select **Currency**, with **2** decimals and **$** symbol. Click **OK** to close the **Format Cells** dialogue box.

(3) The columns may need to be widened to fit the $100 000's.

12. CALCULATING ONGOING SAVINGS DEPOSITS

(1) Type the heading **SAVINGS DEPOSITS** in cell **A20.**

(2) Enter the appropriate formula into cell **B20** and COPY it across that row.

13. CUMULATING INTEREST AND SAVINGS MONTH BY MONTH

(1) Enter the annual interest rate into the top of the worksheet

A4 = "INTEREST RATE="

B4 = 7.25%

A22 = "SAVINGS ACCOUNT"

A23 = "CLOSING BALANCE"

B23 = =B20 (Assume zero beginning balance for January)

C23 = =B23*(1+B4/12) + C20

One reference should be **ABSOLUTE**!!

i.e. February' savings = January balance with interest + February savings

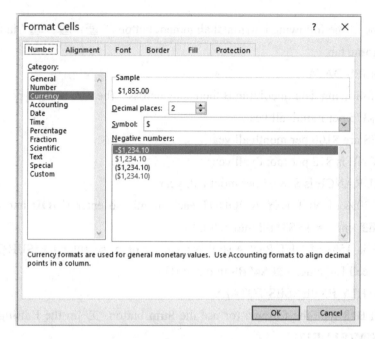

Figure Ⅳ-1-2 Format Cells

(2) Copy this formula across.

(3) Underline the **Closing Balance**.

(If you have it all correct Jan = $263.94, Dec = $3966.81)

Figure Ⅳ-1-3 shows the layout of the spreadsheet:

	A	B	C	D	E	F	G	H	I	J	K	L	M
1	Home Budget by JiangHong												
2													
3	TAX RATE=	23%		FOOD	30%								
4	INTEREST RATE=	7.25%		LEISURE	15%								
5		Jan	Feb	Mar	Apr	May	Jun	Jul	Aug	Sep	Oct	Nov	Dec
6	GROSS INCOME	$1,855.00	$1,855.00	$1,855.00	$1,855.00	$2,055.00	$2,055.00	$2,055.00	$2,055.00	$2,055.00	$2,055.00	$2,055.00	$2,055.00
7	TAX PAYABLE	$426.65	$426.65	$426.65	$426.65	$472.65	$472.65	$472.65	$472.65	$472.65	$472.65	$472.65	$472.65
8	**NET PAY AMOUNT**	$1,428.35	$1,428.35	$1,428.35	$1,428.35	$1,582.35	$1,582.35	$1,582.35	$1,582.35	$1,582.35	$1,582.35	$1,582.35	$1,582.35
9													
10	**EXPENSES**												
11													
12	RENT	$375.00	$375.00	$375.00	$375.00	$375.00	$375.00	$375.00	$375.00	$375.00	$375.00	$375.00	$375.00
13	UTILITIES	$100.00	$100.00	$100.00	$100.00	$100.00	$100.00	$100.00	$100.00	$100.00	$100.00	$100.00	$100.00
14	TELEPHONE	$18.00	$18.00	$18.00	$18.00	$18.00	$18.00	$18.00	$18.00	$18.00	$18.00	$18.00	$18.00
15	LIFE INSURANCE	$28.65	$28.65	$28.65	$28.65	$28.65	$28.65	$28.65	$28.65	$28.65	$28.65	$28.65	$28.65
16	FOOD	$428.51	$428.51	$428.51	$428.51	$474.71	$474.71	$474.71	$474.71	$474.71	$474.71	$474.71	$474.71
17	LEISURE	$214.25	$214.25	$214.25	$214.25	$237.35	$237.35	$237.35	$237.35	$237.35	$237.35	$237.35	$237.35
18	**TOTAL EXPENSES**	$1,164.41	$1,164.41	$1,164.41	$1,164.41	$1,233.71	$1,233.71	$1,233.71	$1,233.71	$1,233.71	$1,233.71	$1,233.71	$1,233.71
19													
20	**SAVINGS DEPOSITS**	$263.94	$263.94	$263.94	$263.94	$348.64	$348.64	$348.64	$348.64	$348.64	$348.64	$348.64	$348.64
21													
22	SAVINGS ACCOUNT												
23	**CLOSING BALANCE**	$263.94	$529.48	$796.62	$1,065.38	$1,420.46	$1,777.68	$2,137.06	$2,498.62	$2,862.35	$3,228.29	$3,596.44	$3,966.81

Figure Ⅳ-1-3 BUDGET Spreadsheet

14. **WHAT...IF**? a few changes are made.

(1) What extra savings would be made if an account offering 12% interest was used? December balance = $____.

(2) The tax rate is increased to 30%. December balance = ____.

(3) Can you afford to spend 25% of income on Leisure?   Yes/No.

(You'd better make alterations so the 25% occurs only once in the model, similar to the interest rate.).

(4) The expense item CAR has been forgotten, which costs $100 per month to run. To add it in you will need to INSERT a new row.

Note: the formulas below automatically alter to accommodate the new row.

15. SAVING YOUR DOCUMENT

Click on the **SAVE** button (or **File→Save**).

Make sure you are saving in *your own drive*. Let's give it the name **BUDGET**.

(BUDGET.XLSX will be created on your disk)

16. QUITTING THE SPREADSHEET

File→Exit.

Lab 2
Orchids Shade House Case Study

1. Before loading Excel, copy the file **ORCHIDS.XLSX** from the files link on the IT Fundamental web site and save it to your own drive. It is a skeleton spreadsheet model for the following problem: (The model is reproduced at the end of this lab)

A flower grower plants a bed of orchids in a shade house at a total cost of $6000. The bed will not start producing during the first year. In the second through eighth years it should produce, respectively, 150, 500, 1900, 2000, 2200, 2000, and 1200 dozen orchid sprays. Fixed annual operating costs are $1200 per bed, while variable expenses during producing years are 25% of sales receipts. Water costs per year are provided in the table in the orchid's spreadsheet. If 2000 or more orchids are produced in a year then an extra staff hand needs to be employed, costing $18 000 for that year. Due to the drought, water costs fluctuate each year: $2000, $3000, $5000, $5000, $5000, $6000, $6000 and $5000 respectively. Orchids currently sell for $15 per dozen, which has currently been increasing at 3.5% per year.

Construct a spreadsheet model that shows a breakdown of anticipated annual income and expenses over the eight years.

2. From within Excel load in the file ORCHIDS.XLSX.

3. Entering the figures into the table:

(1) The orchid sales figures in the calculating area should be read from the table at the top.

   This is an excellent opportunity to obtain practice using the **VLOOKUP** formula:
   In cell B16 enter
   =VLOOKUP(B14,D5:E12,2)

Copy the formulae across the last 7 years (C16:I16) to obtain 150 for year 2 500 for year 3 etc.

- Didn't work properly, did it?

A couple of the cell references in the formula in cell B16 need to be entered as **ABSOLUTE** references. Go back and EDIT B16 then recopy it.

(2) Price/Dozen:
   cell B17 contains =$B$9 (15 dollars)
   Second year is an increase of 3.5% from first year.
   i.e. cell C17 contains =B17*(1+B10)
   (One of these should be an absolute reference also!)

now copy this across to D17...I17

(3) Income from Sales

Cell B18 = B16*B17

Simply copy this across to C18...I18

(4) Cumulative Income

Year 1 contains simply first year income, i.e.: =B18

Year 2 contains first year cum. + second year sales, ie: =B20+C18

Copy this across to D20:I20.

(5) Extra Staff

IF function is used here as extra staff are needed if 2000 or more orchids are produced.

In cell B26 enter **something like**: =IF(B16 >= B7, B8, 0)

(two of the cell references need to be entered as **ABSOLUTE** references if you wish to copy across the row successfully!)

(6) Work out the rest of the model by yourself. Refer to Figure IV-2-1 for correct figures.

	A	B	C	D	E	F	G	H	I
1	ORCHID SHADE HOUSE BUDGET			Modified By Melissa Jiang					
2	************	************	************	************	************	************	************		
3				Anticipated Yearly Sales & Water Costs			*		
4	SETUP COST	$6,000.00		Year	Sales	Water Costs	*		
5	FIXED COSTS	$1,200.00		1	0	$2,000.00	*		
6	VARIABLE COSTS	25%		2	150	$3,000.00	*		
7	SALES FOR MORE STAFF	2000		3	500	$5,000.00	*		
8	Extra Wages/Year	$18,000.00		4	1900	$5,000.00	*		
9	CURRENT PRICE/DOZEN	$15.00		5	2000	$5,000.00	*		
10	PRICE INCREASE	3.5%		6	2200	$6,000.00	*		
11				7	2000	$6,000.00	*		
12				8	1200	$5,000.00	*		
13	************	************	************	************	************	************	************	************	************
14	YEAR	1	2	3	4	5	6	7	8
15	INCOME:								
16	ORCHID SALES	0	150	500	1900	2000	2200	2000	1200
17	PRICE/DOZEN	$ 15.00	$ 15.53	$ 16.07	$ 16.63	$ 17.21	$ 17.82	$ 18.44	$ 19.08
18	INCOME FROM SALES	$ -	$ 2,328.75	$ 8,034.19	$ 31,598.46	$ 34,425.69	$ 39,193.65	$ 36,877.66	$ 22,901.03
19	==========	==========	==========	==========	==========	==========	==========	==========	==========
20	CUMULATIVE INCOME	$ -	$ 2,328.75	$ 10,362.94	$ 41,961.40	$ 76,387.09	$ 115,580.74	$ 152,458.39	$ 175,359.42
21									
22	COSTS:								
23	SETUP	$ 6,000.00							
24	FIXED YEARLY	$ 1,200.00	$ 1,200.00	$ 1,200.00	$ 1,200.00	$ 1,200.00	$ 1,200.00	$ 1,200.00	$ 1,200.00
25	VARIABLE YEARLY	$ -	$ 582.19	$ 2,008.55	$ 7,899.61	$ 8,606.42	$ 9,798.41	$ 9,219.41	$ 5,725.26
26	EXTRA STAFF	$ -	$ -	$ -	$ -	$ 18,000.00	$ 18,000.00	$ 18,000.00	$ -
27	WATER COSTS	$ 2,000.00	$ 3,000.00	$ 5,000.00	$ 5,000.00	$ 5,000.00	$ 6,000.00	$ 6,000.00	$ 5,000.00
28	TOTAL COSTS	$ 9,200.00	$ 4,782.19	$ 8,208.55	$ 14,099.61	$ 32,806.42	$ 34,998.41	$ 34,419.41	$ 11,925.26
29	==========	==========	==========	==========	==========	==========	==========	==========	==========
30	CUMULATIVE COSTS	$ 9,200.00	$ 13,982.19	$ 22,190.73	$ 36,290.35	$ 69,096.77	$ 104,095.18	$ 138,514.60	$ 150,439.86
31	==========	==========	==========	==========	==========	==========	==========	==========	==========
32	PROFIT/LOSS	-$ 9,200.00	-$ 2,453.44	-$ 174.36	$ 17,498.84	$ 1,619.27	$ 4,195.24	$ 2,458.24	$ 10,975.77
33	==========	==========	==========	==========	==========	==========	==========	==========	==========
34	CUMULATIVE BALANCE	-$ 9,200.00	-$ 11,653.44	-$ 11,827.80	$ 5,671.05	$ 7,290.32	$ 11,485.55	$ 13,943.80	$ 24,919.57
35	breakeven production	NO	NO	NO	YES	YES	YES	YES	YES

Figure IV-2-1 ORCHID Shade House Spreadsheet

4. Answer the following:

(1) When will the new bed reach breakeven production (pay for itself)?

A line graph of Cumulative Income against Cumulative Costs would show the exact place.

How does this change if (i.e. What...if ?)

(i) the variable costs were 35%;

(ii) the variable costs were 40%;

(iii) the price increase was 5%.

Make up your own what...if. Compare yours with your neighbor.

(2) Construct a bar graph of annual expenses against annual income. Put your name on the title of the graph.

(3) Construct a stacked bar graph that shows "contributions" from each of the expenses over the 8 years.

(4) Construct a pie graph to show the relative proportion of each of the five expenses for the 6th year. Temporarily add a figure for setup expense in this year to see that the graph automatically compensates.

Assignment 1

The Sharpshooter Snooker Centre is the new craze in town. It is the new local spot where competitors from far and wide gathered to play for the weekly prize money. Henry Thomas, the owner believes the time has come to improve the control and efficiency of his business. Currently, all details are recorded manually but with the increased popularity of the snooker competition, he is convinced that his workload could be somewhat reduced by entering the financial details of his business into a spreadsheet.

The following represents Sharpshooter's current and anticipated income for the next 12 months.

General table hire operates all year round whilst the competition operates only for the first 6 months of the year.

General Takings for table hire - $3500 per month
Extra takings during competition season - $4000 per month
Canteen takings all year - $2600 per month

Due to Sharpshooter's popularity, general takings are increasing by 6% per month.

Donation from an ex-snooker champion who has taken an interest in the new location-$500 at the beginning of each quarter.

Current and anticipated expenses for the next 12 months.

Food supplies for canteen

- 25% of total monthly income (excluding donation) if in competition season
- 16% of total monthly income (excluding donation) if out of competition season

Rent - $2500 per month

Henry employs 2 part-time workers whose number of hours worked differs from month to month. Extra bonuses are given to employees who work for longer hours in the month. This can be shown by the following table. (Assume whole hours only).

Hours worked per month	Bonus paid to employee
0~29	$0
30~64	$30
65~99	$50
100+	$80

Every staff member gets $12 per hour plus extra bonus amounts for longer hours worked.

The number of hours worked per month by each worker is given below.

Month	1	2	3	4	5	6	7	8	9	10	11	12
Worker 1	30	60	105	48	90	25	75	80	58	42	38	50
Worker 2	138	35	40	38	66	80	20	55	74	114	90	46

Maintenance expenses (repair or replace table felt, new chalk, repair or replace billiard cues etc.)—$300 per month (increasing by $8 per month)

Part (a)

Your task is to design a spreadsheet for Henry listing all expenses and income arriving at appropriate totals which show the financial position for the sharpshooter Snooker Centre for each of the 12 months. Allow maximum flexibility so changes can easily be made. Layout of the spreadsheet should be in a neat, easy to follow manner.

Hand in:

(i) An explanation of the general layout of the spreadsheet and how Henry would use the spreadsheet to enter his base figures, and what he should look at to determine profits/losses. (Indicate any lookup tables and sophisticated formulae used).

(ii) A copy of the display of your model. This should be printed with gridlines and row & column headings. To do this, in the **Sheet Options** group on the **Page Layout** tab, click to select **Gridlines:Print** and **Headings:Print**; or click the **Sheet Options** launcher on the **Page Layout** tab. The **Page Setup** dialog box opens with the **Sheet** tab active, click to select **Gridlines & Row and Column Headings** check boxes.

(iii) A copy of the contents of your model (i.e. the formulae you have used—you only need to print a section of your worksheet to show an example of each of the general formulae used (submit at least 2 columns that demonstrate adequate proof of the use of formulae). To do this, in the **Formula Auditing** group on the **Formula** tab, click **Show Formula** button. You may still need to widen some columns.

(iv) Three separate graphs, one showing any profit or loss made over the 12 months, another showing the breakdown of one year's expenses and a third graph of your choice with a description of what it is showing.

Part (b)

By the second half of the year (beginning from month 7), it is believed that a professional snooker player will be performing trick sessions as an added attraction.

His fee is $600 per month which will increase by 4% every month. Include this change into the model whilst maintaining maximum flexibility.

Hand in:

(i) A copy of the **display and formulas** of the revised model with this inclusion.

(ii) A graph spanning 12 months that shows the profits before and after this inclusion.

Lab 3
Creating Your Webpage (1)

1. Creating your first WEB page
Load **Notepad** by selecting **Start → Windows Accessories → Notepad**.
Type in the following: (USE YOUR NAME AND EMAIL: NOT *Pine*'s)

```
<HTML>
<HEAD>
     <TITLE>Lab 1 </TITLE>
</HEAD>
<BODY>
    <H2> Lab 1</H2>

    Welcome to my Web Page.<BR>

    This is the first of many Web pages that I will create.<P>

    <ADDRESS>
        Author : Pine Yu
        <A href="mailto:ITF@teacher.ecnu.edu.cn"> Pine Yu </A>
    </ADDRESS>

    Revision date: today's date

</BODY>
</HTML>
```

(1) Save the file as "**HTML1.html**" in your hard disk drive (e.g., H: drive).
(2) Click on **Computer** on the taskbar. Go to H: drive and double click on HTML1.html.
(3) This should load up **Microsoft Edge** and your web page should appear on the screen.
(4) Every time you make some changes in HTML1.html in Notepad, save the document, go to **Microsoft Edge** window and press **REFRESH**. This will give you the updated version of your web page.

2. Change the font size for the document to be 5 by using the BASEFONT tag in the document header.
(Hint: Add the line **<BASEFONT size = 5>** in the **HEAD** of the web page.)

3. Change the text and background colors of the document to some colors of your choice.

(<u>Hint</u>: Change **<BODY>** to **<BODY bgcolor="blue" text="yellow">**)

4. Add the following lines of text to the document body.

```
hello
   my name is ApplePie

      your name is EggRoll

         welcome to Shanghai!
```

Hint: Type the lines in the body of the document. You will notice that it puts them all in one line even though you type it in as above. Why?

Try adding <PRE> before the new lines and </PRE> after the new lines. Now **SAVE** in Notepad and **REFRESH** in **Microsoft Edge**. Is this what you want? Read up on preformatted text (<PRE>) from the lecture notes; then try to draw the following web page.

NAME	AGE	E-mail	REMARKS
Michael	20	michael@163.com	student
Mary	30	mary@gmail.com	manager
Johnson	40	johnson@ibm.com	CEO

5. Experiment by adding more text to your document and trying out some of the physical and logical appearance markup tags. e.g. bold, italic, strong etc.

Hint: <b> **bold** </b>

<i> *italic* </i>

<u> underline </u>

 start a new line

<p> insert a blank line and start a new paragraph

<hr> horizontal line

<em>*emphasis* </em>

<strong> **strong** </strong>

<cite> *citation* </cite>

<code> fixed-width font </code>

6. Try using a background image for the document.

(1) Copy bg1.gif from the Files link on the IT Fundamental web site into your H: drive and include it into your HTML1.html by changing the code in the **BODY** of the web page to

<BODY background="bg1.gif">

(2) Now **SAVE** in Notepad and **REFRESH** in **Microsoft Edge**. You should see a new background of your web page.

7. Write the full HTML code (be sure to save it in the file of **HTML2.htm** in your hard disk drive (e.g., H: drive) to display the following web page (ref. Figure Ⅳ-3-1). Note: Web page

font size is 4, font color is blue, and background color is pink. "Great Wall's Long History" has the H1 format (green and center-alignment), "Specials information about China" has the H2 format. "Olympic Games" and "Expos" are in red and font sizes are 2. Be sure that the web page has the title of "Great Wall".

Figure IV-3-1　Web Page for HTML2.htm

Lab 4
Creating Your Webpage (2)

1. Creating your Web page.

(1) Load **Notepad** by selecting **Start → Windows Accessories → Notepad**.

(2) Copy the file ***basketball.htm***, ***English.html***, ***ITF.html***, ***sports.html*** and ***basket.jpg***, ***ski.jpg***, ***happy.gif***, ***bg2.gif*** from the **Files** link on the IT Fundamental web site and save them into your own hard disk (e.g., H: drive).

(3) Type in the following.

```
<HTML>
<HEAD>
    <TITLE>Welcome to my website</TITLE>
</HEAD>
<BODY>
    <H1>Your Name's curriculum </H1>
    <H2>Bachelor/Master of Business/Computing </H2>
    <B>Subjects studying in 2011:</B>

    <P>Semester 1:
    <UL>
        <LI>IT Fundamentals
        <LI>Advanced English
        <LI>Management Information System
    </UL>

    My Interests:
    <OL>
        <LI>Tennis
        <LI>Basketball
        <LI>Introduction to sports
    </OL>

    Please email me here: Melissa Jiang
</BODY>
</HTML>
```

(1) Save the file as **HTML3.html** in your hard disk drive (e.g., H: drive).

(2) Click on **Computer** on the taskbar. Go to H: drive and double click on HTML3.html.

(3) This should load up **Microsoft Edge** and your web page should appear on the screen (ref. Figure Ⅳ-4-1):

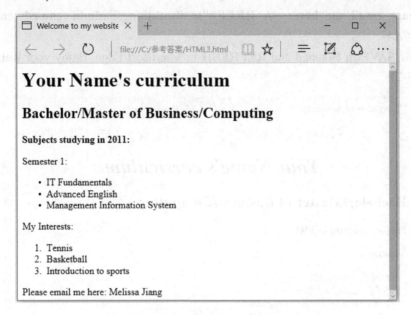

Figure Ⅳ-4-1　Web Page for HTML3.htm

2. You have to make the following changes to your web page.

Note: Every time you make a change in HTML3.html in Notepad, you have to **SAVE** in Notepad and **REFRESH** in **Microsoft Edge** to see the new changes you have made. You do not need to close HTML3.html until you have finished making all the changes.

(1) Put your name at the top of your web page, and make sure your name is bold and italic and center-alignment.

(2) Change the text "Semester 1:" and "My Interests:" to be RED and BOLD.

　　Hint: The code should look like something like the following:

　　　　　<P><B><FONT COLOR="red">Semester 1:</B></FONT>

(3) Link the three subjects to their web links.

subject	web link
IT Fundamentals	ITF.html
Advanced English	English.html
Management Information System	http://www.scsite.com/dc2007

(4) Link the three interests to their web links:

Interest	web link
Tennis	http://www.echinasport.com/
Image: basket.jpg, top-alignment	basketball.htm
Image: ski.jpg, middle-alignment, ALT= Introduction to sports	http://www.sportschina.com/

(5) Add a horizontal line and then a blank line before the email information.

(6) Change the email address to your email address.

Hint: The code should look something like the following.

Please email me here: `<A HREF="mailto:hjiang@admin.ecnu.edu.cn">Melissa Jiang</A>`

(7) Now **SAVE** in Notepad and **REFRESH** in Microsoft Edge. Enjoy your web page (ref. Figure Ⅳ-4-2):

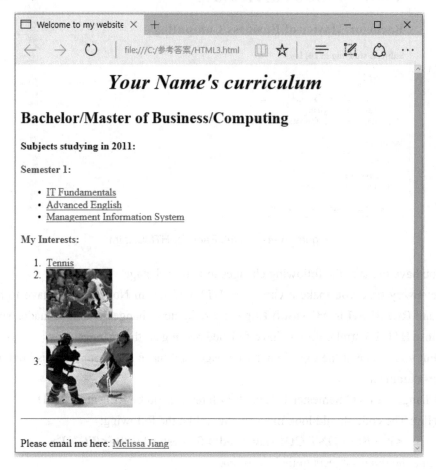

Figure Ⅳ-4-2 Web Page for Updated HTML3.htm

3. Creating your Web page with a table into it:

(1) In notepad, type the following code into another web document named **HTML4.html**.

```
<HTML>
<BODY>
<TABLE>
<CAPTION> LONGLIVE FOOTBALL CLUB </CAPTION>
        <TR>
            <TH> First Column </TH>
```

```
                <TH> Second Column </TH>
                <TH> Third Column </TH>
            </TR>
            <TR>
                <TD>Cell 11.</TD>
                <TD>Cell 12.</TD>
                <TD>Cell 13.</TD>
            </TR>
            <TR>
                <TD>Cell 21.</TD>
                <TD>Cell 22.</TD>
                <TD>Cell 23.</TD>
            </TR>
        </TABLE>
    </BODY>
</HTML>
```

(2) Save it and have a look at it in Microsoft Edge. It should look like a table without any borders (ref. Figure Ⅳ-4-3).

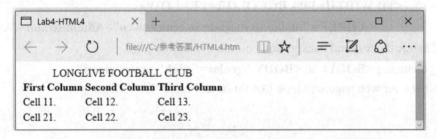

Figure Ⅳ-4-3　Web Page for HTML4.htm

Now add some table attributes into your web page.

(3) Change **<TABLE>** to

<TABLE WIDTH=75% BGCOLOR="orange" BORDER=3 CELLPADDING=10>

Have a look at the result in Microsoft Edge.

(4) You should now experiment with some of the other table attributes.

　　-colspan, bgcolor, align, width, cellpadding

　　For example, replace the following lines

　　　　<TH> First Column　</TH>

　　　　<TH> Second Column </TH>

　　　　<TH> Third Column　</TH>

　　with:

　　　　<TH colspan=3 bgcolor=cyan align="center">

　　　　　<FONT COLOR="red">First Row</FONT></TH>

(5) To link the text in the first row and second column with a suitable web site replace the following line. Try it.

    `<TD>Cell 12.</TD>`

  with

    `<TD ALIGN="center">`

     `<A HREF="http://www.ecnu.edu.cn">Our University</A></TD>`

 (6) Change the text in the second row and second column to an image. Be sure the width of this cell is 50% of the table, and the content of this cell is horizontal center-alignment and vertical middle-alignment. The image's alternative text is "beautiful campus".

 Hint: replace `<TD>Cell 22.</TD>`

  with

    `<TD WIDTH=50% ALIGN=center VALIGN=middle>`

     `<IMG SRC="happy.gif " ALT="beautiful campus"></TD>`

 (7) Change the text in the second row and third column to the Administrator's email address. Be sure the width of this cell is 30% of the table, and the background color of this cell is YELLOW. Administrator's email address is qsyu@admin.ecnu.edu.cn.

 Hint: replace `<TD>Cell 23.</TD>`

  with

    `<TD WIDTH=30% BGCOLOR=YELLOW>`

     `<A HREF= "mailto:qsyu@admin.ecnu.edu.cn">Administrator</A></TD>`

 (8) Change the text and background colors of your web page.

 Hint: changing `<BODY>` to `<BODY bgcolor="pink" text="green">`.

 Now your web page will look like this (ref. Figure Ⅳ-4-4).

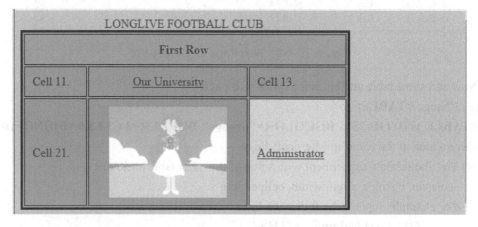

Figure Ⅳ-4-4 Web Page for Updated HTML4.htm(1)

 (9) Try using a background image for your web page.

 Hint: Copy bg2.gif from the Files link on the IT Fundamental web site into your H: drive and include it into your **HTML4.html** by changing the code in the **BODY** of the web page to

  `<BODY background="bg2.gif">`

 Now your web page will look like this (ref. Figure Ⅳ-4-5).

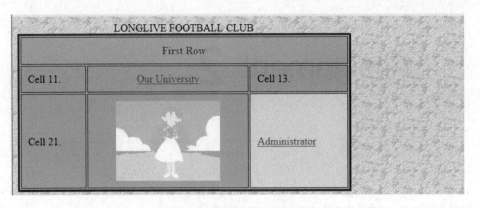

Figure Ⅳ-4-5　Web Page for Updated HTML4.htm(2)

Assignment 2

Your job is to make a web site on your own choice. You are to make the web site as interesting as possible and include features you have learnt in this module.

Your home page should include.

(1) your own personal information (student ID, student name, etc.) on your home page (whose file name must be *index.htm* or *index.html*).

(2) ordered lists and unordered lists.

(3) suitable page font size, page font color, as well as words color, words size and words style, etc.

(4) interesting background style or color. You must try both of them in your home page or the linked web pages.

(5) text links and image links:

- one link to another web page (which is set up by you) which contains a table. Following table elements and attributes must be included in your table: CAPTION, TR, TH, TD, WIDTH, BORDER, CELLPADDING, ALIGN, COLSPAN, BGCOLOR, etc.;
- links to a few existing web sites (at least 3 web sites);
- links to a few web pages set up by you (at least 3 HTML files) or downloaded from the Internet (at least 3 HTML files);
- links to your own email address.

(6) graphics downloaded from the Internet (at least 3 graphics, which can be .GIF or .JPG). Following image attributes must be included in your web pages: WIDTH, HEIGHT, ALT, ALIGN, etc.

(7) any other features that will enhance your web page.

Note: Highest marks will be given to web pages that are attractively presented and use many different HTML tags effectively.

Lab 5
Creating a New Datafile

1. Load Access by selecting **Start → Access 2016**.
2. Creating a new database.
(1) Click **Blank desktop database**.
(2) In the popup dialog, type **EMPLOYEE.accdb** in the **File Name** box. To change the location of the file from the default, click **Browse** ( ) for a location on your own disk to put your database, browse to the new location, and then click **OK**.
(3) Click **Create**. Access creates the database with an empty table named **Table1**, and then opens Table1 in Datasheet view. The cursor is placed in the first empty cell in the **Click to Add** column.
3. Creating a new data Table.
(1) On the **File** tab, click **Save**, or click **Save** button on **Quick Access Toolbar**, save the new table as **EmployeeTBL**.
(2) On the **Fields** tab, in the **Views** group, click **Design View** . The following screen (ref. Figure Ⅳ-5-1, blank initially of course!) will appear.

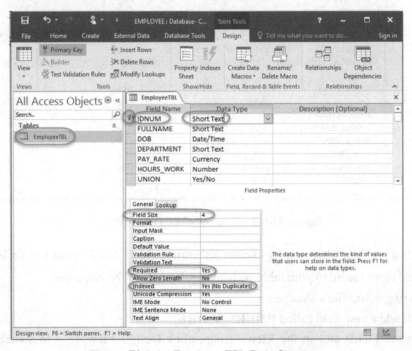

Figure Ⅳ-5-1 EmployeeTBL Data Structure

(3) Enter the following data structure information according to the following table.

Field name	Data Type	Field Size	Format	Decimal Places	Required	Indexed
IDNUM	Short Text	4			Yes	Yes (No Duplicates)
FULLNAME	Short Text	20			Yes	Yes (Duplicates OK)
DOB	Date/Time		Short Date		No	No
DEPARTMENT	Short Text	15			No	No
PAY_RATE	Currency		Currency	2	No	No
UNION	Yes/No		Yes/No			No

(4) When finished, make sure that there is a key indicator on the left of the **IDNUM** field. Otherwise, on the **Design** tab, in the **Tools** group, click **Primary Key** to set the **IDNUM** field as Primary Key.

(5) Click **Save** button on **Quick Access Toolbar** to save the **EmployeeTBL** table.

4. Entry of Records.

(1) On the **Design** tab, in the **Views** group, click **Datasheet View** to switch to Datasheet view.

(2) Enter the following records as shown in Figure Ⅳ-5-2.

IDNUM	FULLNAME	DOB	DEPARTMEN	PAY_RATE	HOURS_WO	UNION
1011	Graham, Ben	1/25/1978	Back	$6.00	30	
1012	Hocking, Garry	10/8/1968	Utility	$8.75	12	✓
1013	Harley,Tom	7/8/1979	Back	$9.00	35	
1016	Houlihan,Adam	5/30/1977	Forward	$8.75	30	
1017	Burns,Ronnie	11/8/1974	Forward	$8.75	23	✓
1012	Kilpatrick,Glen	9/17/1973	Centre	$8.60	20	✓
1020	King, Steven	9/25/1978	Utility	$6.00	35	
1022	Clarke, David	7/19/1980	Centre	$9.10	32	✓
1025	Corey, Joel	2/9/1979	Utility	$8.00	28	
1026	Riccardi, Peter	12/17/1972	Centre	$6.75	30	✓
1027	Spriggs, David	8/2/1981	Utility	$9.00	12	
1028	Bizzell, Clint	3/30/1978	Forward	$8.75	18	
1029	Ling, Cameron	9/9/1981	Forward	$9.00	20	✓
1030	Scarlett, Matthew	9/10/1979	Back	$8.60	40	✓
1037	Mensh, David	6/7/1975	Forward	$11.00	30	

Figure Ⅳ-5-2 EmployeeTBL Data Records

There is a problem when you try to enter Kilpatrick,Glen's details—can you fix it?

Add one more record to your table with YOUR NAME as the employee.

5. Altering a data file's structure.

We will add a new field called **HOURS_WORK**.

(1) On the **Fields** tab, in the **Views** group, click **Design View** to switch to design

view.

(2) Right click on the **Union** field, and then select shortcut command option "**Insert Rows**".

(3) Enter the new field **HOURS_WORK** just before the **Union** field. Make it type **Number** (Field Size: **Single**), Required – **No**, Indexed – **No**.

(4) Click on the **Save** button to save the new design.

(5) Change back to the datasheet screen by click **Datasheet View** in the **Views** group on the **Design** tab.

(6) Add data to the new field **HOURS_WORK** for each record, keeping the hours worked somewhere between 0 and 40 hours.

6. Create another table in Design view.

We will add a new table called **BonusTBL**.

(1) On the **Create** tab, in the **Tables** group, click **Table Design**.

(2) Enter the following data structure information according to the following table.

Field name	Data Type	Field Size	Format	Decimal Places	Required	Indexed
IDNUM	Short Text	4			Yes	Yes (Duplicates OK)
WORKMONTH	Number	Integer			Yes	Yes (Duplicates OK)
BONUS	Currency		Currency	2	No	No

(3) When finished, hold down **Ctrl**, and then click the row selectors of **IDNUM** and **WORKMONTH** to select these two fields.

(4) On the **Design** tab, in the **Tools** group, click **Primary Key**.

(5) On the **Design** tab, in the **Views** group, click **Datasheet View** to switch to Datasheet view.

(6) Enter the following records as shown in Figure Ⅳ-5-3.

IDNUM	WORKMONTH	BONUS
1012	1	$5.00
1012	2	$4.58
1012	3	$6.78
1012	4	$5.55
1018	1	$6.50
1018	2	$4.25
1018	3	$6.21
1020	1	$5.66
1020	2	$3.89
1025	4	$7.94
1037	2	$5.88
1037	3	$8.89

Figure Ⅳ-5-3 BonusTBL Data Records

(7) Click **Save** button on **Quick Access Toolbar** to save the new table as **BonusTBL**.

7. Create a relationship between two tables

A relationship is a link or connection between two tables sharing a common field, which must be of same data type and the same size. When a common field is a primary key in one table, it becomes a foreign key in the other table. A foreign key is a field that links to a primary key field in the related table.

(1) On the **Database Tools** tab, in the **Relationships** group, click **Relationships** "▣".

(2) On the **Design** tab, in the **Relationships** group, click **Show Table** "▣". The **Show Table** dialog box lists the tables and queries that are in the database.

(3) In the **Show Table** dialog box, on the **Tables** tab, double-click the **EmployeeTBL** table and **BonusTBL** table.

(4) Close the **Show Table** dialog box.

(5) Click and drag the **IDNUM** field from the **EmployeeTBL** field list to the **IDNUM** field in the **BonusTBL** field list.

(6) The **Edit Relationships** dialog box opens (ref. Figure Ⅳ-5-4).

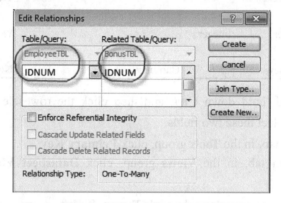

Figure Ⅳ-5-4　Edit Relationships Dialog Box

(7) Click **Create**. A join line links the common field **IDNUM** (ref. Figure Ⅳ-5-5).

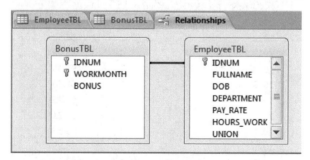

Figure Ⅳ-5-5　A Relationship between Two Tables

(8) Click **Save** button on **Quick Access Toolbar** to save the relationship between **EmployeeTBL** table and **BonusTBL** table.

(9) On the **Design** tab, in the **Relationships** group, click **Close** command ▣.

Lab 6

Access Queries

Load Access and open the database **EMPLOYEE** you created in Lab 5.

Queries provide retrieval of records according to given criteria, and allow the matching records to be listed in a particular order.

QUERY 1

List the IDNUM, FullName and Pay_rate of all employees in the Back department. Save the query as **EmpQRY1**.

(1) Starting a new query:

On the **Create** tab, in the **Queries** group, click **Query Design**.

In the **Show Table** dialog box, on the **Tables** tab, double-click the **EmployeeTBL** table.

Close the **Show Table** dialog box.

(2) Selecting the fields to be included in the query.

This query only requires data from the three fields IDNUM, FullName and Pay_rate.

Drag the **IDNUM** field from the **EmployeeTBL** list to the first column of the Query by Example (QBE) grid.

Drag the fields of **FullName** and **Pay_rate** into the next two columns seperately.

(3) Generating a response to the query:

On the **Design** tab, in the **Results** group, click **Run** or **Datasheet View**.

Does the output match the requirements for the query? Why?

(4) Transferring back to Query Design:

On the **Home** tab, in the **Views** group, click **Design View**.

(5) Selecting the records to be displayed:

Add the **Department** field to your QBE (Query by Example) grid.

Check the **Datasheet View** again to verify the current output.

Back to Query Design, in the **Criteria** row of the Department field type **"Back"**

Again check the result in **Datasheet View**.

(6) Hiding columns:

The Department field is not required by this query, so let's hide it.

Click the **Show** box of the **Department** column so it is not checked.

Again view the results (ref. Figure IV-6-1) of your query.

IDNUM	FULLNAME	PAY_RATE
1011	Graham, Ben	$6.00
1013	Harley,Tom	$9.00
1030	Scarlett, Matthew	$8.60

Figure Ⅳ-6-1 Output of Query1

(7) Saving a query:

From within the **Design View** screen select **File→Save**(or press the **Save** button):

Give it the name **EmpQRY1**.

(Select **File→Save Object As** if you wish to save it with a new name.)

QUERY 2

List the FullName, Pay_Rate and Date of Birth of employees born after 17th December 1980. Save the query as **EmpQRY2**.

(1) First step:

Create a new query based on the EmployeeTBL table as in Query 1 above.

Add the three required fields (FullName, Pay_Rate, and DOB) to the QBE grid.

(2) Date fields:

In the Criteria row of DOB field type

> **> 80/12/17**

(Access stores this as>#12/17/1980#. The # represents a date conversion to Access.)

(3) Generating a response to the query:

Click on the "**Datasheet View**" button or click on the "**Run**" query button to generate a response to the query. View the results (ref. Figure Ⅳ-6-2) of your query.

FULLNAME	PAY_RATE	DOB
Spriggs, David	$9.00	8/2/1981
Ling, Cameron	$9.00	9/9/1981

Figure Ⅳ-6-2 Output of Query2

(4) Saving a query:

Transfer back to the **Query Design** screen and save it with the name **EmpQRY2**.

QUERY 3

List the Department, Fullname, Date of Birth and Union of employees born on or after 17th Dec. 1972 AND who are also union members. Display the records in alphabetical order by FullName within each Department. Save the query as **EmpQRY3**.

(1) First step:

Create a new query and add the fields of Department, Fullname, DOB and Union to the QBE grid.

(2) Combining Record Conditions:

We only want employees who are union members, i.e., **UNION** field contains **Yes**, so add this to the Criteria row of the Union column.

Similarly add the **DOB** condition as outlined in query 2. (Use>=for "on or after")

Conditions on the same Criteria line are joined with an **AND**. Records will only be displayed if both conditions are met.

i.e., DOB must be >=1972/12/17 and UNION must be Yes

(3) Setting the order in which records should be displayed:

Select **Ascending** from the **Sort** row in both the **DEPARTMENT** and **FULLNAME** fields.

Note: to generate the correct sort order, the **DEPARTMENT** field must be placed before the **FULLNAME** field in the QBE grid.

(4) Generating the response to Query 3:

View the results (ref. Figure Ⅳ-6-3) of your query.

DEPARTMENT	FULLNAME	DOB	UNION
Back	Scarlett, Matthew	9/10/1979	☑
Centre	Clarke, David	7/19/1980	☑
Centre	Kilpatrick, Glen	9/17/1973	☑
Centre	Riccardi, Peter	12/17/1972	☑
Forward	Burns, Ronnie	11/8/1974	☑
Forward	Ling, Cameron	9/9/1981	☑

Figure Ⅳ-6-3　Output of Query3

(5) Transferring back to the **Design View** screen and save this query with the name **EmpQRY3**.

QUERY 4

Display all details and the WEEKS PAY of all employees. Save the query as **EmpQRY4**.

(1) First step:

Create a new query and add all fields from EmployeeTBL (double click on the top "*" field).

(2) Adding a calculated field:

WEEKS_PAY is the result of a small calculation between two other fields.

In the Field row of a new column enter

　　Weeks_Pay:　[PAY_RATE] * [HOURS_WORK]

or

Right-click the mouse on the Field row of a new column and choose the **Build** option on the menu (ref. Figure Ⅳ-6-4), an Expression Builder window will be opened. Double-click **PAY_RATE** field name, enter "*", and double-click **HOURS_WORK** field name, then the calculated expression will be displayed automatically in the Expression Builder window (ref. Figure Ⅳ-6-5). Click **OK** button to close the Expression Builder window. In the query design view, rename "Expr1" as "**Weeks_Pay**" (ref. Figure Ⅳ-6-6).

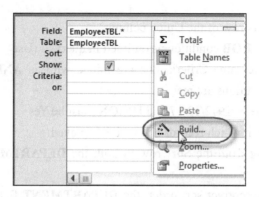

Figure Ⅳ-6-4　Choose the Build Option

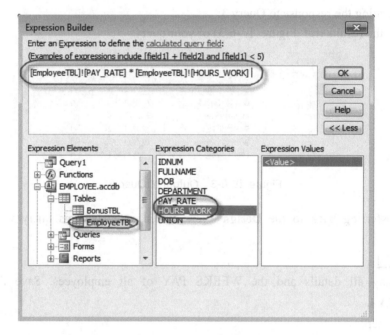

Figure Ⅳ-6-5　Expression Builder Window

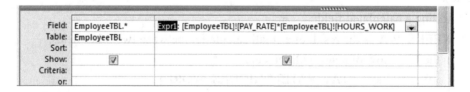

Figure Ⅳ-6-6　Editing the Calculation Field

(3) Formatting the calculated field to Currency format:

Still in design view, right click on the new calculated field, Weeks_Pay, and then click **Properties** to open **Property Sheet** table. In the **Property Sheet**, click the **General** tab, and then select **Currency** from **Format** property.

(4) Generating the response to Query 4.

View the results (ref. Figure Ⅳ-6-7) of your query.

IDNUM	FULLNAME	DOB	DEPARTMEN	PAY_RATE	HOURS_WOI	UNION	Weeks_Pay
1011	Graham, Ben	1/25/1978	Back	$6.00	30	☐	$180.00
1012	Hocking, Garry	10/8/1968	Utility	$8.75	12	☑	$105.00
1013	Harley,Tom	7/8/1979	Back	$9.00	35	☐	$315.00
1016	Houlihan,Adam	5/30/1977	Forward	$8.75	30	☐	$262.50
1017	Burns,Ronnie	11/8/1974	Forward	$8.75	23	☑	$201.25
1018	Kilpatrick,Glen	9/17/1973	Centre	$8.60	20	☑	$172.00
1020	King, Steven	9/25/1978	Utility	$6.00	35	☐	$210.00
1022	Clarke, David	7/19/1980	Centre	$9.10	32	☑	$291.20
1025	Corey, Joel	2/9/1979	Utility	$8.00	28	☐	$224.00
1026	Riccardi, Peter	12/17/1972	Centre	$6.75	30	☑	$202.50
1027	Spriggs, David	8/2/1981	Utility	$9.00	12	☐	$108.00
1028	Bizzell, Clint	3/30/1978	Forward	$8.75	18	☐	$157.50
1029	Ling, Cameron	9/9/1981	Forward	$9.00	20	☑	$180.00
1030	Scarlett, Matthew	9/10/1979	Back	$8.60	40	☑	$344.00
1037	Mensh, David	6/7/1975	Forward	$11.00	30	☐	$330.00

Figure Ⅳ-6-7 Output of Query4

(5) Saving the query as **EmpQRY4**.

(6) **File → Save As → Save Object As→ Save As**, save the query again as **WeeksPayQRY** (ref. Figure Ⅳ-6-8, and WeeksPayQRY will be required in the next lab).

Figure Ⅳ-6-8 Save Query as Another

QUERY 5

Display the IDNUM, FullName, Department and the WEEKS PAY of employees who either have a Pay_Rate between $9.00 and $10.00 OR are in the Union. Display the records in order from the highest pay_rate to the lowest. Save the query as **EmpQRY5**.

(1) First step:

Create a new query and add the three fields (IDNUM, FULLNAME, and DEPARTMENT) from EmployeeTBL to be displayed plus the PAY_RATE and UNION fields.

(2) Numeric fields:

In the first Criteria row of the PAY_RATE field enter

>= 9 and <=10

(Note: no quotes needed for Numeric data)

(3) Sort descending on **PAY_RATE** field.

(4) OR conditions:

In the second Criteria row of the Union field enter **Yes**.

Check the output of this criteria. The two conditions in Pay_Rate and Union being on separate lines means only one has to be true for a record to be selected.

(5) Adding a calculated field:

WEEKS_PAY is the result of a small calculation between two other fields.

In the Field row of a new column enter (or create the calculated expression automatically in the Expression Builder window):

Weeks_Pay: [PAY_RATE] * [HOURS_WORK]

(6) Formatting the calculated field to Currency format:

Still in Design view, right click on the new calculated field, Weeks_Pay, and then click **Properties** to open **Property Sheet** table. In the **Property Sheet**, click the **General** tab, and then select **Currency** from **Format** property.

(7) Generating the response to Query 5.

View the results (ref. Figure Ⅳ-6-9) of your query.

IDNUM	FULLNAME	DEPARTMENT	Weeks_Pay
1022	Clarke, David	Centre	$291.20
1029	Ling, Cameron	Forward	$180.00
1027	Spriggs, David	Utility	$108.00
1013	Harley, Tom	Back	$315.00
1017	Burns, Ronnie	Forward	$201.25
1012	Hocking, Garry	Utility	$105.00
1030	Scarlett, Matthew	Back	$344.00
1018	Kilpatrick, Glen	Centre	$172.00
1026	Riccardi, Peter	Centre	$202.50

Figure Ⅳ-6-9 Output of Query5

(8) Saving the query as **EmpQRY5**.

QUERY 6

Display the Department, IDNUM, and FullName of employees with a Pay_Rate at least $8.00 from the Utility or Back department. Display the records in alphabetical order by employee IDNUM within each Department. Save the query as **EmpQRY6**. Figure Ⅳ-6-10 shows the output of your query.

DEPARTMEN	IDNUM	FULLNAME
Back	1013	Harley, Tom
Back	1030	Scarlett, Matthew
Utility	1012	Hocking, Garry
Utility	1025	Corey, Joel
Utility	1027	Spriggs, David

Figure Ⅳ-6-10 Output of Query6

QUERY 7

Display alphabetically by name within each department all details of employees who are not union members, and who were born before 1st April 1978. Display the records in alphabetical order by employee FullName within each Department. Save the query as EmpQRY7. Figure Ⅳ-6-11 shows the output of your query.

EmpQRY7		
DEPARTMEN	FULLNAME	DOB
Back	Graham, Ben	1/25/1978
Forward	Bizzell, Clint	3/30/1978
Forward	Houlihan, Adam	5/30/1977
Forward	Mensh, David	6/7/1975

Figure Ⅳ-6-11   Output of Query7

QUERY 8

Display the Department and the number of employees, average Pay_Rate, maximum Pay_Rate, and minimum Pay_Rate within each department. Display the records in order from the least employee number to the most. Save the query as **EmpQRY8**.

(1) First step:

Create a new query and add the five fields (DEPARTMENT, IDNUM, Pay_Rate, Pay_Rate, and Pay_Rate) from EmployeeTBL to the QBE grid.

(2) Showing **Total** row on the QBE grid:

Now click the **Totals** button ∑ in the **Show/Hide** group on the **Design** tab, **Total** row will appear on the QBE grid. Note: each column in the query will have the words "**Group By**" in this **Total** row.

(3) Summarising Data (ref. Figure Ⅳ-6-12):

Choosing a summary function for each column you want to summaries.

Click in the Total row under the **IDNUM** column. Click on the button with the down arrow and select **Count** function.

Click in the Total row under the first **Pay_Rate** column. Click on the button with the down arrow and select **Avg** function.

Click in the Total row under the second **Pay_Rate** column. Click on the button with the down arrow and select **Max** function.

Click in the Total row under the third **Pay_Rate** column. Click on the button with the down arrow and select **Min** function.

Field:	DEPARTMENT	IDNUM	PAY_RATE	PAY_RATE	PAY_RATE
Table:	EmployeeTBL	EmployeeTBL	EmployeeTBL	EmployeeTBL	EmployeeTBL
Total:	Group By	Count	Avg	Max	Min
Sort:		Ascending			
Show:	✓	✓	✓	✓	✓
Criteria:					

Figure Ⅳ-6-12   Summarising Data

(4) Renaming the column heading for the summarized columns (ref. Figure Ⅳ-6-13):

Click separately in front of each name of the column (i.e., IDNUM, Pay_Rate, Pay_Rate, and Pay_Rate) in the QBE grid;

Type separately the new column heading name (i.e., HeadCount, AvgPay, MaxPay, and MinPay) followed by a colon in front of each field name in the QBE grid.

(5) Setting the order in which records should be displayed:

Select **Ascending** from the **Sort** row in the **HeadCount** field.

(6) Generating the response to Query 8.

View the results (ref. Figure Ⅳ-6-13) of your query.

DEPARTMEN	HeadCount	AvgPay	MaxPay	MinPay
Centre	3	$8.15	$9.10	$6.75
Back	3	$7.87	$9.00	$6.00
Utility	4	$7.94	$9.00	$6.00
Forward	5	$9.25	$11.00	$8.75

Figure Ⅳ-6-13 Output of Query8

(7) Saving the query as **EmpQRY8**.

(8) Try to use another method (ref. Figure Ⅳ-6-14) to accomplish this query and save it as **EmpQRY8-2**.

Field:	DEPARTMENT	HeadCount: Count(*)	AvgPay: Avg([EmployeeTBL]![PAY_RATE])	MaxPay: Max([EmployeeTBL]![PAY_RATE]	MinPay: Min([EmployeeTBL]![PAY_RA
Table:	EmployeeTBL				
Total:	Group By	Expression	Expression	Expression	Expression
Sort:		Ascending			
Show:	✓	✓	✓	✓	✓
Criteria:					

Figure Ⅳ-6-14 Constructing Query8 (Another Method)

QUERY 9

Display the bonus each month for each employee. The query should list the FullName, WorkMonth, and Bonus, in descending FullName followed by ascending WorkMonth order. Save the query as **EmpQRY9**.

(1) Starting a new query:

On the **Create** tab, in the **Queries** group, click **Query Design** .

In the **Show Table** dialog box, on the **Tables** tab, double-click the **EmployeeTBL** table and **BonusTBL** table.

Close the **Show Table** dialog box.

(2) Selecting the fields to be included in the query:

Drag the **FullName** field from the **EmployeeTBL** list to the first column of the QBE grid.

Drag the **WorkMonth** field and the **Bonus** field from the **BonusTBL** list into the next two columns.

(3) Setting the order in which records should be displayed:

Select **Descending** from the **Sort** row in the **FullName** field and **Ascending** from the **Sort** row in the **WorkMonth** field.

(4) Generating the response to Query 9.

View the results (ref. Figure Ⅳ-6-15) of your query.

FULLNAME	WORKMONTH	BONUS
Mensh, David	2	$5.88
Mensh, David	3	$8.89
King, Steven	1	$5.66
King, Steven	2	$3.89
Kilpatrick, Glen	1	$6.50
Kilpatrick, Glen	2	$4.25
Kilpatrick, Glen	3	$6.21
Hocking, Garry	1	$5.00
Hocking, Garry	2	$4.58
Hocking, Garry	3	$6.78
Hocking, Garry	4	$5.55
Corey, Joel	4	$7.94

Figure Ⅳ-6-15　Output of Query9

(5) Saving the query as **EmpQRY9**.

QUERY 10

Display the average bonus, maximum bonus, and minimum bonus for each employee, in order from the highest average bonus to the lowest. Save the query as **EmpQRY10**.

(1) First step:

Create a new query and add the four fields (FullName, Bonus, Bonus, and Bonus) from the **EmployeeTBL** table and **BonusTBL** table respectively to the QBE grid.

(2) Showing **Total** row on the QBE grid:

Click the **Totals** button ∑ in the **Show/Hide** group on the **Design** tab.

(3) Summarising Data (ref. Figure Ⅳ-6-16):

Choosing a summary function for each column you want to summaries.

Click in the **Total** row under the first **Bonus** column. Click on the button with the down arrow and select **Avg** function.

Click in the **Total** row under the second **Bonus** column. Click on the button with the down arrow and select **Max** function.

Click in the **Total** row under the third **Bonus** column. Click on the button with the down arrow and select **Min** function.

(4) Renaming the column heading for the summarized columns (ref. Figure Ⅳ-6-16):

Click separately in front of each name of the three **Bonus** columns in the QBE grid;

Type separately the new column heading name (i.e., AvgBonus, MaxBonus, and MinBonus) followed by a colon in front of each field name in the QBE grid.

(5) Setting the order in which records should be displayed:
Select **Descending** from the **Sort** row in the **AvgBonus** field.

Field:	FULLNAME	AvgBonus: BONUS	MaxBonus: BONUS	MinBonus: BONUS
Table:	EmployeeTBL	BonusTBL	BonusTBL	BonusTBL
Total:	Group By	Avg	Max	Min
Sort:		Descending		
Show:	✓	✓	✓	✓

Figure Ⅳ-6-16　Summarising Data

(6) Generating the response to Query 10.
　　View the results (ref. Figure Ⅳ-6-17) of your query.

FULLNAME	AvgBonus	MaxBonus	MinBonus
Corey, Joel	$7.94	$7.94	$7.94
Mensh, David	$7.39	$8.89	$5.88
Kilpatrick, Glen	$5.65	$6.50	$4.25
Hocking, Garry	$5.48	$6.78	$4.58
King, Steven	$4.78	$5.66	$3.89

Figure Ⅳ-6-17　Output of Query10

(7) Saving the query as **EmpQRY10.**

(8) Try to use another method (ref. Figure Ⅳ-6-18) to accomplish this query and save it as **EmpQRY10-2.**

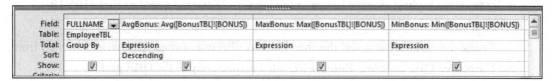

Figure Ⅳ-6-18　Constructing Query10 (Another Method)

Lab 7
Access Report Generation

1. This lab creates a grouped report by using a Wizard. The Report Wizard is the quickest method for creating a grouped report. The report view of this lab is shown in Figure IV-7-1.

DEPARTMENT	IDNUM	FULLNAME	DOB	PAY_RATE	HOURS_WORK	UNION	Weeks_Pay
Back							
	1011	Graham, Ben	1/25/1978	$6.00	30	☐	$180.00
	1013	Harley,Tom	7/8/1979	$9.00	35	☐	$315.00
	1030	Scarlett, Matthew	9/10/1979	$8.60	40	☑	$344.00
Summary for 'DEPARTMENT' = Back (3 detail records)							
Sum of Department							$839.00
Centre							
	1018	Kilpatrick,Glen	9/17/1973	$8.60	20	☑	$172.00
	1022	Clarke, David	7/19/1980	$9.10	32	☑	$291.20
	1026	Riccardi, Peter	12/17/1972	$6.75	30	☑	$202.50
Summary for 'DEPARTMENT' = Centre (3 detail records)							
Sum of Department							$665.70
Forward							
	1016	Houlihan,Adam	5/30/1977	$8.75	30	☐	$262.50
	1017	Burns,Ronnie	11/8/1974	$8.75	23	☑	$201.25
	1028	Bizzell, Clint	3/30/1978	$8.75	18	☐	$157.50
	1029	Ling, Cameron	9/9/1981	$9.00	20	☑	$180.00
	1037	Mensh, David	6/7/1975	$11.00	30	☐	$330.00
Summary for 'DEPARTMENT' = Forward (5 detail records)							
Sum of Department							$1,131.25
Utility							
	1012	Hocking, Garry	10/8/1968	$8.75	12	☑	$105.00
	1020	King, Steven	9/25/1978	$6.00	35	☐	$210.00
	1025	Corey, Joel	2/9/1979	$8.00	28	☐	$224.00
	1027	Spriggs, David	8/2/1981	$9.00	12	☐	$108.00
Summary for 'DEPARTMENT' = Utility (4 detail records)							
Sum of Department							$647.00
Grand Total							$3,282.95

Saturday, December 31, 2016

Figure IV-7-1   Weeks Pay Report

2. Creating a new report by using a Wizard.
(1) On the **Create** tab, in the **Reports** group, click **Report Wizard** 🔍.
(2) In the **Tables/Queries** drop-down box, choose **Query:WeeksPayQRY** which was

created in Lab 6.

(3) Click the **Add All** button [>>]. Click **Next** (ref. Figure Ⅳ-7-2).

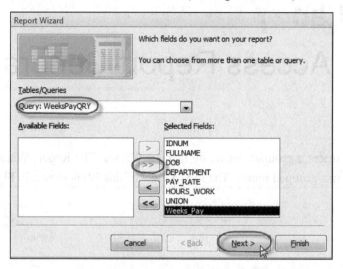

Figure Ⅳ-7-2　Starting the Weeks Pay Report

3. Grouping the report by Department:

(1) Click **DEPARTMENT** and click the **Add One** button [>].

(2) Click **Next** (ref. Figure Ⅳ-7-3).

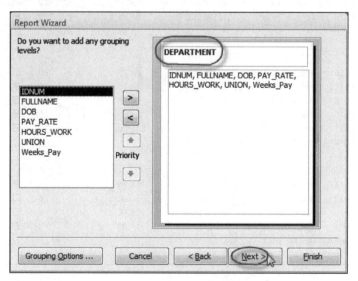

Figure Ⅳ-7-3　Grouping the report by Department

4. Sorting the report by IDNUM:

(1) Click the first combo box drop-down arrow and select **IDNUM**.

(2) Click **Summary Options** (ref. Figure Ⅳ-7-4).

5. Aggregating the report:

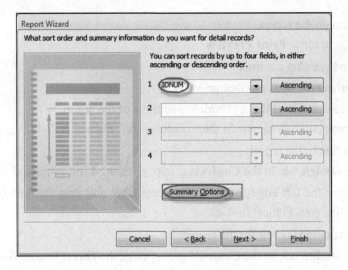

Figure Ⅳ-7-4　Sorting the Report by IDNUM

(1) Click the check box for **Sum** of the Weeks_Pay field.
(2) Click **OK** (ref. Figure Ⅳ-7-5).
(3) Click **Next**.

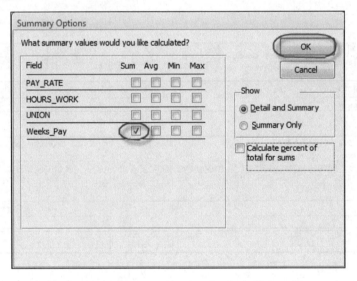

Figure Ⅳ-7-5　Aggregating the Report

6. Laying out the report:
(1) In the Layout section, select **Stepped** and in the **Orientation** section, select **Portrait**.
(2) Click **Next**.
7. Finishing the Report Wizard:
(1) Modify the title of the report to **WeeksPayReport**.
(2) Select **Preview the report** and click **Finish**.
8. Preview the report:

(1) Scroll through the report.

(2) Click ☒ to close **Print Preview**.

9. Moving and resizing controls:

(1) The report is now opens in Design View.

(2) Drag some controls to the appropriate positions of the report.

(3) Resize some controls to best fit their contents (ref. Figure Ⅳ-7-1).

10. Adding a line to the report:

(1) From the **Design** tab, in the **Controls** group, click the **Line** command ＼.

(2) Click above the left edge of the Grand Total label and drag the line control across the page in a straight line (ref. Figure Ⅳ-7-6).

11. Adding the company log for the report:

(1) From the **Design** tab, in the **Controls** group, click the **Insert Image-Browse** command.

(2) Select the company log picture file in the **Insert Picture** dialog.

(3) Click beside the Report Title in the Report Header section and drag the picture control to show the log.

12. The design view of the report is shown in Figure Ⅳ-7-6.

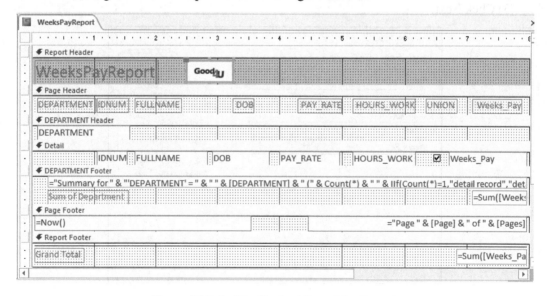

Figure Ⅳ-7-6　The Design View of the Report

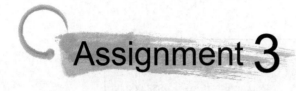

Assignment 3

The following two tables represent respectively 10 records from the **EMPLOYEE** table and the **Bonus** table of a MS-Access database.

EMPLOYEE					
EmployeeName	Department	Gender	EmployedDate	Salary	
Cook Conrad	Personnel	F	9/19/1987	$26,800	
Johnson Frank	Personnel	M	7/18/1988	$46,800	
Jones David	Finance	M	11/24/1990	$33,900	
Knat Michael	Administration	M	11/12/1985	$41,600	
Martins Prue	Personnel	F	7/10/1983	$26,500	
Nichols Sandy	Administration	F	4/16/1994	$39,000	
Pratt John	Finance	M	6/22/1982	$24,940	
Sill Sally	Finance	F	7/12/1988	$45,000	
Smith Paul	Administration	M	5/13/1990	$32,900	
Smoler Ellen	Personnel	F	9/15/1984	$36,888	

BONUS	
EmployeeName	Bonus
Cook Conrad	$569.78
Johnson Frank	$459.23
Jones David	$569.99
Knat Michael	$800.89
Martins Prue	$689.87
Nichols Sandy	$987.67
Pratt John	$779.93
Sill Sally	$498.65
Smith Paul	$798.56
Smoler Ellen	$618.34

1. Create this MS-Access database file and save it with the filename "Assignment3.accdb".

2. Access queries.

(1) Construct a query that lists all employees with the salary exceeding $30 000. The query should list the EmployeeName, Department, and Salary, in ascending Department followed by ascending Salary order. Save this query with the name of Query1. The output of the query should be as follows.

Query1		
EmployeeName	Department	Salary
Smith Paul	Administration	$32,900
Nichols Sandy	Administration	$39,000
Knat Michael	Administration	$41,600
Jones David	Finance	$33,900
Sill Sally	Finance	$45,000
Smoler Ellen	Personnel	$36,888
Johnson Frank	Personnel	$46,800

(2) Construct a query that lists the EmployeeName, Department, and Double of the Salary of all employees in Personnel or Administration department. Display the records in order from the lowest Double salary to the highest. Save this query with the name of Query2. The output of

the query should be as follows.

EmployeeName	Department	Double
Martins Prue	Personnel	$53,000.00
Cook Conrad	Personnel	$53,600.00
Smith Paul	Administration	$65,800.00
Smoler Ellen	Personnel	$73,776.00
Nichols Sandy	Administration	$78,000.00
Knat Michael	Administration	$83,200.00
Johnson Frank	Personnel	$93,600.00

(3) Construct a query that lists all employees who are female or employed after the year 1988. The query should list the EmployeeName, Department, Gender, and EmployedDate, in descending Department followed by ascending EmployedDate order. Save this query with the name of Query3. The output of the query should be as follows.

EmployeeName	Department	Gend	EmployedDate
Martins Prue	Personnel	F	7/10/1983
Smoler Ellen	Personnel	F	9/15/1984
Cook Conrad	Personnel	F	9/19/1987
Sill Sally	Finance	F	7/12/1988
Jones David	Finance	M	11/24/1990
Smith Paul	Administration	M	5/13/1990
Nichols Sandy	Administration	F	4/16/1994

(4) Construct a query that lists all Finance or Personnel employees and who have a salary over 30 000. The query should list the EmployeeName, Gender, and Salary, in order from the highest salary to the lowest. Save this query with the name of Query4. The output of the query should be as follows.

EmployeeName	Gende	Salary
Johnson Frank	M	$46,800
Sill Sally	F	$45,000
Smoler Ellen	F	$36,888
Jones David	M	$33,900

(5) Construct a query that lists all male employees and who are employed not before the year 1988. The query should list the EmployeeName, Salary, and EmployedDate, in order from the earliest employed date to the newest. Save this query with the name of Query5. The output of the query should be as follows.

Query5		
EmployeeName	Salary	EmployedDate
Johnson Frank	$46,800	7/18/1988
Smith Paul	$32,900	5/13/1990
Jones David	$33,900	11/24/1990

(6) Construct a query that lists all Administration employees and who are employed before the year 1991. The query should list the EmployeeName, EmployedDate and NewSalary (increased by 10%). Save this query with the name of Query6. The output of the query should be as follows.

Query6		
EmployeeName	EmployedDate	NewSalary
Knat Michael	11/12/1985	$45,760.00
Smith Paul	5/13/1990	$36,190.00

(7) Construct a query that lists all Administration employees or who have a salary between 40 000 and 45 000. The query should list the Department, EmployeeName, and YearsEmployed, in descending YearsEmployed order within each Department. Save this query with the name of Query7. The output of the query should be as follows.

Query7		
Department	EmployeeName	YearsEmployed
Administration	Knat Michael	26
Administration	Smith Paul	21
Administration	Nichols Sandy	17
Finance	Sill Sally	23

(8) Construct a query that lists the number of the male and female employees within each department. The query should list the Department, Gender, and number of employees (shown as HeadCount), in order from the most number of employees to the least within each Department. Save this query with the name of Query8. The output of the query should be as follows.

Query8		
Department	Gender	HeadCount
Administration	M	2
Administration	F	1
Finance	M	2
Finance	F	1
Personnel	F	3
Personnel	M	1

(9) Construct a query that lists all departments whose average salary exceeds $34 500. The query should list the Department, number of employees (shown as HeadCount), and average salary (shown as AvgSalary). Save this query with the name of Query9. The output of the query

should be as follows.

Query9		
Department	HeadCount	AvgSalary
Administration	3	$37,833.33
Finance	3	$34,613.33

(10) Construct a query that lists all departments which have not less than two employees whose salary more than $35 000. The query should list the Department, number of employees (shown as HeadCount), and average salary (shown as AvgSalary). Save this query with the name of Query10. The output of the query should be as follows.

Query10		
Department	HeadCount	AvgSalary
Administration	2	$40,300.00
Personnel	2	$41,844.00

(11) Construct a query that lists the Salary and Bonus for each employee. The query should list the EmployeeName, Salary, and Bonus, in order from the lowest bonus to the highest. Save this query with the name of Query11. The output of the query should be as follows.

Query11		
EmployeeName	Salary	Bonus
Johnson Frank	$46,800	$459.23
Sill Sally	$45,000	$498.65
Cook Conrad	$26,800	$569.78
Jones David	$33,900	$569.99
Smoler Ellen	$36,888	$618.34
Martins Prue	$26,500	$689.87
Pratt John	$24,940	$779.93
Smith Paul	$32,900	$798.56
Knat Michael	$41,600	$800.89
Nichols Sandy	$39,000	$987.67

(12) Construct a query that lists number of employees, average salary, and average bonus within each department. The query should list the Department, number of employees (shown as HeadCount), average salary (shown as AvgSalary), and average bonus (shown as AvgBonus). Display the records in order from the most number of employees to the least. Save this query with the name of Query12. The output of the query should be as follows.

Query12			
Department	HeadCount	AvgSalary	AvgBonus
Personnel	4	$34,247.00	$584.31
Finance	3	$34,613.33	$616.19
Administration	3	$37,833.33	$862.37

3. Create the following report (saved as QueryReport) grouped by department and sorted by EmployeeName with aggregation of average salary and average bonus in each department.

QueryReport

Monday, May 16, 2011

Department	EmployeeName	Gender	EmployedDate	Salary	Bonus
Administration					
	Knat Michael	M	11/12/1985	$41,600	$800.89
	Nichols Sandy	F	4/16/1994	$39,000	$987.67
	Smith Paul	M	5/13/1990	$32,900	$798.56
Summary for 'Department' = Administration (3 detail records)					
Average				$37,833	$862.37
Finance					
	Jones David	M	11/24/1990	$33,900	$569.99
	Pratt John	M	6/22/1982	$24,940	$779.93
	Sill Sally	F	7/12/1988	$45,000	$498.65
Summary for 'Department' = Finance (3 detail records)					
Average				$34,613	$616.19
Personnel					
	Cook Conrad	F	9/19/1987	$26,800	$569.78
	Johnson Frank	M	7/18/1988	$46,800	$459.23
	Martins Prue	F	7/10/1983	$26,500	$689.87
	Smoler Ellen	F	9/15/1984	$36,888	$618.34
Summary for 'Department' = Personnel (4 detail records)					
Average				$34,247	$584.31

Part V Revisions

（第五部分 复习题）

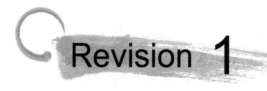

Revision 1

SECTION A

1. How much memory is required for the following display? An IBM PC with a display of 128 colors and 800 × 600 pixels_____.
   A. 420 000 bytes
   B. 800 000 bytes
   C. 3 360 000 bytes
   D. 61 440 000 bytes

2. A personal computer attached to a LAN is referred to as a_____.
   A. host
   B. router
   C. gateway
   D. node

3. A company wishes to store records on its 1500 customers and 200 stock items. Each customer requires 25KB, each stock item 20KB and each sales order 30KB. Each customer has on average 20 sale orders per year. Estimate the size of the files over a one-year period_____.
   A. 0.8GB
   B. 1.8GB
   C. 0.9GB
   D. 9GB

4. A 3 1/2 inch floppy disk contains 80 tracks and is divided into 18 sectors. Each sector in each track holds 512 characters. The disk is double sided. What is the capacity of the disk?_____
   A. 368 640 bytes
   B. 512 000 bytes
   C. 737 280 bytes
   D. 1 474 560 bytes

5. _____is the processing of encoding data and information into unreadable form.
   A. Encryption
   B. Formatting
   C. Transmission
   D. Modulation

6. Modulation converts a signal from_____.
   A. analog to digital
   B. digital to synchronous
   C. digital to analog
   D. synchronous to digital

7. What is the average access time of a hard drive which has a seek time of 10 milliseconds and rotates at 3600 rpm? Assume that the data transfer time is negligible._____
   A. 14.17ms
   B. 18.33ms
   C. 22.33ms
   D. 26.67ms

8. Given the following table in Excel, what will be displayed in the cell with the following formula = VLOOKUP(19,$A$2:$C$7,3)?_____

	A	B	C
1	Hours	Bonus	Payrate
2	0	0	$8.00
3	5	$20	$10.00
4	10	$30	$11.00
5	20	$50	$12.00
6	25	$60	$13.50
7	40	$70	$15.00

   A. $30
   B. $70
   C. $11.00
   D. $12.00

9. By default, Excel positions number_____the cell.
   A. left-aligned, which means the text is positioned at the far left in
   B. right-aligned, which means the text is positioned at the far right in
   C. centered, which means the text is positioned in the middle of
   D. justified, which means the text is spread across the width of

10. To create a link to your email address on your Web Page, the following code should be used_____.
   A. <A HREF=bobby@student.ecnu.edu.cn>My email</A>
   B. <A MAIL TO=bobby@student.ecnu.edu.cn>My email</A>
   C. <A HREF=mailto:bobby@student.ecnu.edu.cn>My email</A>
   D. <A HREF=mail to:bobby@student.ecnu.edu.cn>My email</A>

11. A network on which a cable forms a closed loop with all computers and devices arranged along the loop is known as a_____.
   A. bus network
   B. star network
   C. token ring
   D. ring network

12. Which of the following pieces of code will show the entire text on your web page in large font?_____
    A. <TITLE><BASEFONT SIZE=7></TITLE>
    B. <BODY ><BASEFONT SIZE=7></BODY>
    C. <HEAD><BASEFONT SIZE=3></HEAD>
    D. <HEAD><BASEFONT SIZE=7></HEAD>

13. The HTML code to include an image (image.gif) on top right corner of the web page is_____.
    A. <A IMG SRC="image.gif " align=top>
    B. <IMG SRC="image.gif " align=right>
    C. <IMG SCR="image.gif " align=right>
    D. <A HREF="image.gif " align=right>

14. Which is not true of a "primary key" in MS Access?_____
    A. it is good practice to set up a primary key in each tale of a database
    B. there cannot be two records with the same primary key
    C. it maintains the integrity of the database
    D. it restricts access to the table

15. _____is NOT a feature of Access Database Management System.
    A. Form
    B. Query
    C. Report
    D. Label

16. _____is NOT a valid field type in Access databases.
    A. character
    B. money
    C. currency
    D. numeric

17. A cylinder is a term used to describe how data is stored on a_____.
    A. floppy disk
    B. hard disk
    C. DVD
    D. CD ROM

18. The extensions .gov, .edu, .mil, and, .net are called_____.
    A. E-mail targets
    B. domain extensions
    C. domain name system
    D. top-level domains

19. For a spreadsheet set up with the percentages located in the key values table, a formula to decrease the value in cell B20 by a percentage contained in cell B9 would look like_____.
    A. B20*B9

B. B20*(1–B9)

C. B20*(1–$B$9)

D. $B$20*(1–$B$9)

20. In a MS Access query, what would be the criteria for displaying all books published before the year 2000 OR which cost more than $30.00?_____

A.
Field:	Year	Price
Criteria:	>=2000	>30.00
Or:		

B.
Field:	Year	Price
Criteria:	>2000	
Or:		>=30.00

C.
Field:	Year	Price
Criteria:	<2000	>30.00
Or:		

D.
Field:	Year	Price
Criteria:	<2000	
Or:		>30.00

SECTION B

1. Briefly explain the following computing terms: RISC; ROM; and DVD.

2. Write short notes on the following terms: fibre optics; EDI; and OS.

3. Explain what is meant by the following terms in the context of web authoring: shopping cart, browser; and anchor tag.

4. Describe three ways in which E-Commerce can be beneficial to a small business enterprise.

5. What sort of transmission (simplex, half-duplex or full-duplex) is required for the following situations?_____

    (a) CB radio

    (b) Mobile phone

    (c) Television

6. Describe the characteristics and uses of the following storage media: CD ROM & USB Flash Drive.

7. Before the computer can do useful work, the operating system must be "booted". What are the steps involved in booting a simple operating system?

8. What are the benefits of having a Local Area Network for a medium sized business which consisted of 3 staff members in the Accounts department, 2 in the IT department, 5 in managerial

positions, 10 in the sales department and 3 administrative staff? What would be the benefits to the business if a Wide Area Network is set up for this company?

9. Compare the 3 different types of monitors (CRT, LCD and Gas Plasma) with respect to quality of picture, cost and the technology they use.

10. This question comprises four parts.

    (a) Convert 15 000 bits to KB.

    (b) Convert 5.2GB to KB.

    (c) Suppose you send a block of 2000 bytes of data using asynchronous data transmission at 7200 bps. The asynchronous format uses 1 start bit, 2 parity bits and 2 stop bits for each byte of data. How long (in seconds with two decimals) will it take to send the data?

    (d) Calculate the average access time (in ms with two decimals) of a hard drive which has a seek time of 35 milliseconds and rotates at 7200 rpm. Assume negligible data transfer time.

SECTION C

1. Windows Operations

You are assigned as a DB Administrator for Multimedia Technology Co., Ltd. In the era of Big Data, where companies are able to quickly make sense of larger quantities of data than ever, information is finally recognized as a critical business asset.

You have been given the responsibility for the collection and distribution of information inside and outside of the company.

(a) Your company provides many kinds of multimedia data for different purposes: word processing, presentation, and web-authoring. Design and implement the suitable directory structure and files as shown below that will be capable of storing the files for each purpose. In a well-organised and efficient way. Print a screenshot of the directory structure.

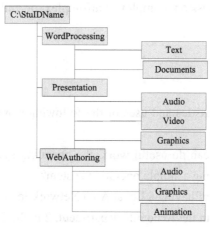

(b) Your lecturer will provide you with the following files:
- .mp3, .mp4, .wav, .jpg, and .avi files for Presentation folder.
- .swf, .jpg, .wav, and.gif files for WebAuthoring folder.
- .txt, and.docx files for WordProcessing folder.

Copy each of the files provided into the appropriate folder in the directory structure you created. Print a screen shot of the contents of the Graphics-WebAuthoring folder.

(c) Try to copy win.ini in drive C into the WebAuthoring folder, and rename it as winsys.cfg.

(d) Display the detailed information and the file extensions for the files in file explorer. Print a screenshot for the option settings.

(e) Compress the files ITF_WebAuthoring_snow.jpg, ITF_WebAuthoring_egg.jpg and ITF_Presentation_birds.jpg your lecturer provided for you in Question (b) into the WebAuthoring folder, named as pictures.rar with the password 123456. Take a screenshot of the password set up screen, and a screenshot of the contents of the WebAuthoring folder.

(f) Carry out a virus scan of Drive C. Take screenshots of the virus scan set up and scanning screen.

2. Word Operations

Recall the file saved as "Go Abroad China Special Offer Report.docx" and edit it as follows:

(a) Insert automatic page number at the bottom of the report.

(b) Insert "Go Abroad International Programs Agency" as a header.

(c) Insert your StuID, name and today's date (update automatically) as a footer.

(d) Copy the information from the spreadsheet file called scholarship.xlsx to the Go Abroad China Chinese Program Scholarship section.

(e) Insert an automatic table of contents (on a separate page).

(f) Insert a front cover and include an appropriate report title and the Agency logo.

(g) Tide up the report as required.

3. PowerPoint Operations

Using suitable websites for reference create a presentation on Porter's three generic strategies. Each slide should contain the Agency logo. On completion save the presentation with the name of StuIDNamePPT.pptx, and print it (6 slides per page) to StuIDNamePPT.pdf.

Your presentation should include the following slides:

Slide 1: Title slide giving your StuID, name and presentation title.

Slide 2: Explanation of what Porter's three generic strategies is.

Slide 3: Provide an example of a company that is using each strategy.

Slide 4: Copy results of Comparison of Sales and display as a graph.

Slide 5: Three hyperlinks to websites giving further information on Porter's three generic strategies.

Slide 6: Closing image on Porter's three generic strategies.

4. Spreadsheet Operations

A spreadsheet has been developed for Harry's Bed Shop to estimate his sales and expenses over 5 years. The estimated sales are $15 000 per year for single beds increasing by $8000 per year, $45 000 per year for double beds increasing by $12 000 per year, $18000 for bunks decreasing by 8% per year and $30000 for mattresses increasing by 7% per year. The estimated expenses are $50 000 per year for stock increasing by 16% each year, $45 000 for wages per year, and $5500 for other expenses per year. Rent will vary per year as indicated by the table included in the spreadsheet.

	A	B	C	D	E	F
1		**Harry's Bed Shop**				
2	**Sales**					
3	Single Beds	$15,000.00	per year			
4	increasing by	$8,000.00	per year			
5	Double Beds	$45,000.00	per year			
6	increasing by	$12,000.00	per year			
7	Bunks	$18,000.00	per year			
8	% decrease	8%	per year			
9	Mattresses	$30,000	per year			
10	% increase	7%	per year			
11						
12	**Expenses**					
13	Cost of stock	$50,000	per year			
14	increasing by	16%	per year			
15	Wages	$45,000.00	per year			
16	Other expenses	$5,500.00	per year			
17	Rent	Year	Rent			
18		1	$10,000.00			
19		2	$12,000.00			
20		3	$15,000.00			
21		4	$18,000.00			
22		5	$22,000.00			
23		1	2	3	4	5
24	**Sales**					
25	Single Beds	$15,000.00	$23,000.00	$31,000.00	$39,000.00	$47,000.00
26	Double Beds	$45,000.00	$57,000.00	$69,000.00	$81,000.00	$93,000.00
27	Bunks	$18,000.00	$16,560.00	$15,235.20	$14,016.38	$12,895.07
28	Mattresses	$30,000.00	$32,100.00	$34,347.00	$36,751.29	$39,323.88
29	TOTAL	$108,000.00	$128,660.00	$149,582.20	$170,767.67	$192,218.95
30						
31	**EXPENSES**					
32	Cost of stock	$50,000.00	$58,000.00	$67,280.00	$78,044.80	$90,531.97
33	Wages	$45,000.00	$45,000.00	$45,000.00	$45,000.00	$45,000.00
34	Other	$5,500.00	$5,500.00	$5,500.00	$5,500.00	$5,500.00
35	Rent	$10,000.00	$12,000.00	$15,000.00	$18,000.00	$22,000.00
36	TOTAL	$110,500.00	$120,500.00	$132,780.00	$146,544.80	$163,031.97
37						
38	LOSS/PROFIT	-$2,500.00	$8,160.00	$16,802.20	$24,222.87	$29,186.99
39		LOSS	PROFIT	PROFIT	PROFIT	PROFIT

The following cells have formulae that help to generate the desired results for the five years. Write down the formula that would appear in the cells given.

C26		C33	
C27		C35	
C28		C36	
C29		C38	
C32		C39	

5. HTML Operations

   (a) Draw the web page created by the following HTML code.

```
<HTML>
<HEAD>
    <TITLE> Table Example</TITLE>
</HEAD>
<TABLE WIDTH=75% BGCOLOR="#FFFF66" BORDER=3 CELLPADDING=10>
<CAPTION> <H1>JUST 4 CELLS </H1></CAPTION>
    <TR>
        <TH WIDTH=30% COLSPAN=3 ALIGN="center">This is cell 1.</TH>
    </TR>
    <TR>
        <TD WIDTH=20% BGCOLOR="#FFCC66">This is cell 2.</TD>
        <TD>This is cell 3.</TD>
        <TD WIDTH=50% BGCOLOR="#99FF66">This is cell 4.</TD>
    </TR>
</TABLE>
</HTML>
```

   (b) Write the complete HTML code to create the following center-aligned table (whose border is 3 and cellpadding is 10 and width is the 80% of the window) with 3 rows and 2 columns, with the picture "family.jpg" spanned across both columns in the first row, "Bob" and "Jane" in the columns of the second row and "Tom" and "Andrea" in the columns of the third row.

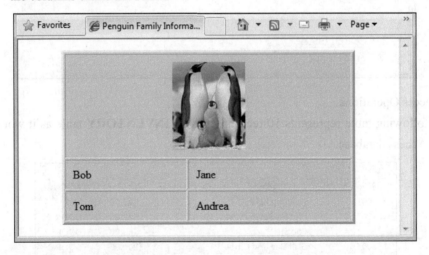

   (c) Draw the web page created by the following HTML code.

```
<HTML>
<HEAD>
  <TITLE> Flora Hill Golf Club</TITLE>
</HEAD>
```

```
<BODY>
  <H1 ALIGN=CENTER><B> President: Gary Counchly</B></H1>
  <IMG SRC="sports.gif">
  <UL> <LI><A HREF="interesting.html">Our common interests</A>
       <LI><A HREF="sports.html">The sports information</A>
  </UL>
</BODY>
</HTML>
```

(d) Write the full HTML code to display the following web page. A picture (heart.gif) should appear in your web page with the width of 100 and height of 120. Assume that the first hyperlink refers to the following URL "http://www.163.com" and the second hyperlink refers to the email address of "friend@ecnu.edu.cn". Note: "How Beautiful Our Life Is" has the H2 format. Be sure that the web page has the title of "Table Tennis Club".

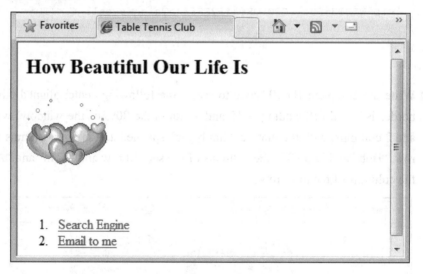

6. Access Operations

The following table represents 10 records from an **INVENTORY** table as it would appear in an MS-Access database.

ITEM_NUM	ITEMDESC	QOH	ITEM_PRICE	LOCATION
1001	Bowling Ball	3	$70.00	INDOOR
1002	Cricket Bat	2	$80.00	OUTDOOR
1003	Football	3	$60.00	OUTDOOR
1004	Golf Bag and Clubs	2	$200.00	OUTDOOR
1005	Tennis Racket	4	$100.00	BOTH
1006	Basketball	2	$50.00	BOTH
1007	Bike Helmet	2	$40.00	OUTDOOR
1008	Fishing Rod	4	$60.00	OUTDOOR
1009	Bench Press	2	$200.00	INDOOR
1010	Baseball Bat	3	$80.00	OUTDOOR

(a) Write down the data displayed for the following query.

Field:	ITEM_NUM	ITEMDESC	LOCATION	ITEM_PRICE	QOH
Table:	INVENTORY	INVENTORY	INVENTORY	INVENTORY	INVENTORY
Sort:			Ascending	Ascending	
Show:	☑	☑	☑	☑	☐
Criteria:					>=4
or:				<=200 And >=100	

(b) Construct a query that lists all items that have a total value exceeding $200 and locate in OUTDOOR. The query should list the ITEM_NUM, ITEMDESC and TOTVALUE, in a descending TOTVALUE order. Save this query with the name of Query1.

(c) Write down the output of the query in 6(b).

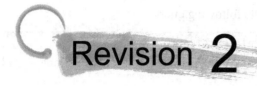

Revision 2

SECTION A

1. How much memory is required for the following display? An IBM PC with a display of 512 colors and 1024 × 768 pixels._____
   A. 884 736 bytes
   B. 720 000 bytes
   C. 786 432 bytes
   D. 1 769 472 bytes

2. The technology used by most desktop computer screens is_____.
   A. POS
   B. OMR
   C. OCR
   D. LCD

3. A 3 1/2 inch floppy disk contains 80 tracks and is divided into 9 sectors. Each sector in each track holds 512 characters. The disk is double sided. What is the capacity of the disk?_____
   A. 368 640 bytes
   B. 512 000 bytes
   C. 737 280 bytes
   D. 1.44 MB

4. A(n)_____port can connect up to 127 different peripherals together with a single connector type.
   A. serial
   B. Bluetooth
   C. USB
   D. IrDA

5. In a_____network, all of the devices in the network connect to a central computer.
   A. bus
   B. star

C. ring

D. peer-to-peer

6. A special computer that directs communications traffic among several linked networks is a_____.

A. router

B. client

C. host

D. node

7. The two types of system software are operating systems and_____.

A. file viewers

B. virus checkers

C. utility programs

D. compression programs

8. The transmission that sends data as distinct pulses, either on or off is called_____.

A. analog

B. data

C. binary

D. digital

9. By default, Excel positions text____the cell.

A. left-aligned, which means the text is positioned at the far left in

B. right-aligned, which means the text is positioned at the far right in

C. centered, which means the text is positioned in the middle of

D. justified, which means the text is spread across the width of

10. The following formula is considered to be a relative reference_____.

A. B17

B. $B17

C. B$17

D. $B$17

11. For the cell values: A1 = 6, B7 = 12 and A3 = 9. What is displayed in the cell with the following formula? IF(B7>A1+A3,A1,A3)_____

A. 3

B. 6

C. 9

D. 15

12. WINRAR is an example of a(n)_____.

A. antivirus utility

B. file compression utility

C. personal firewall

D. personal computer maintenance utility

13. _____ is a security protocol commonly used on the internet.

    A. Common Gateway Interface (CGI)

    B. TCP/IP

    C. Shopping Carts

    D. Secure Sockets Layer (SSL)

14. Anne is looking for some information on the Internet. She types www.baidu.com on the browser. Once the browser is connected, a document file is sent back to Anne's computer. What does this document contain?_____

    A. Location or address of the resources

    B. Hypertext Markup Language

    C. Web page

    D. Domain name

15. Demodulation converts a signal from_____.

    A. analog to digital

    B. digital to synchronous

    C. digital to analog

    D. synchronous to digital

16. _____ is NOT a valid field type in Access databases.

    A. character

    B. logical

    C. operator

    D. numeric

17. _____ is the repetition of the same data values in multiple files.

    A. Data redundancy

    B. Data sharing

    C. Data integrity

    D. Data maintenance

18. The two basic parts of a URL are_____.

    A. the protocol and the domain name

    B. destination and device

    C. TCP/IP and ISP

    D. TCP and IP

19. A company wants to transfer a file from one computer to another using a data link. The file is 2MB in length, and must be transferred in 30 minutes. What bit rate is this?_____

    A. 4000 bps

B. 50 000 bps
C. 30 000 bps
D. 10 000 bps

20. _____ is NOT a feature of Access Database Management System.
A. Form
B. Menu
C. Report
D. Query

SECTION B

1. Name three input or output devices used for physically challenged users.

2. Briefly explain the following computing terms: RAID Storage, pipelining and multifunction device.

3. Write short notes on the following terms: TCP/IP, firewall and plug and play.

4. What are cookies used for and where are they stored? What are the positives and negatives that result from the use of cookies?

5. Discuss the difference between a client-server network and a peer-to-peer network.

6. What is meant by full-duplex data transmission? Give an example of this type of data transmission.

7. Five PCs in a Purchasing Department of a large consumer-goods manufacturer are used primarily for word processing and database applications. What would be the benefits associated with connecting the PCs in a local area network?

8. Briefly describe the following WWW terms: E-commerce, a server and a search engine.

9. How can a spreadsheet be designed to allow a "What IF" analysis to be performed? What is the purpose of a "What IF" analysis?

10. This question comprises four parts.
    (a) How many bits in 12KB?
    (b) Calculate the average access time (in ms with two decimals) of a hard drive which has a seek time of 25 milliseconds and rotates at 3600 rpm. Assume negligible data transfer time.
    (c) A company wishes to store records on its 5000 customers and 100 stock items. Each customer requires 5KB, each stock item 10KB and each sales order 30KB. Each customer has on average 50 sale orders per year. Calculate the size of the files over a one-year period (in GB with two decimals).
    (d) Suppose you send a block of 3000 bytes of data using synchronous data transm-

ission at 9600 bps. The synchronous data block begins with 2 SYN characters, 12 bytes of control information, and then the 3000 bytes of data and finishes with 8 bytes of error checking. How long (in seconds with two decimals) will it take to send the data?

SECTION C

1. Windows Operations

You are employed as a Marketing Assistant with Synergie Fitness Club. The company has decided to expand the business by issuing franchise licenses. You have been given the responsibility for the collection and distribution of information to and from franchisees.

(a) You are involved with the setting up of three of the new Synergie franchises. You will need to be able access relevant files quickly. In order to do this, design and implement a suitable directory structure as shown below that will be capable of storing your files for each of the franchises in a well-organised and efficient way. Print a screenshot of the directory structure.

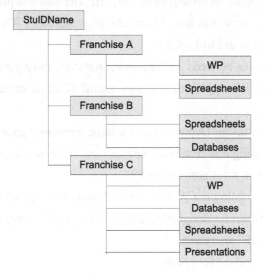

(b) Your lecturer will provide you with the following files:
- Franchise A Report
- Franchise A Cash Budget
- Franchise B Staff
- Franchise B Cash Budget
- Franchise C Cash Budget
- Franchise C Report
- Franchise C Presentation

Copy each of the files provided into the appropriate folder in the directory structure you created. Print a screen shot of the contents of the Franchise B—Database folder.

(c) Let the author of ITF_Franchise_A_Report.doc be your name. Print a screenshot of the author editing dialog.

(d) Try to copy State.ini in drive C into the Franchise C folder, and rename it as Status.cfg. Print a screenshot of the contents of the Franchise C folder.

(e) Compress the database files ITF_Franchise_B_Staff.accdb and ITF_Franchise_C_Employee.mdb your lecturer provided for you in Question (b) into the Franchise C folder, named MYDBs.rar with the password 123. Take a screenshot of the password set up screen, a screenshot of the contents of MYDBs.rar.

2. Word Operations

Recall the report "Synergie Marketing Report" and edit it as follows:

(a) Insert automatic page numbers at the bottom of the report.

(b) Insert "Synergie Fitness Club" as a header.

(c) Insert your StuID, name and today's date (update automatically) as a footer.

(d) Copy the information from the spreadsheet file called Competitors.xls to the competitor analysis section.

(e) Insert an automatic table of contents (on a separate page).

(f) Insert a front cover and include an appropriate report title and the Synergie logo.

(g) Tidy up the report as required.

3. PowerPoint Operations

Using suitable websites for reference create a presentation on Market Segmentation. Each slide should contain the Synergie logo which is saved in the image file called Synergie Logo. On completion save the presentation with the name of StuIDNamePPT.pptx, and print it (6 slides per page) to StuIDNamePPT.pdf.

Your presentation should contain the following slides:

Slide 1 Title slide giving your name and presentation title.

Slide 2 Explanation of what market segmentation is.

Slide 3 Benefits of market segmentation.

Slide 4 Methods of market segmentation.

Slides 5,6 and 7 Give three example segments from Sport England's segmentation of the health and fitness market.

Slide 8 Copy results of Members' Survey and display as a graph.

Slide 9 Two hyperlinks to websites giving further information on Market Segmentation.

Slide 10 Closing image on Market Segmentation.

4. Spreadsheet Operations

The spreadsheet has been developed for Lilian's Golf Supplies to project the sales and expenses over a six-year period. Lilian plans to introduce golf lessons that will be taken by the existing staff. The estimated number of golf lessons per year is shown in the spreadsheet. Golf lessons cost $50 each. The golf shop has sales of $148 000 per year and three staff are employed at $35000 each. Insurance costs are $3500 increasing by 7% per year. The rental is $8000 increasing by $1000 per year. The cost of stock is $50 000 per year.

	A	B	C	D	E	F	G
1			**Lilian's Golf Supplies**				
2							
3	Sales	$148,000.00	per year		YEAR	LESSONS	
4	Lesson Cost	$50.00	per lesson		1	350	
5					2	400	
6	Staff	$35,000.00	each		3	600	
7	no of staff	3			4	650	
8	Insurance	$3,500.00	per year		5	850	
9	increasing by	7%	per year		6	1200	
10	Rent	$8,000	per year				
11	increasing by	$1,000.00	per year				
12	Stock	$50,000.00	per year				
13							
14		1	2	3	4	5	6
15	INCOME						
16	Sales	$148,000.00	$148,000.00	$148,000.00	$148,000.00	$148,000.00	$148,000.00
17	Lessons	$17,500.00	$20,000.00	$30,000.00	$32,500.00	$42,500.00	$60,000.00
18	Total income	$165,500.00	$168,000.00	$178,000.00	$180,500.00	$190,500.00	$208,000.00
19							
20	EXPENSES						
21	Staff	$105,000.00	$105,000.00	$105,000.00	$105,000.00	$105,000.00	$105,000.00
22	Insurance	$3,500.00	$3,745.00	$4,007.15	$4,287.65	$4,587.79	$4,908.93
23	Rent	$8,000.00	$9,000.00	$10,000.00	$11,000.00	$12,000.00	$13,000.00
24	Stock	$50,000.00	$50,000.00	$50,000.00	$50,000.00	$50,000.00	$50,000.00
25	Total Expenses	$166,500.00	$167,745.00	$169,007.15	$170,287.65	$171,587.79	$172,908.93
26							
27	Profit/Loss	-$1,000.00	$255.00	$8,992.85	$10,212.35	$18,912.21	$35,091.07
28		LOSS	PROFIT	PROFIT	PROFIT	PROFIT	PROFIT

The following cells have formulae that help to generate the desired results for the six years. Write down the formula that would appear in each of the ten cells listed below.

B16		B23	
B17		C23	
B18		B25	
B22		B27	
C22		B28	

5. HTML Operations

(a) Draw the web page that will be shown in a browser as a result of the following HTML code.

```
<HTML>
<HEAD>
  <TITLE> Young Kids Academic Club</TITLE>
</HEAD>
<BODY>
  <UL>  <LI><A HREF="http://www.163.com">The Web Page</A>
        <LI><A HREF="http://www.cc.ecnu.edu.cn/">ITF Page</A>
  </UL>
```

```
<H1>  Ring 12345678 for more details  </H1>
</BODY>
</HTML>
```

(b) Write the full HTML code to display the following web page. Assume that the width of the table is 100% of the window, the border width is 3, and cellpadding is 5. The caption of the table has the H1 format. The width of "First" cell is half of the table. Your web page should consider the color of each cell. Be sure that the web page has the title of "Swimming Club".

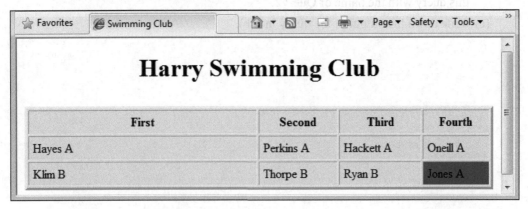

(c) Write the complete HTML code segment so that the words my other page will appear on your web page and when these words are clicked on, a page that you have created called other.html, will appear.

(d) Write the complete HTML code so that only a picture of yourself ("me.gif") will appear right-aligned on the web page with the words "This is me" (with H1 format) above it.

6. Access Operations

The following table represents 10 records from the **DOG** table of a MS-Access database set up for a veterinarian.

License No	Dog Name	Dog Type	Color	Gender	Amount Owing
1001	Curly	Boxer	Gold	Male	$180.00
1003	Lassie	Boxer	White	Female	$50.00
1012	Rex	Labrador	Tan	Female	$50.00
1111	Gypsy	Poodle	White	Male	$120.00
1118	Sam	Boxer	Tan	Male	$150.00
1136	Viper	Poodle	Black	Male	$60.00
1199	Toto	Pomeranian	Black	Male	$80.00
1291	Misty	Boxer	Gold	Female	$60.00
1330	Spec	Poodle	White	Female	$200.00
1555	Pickles	Labrador	Black	Male	$180.00

(a) Write down the data displayed for the following query.

Field:	License No	Dog Name	Amount Owing	Dog Type	Color
Table:	DOG	DOG	DOG	DOG	DOG
Sort:					
Show:	✓	✓	✓	☐	☐
Criteria:				"Poodle"	"White"

(b) Construct a query that lists all Black or White color dogs and whose owners owe not more than $160.00. The query should list the Dog Type., Dog Name and Amount Owing, in ascending Dog Type followed by descending Amount Owing order. Save this query with the name of Query2.

(c) Write down the output of the query in 6(b).

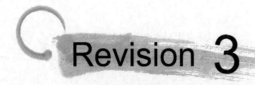

Revision 3

SECTION A

1. The most widely used coding system to represent data is the_____scheme.
   A. Extended Binary Coded Decimal Interchange Code (EBCDIC)
   B. American Standard Code for Information Interchange (ASCII)
   C. Unicode
   D. Infrared Data Association (IrDA)

2. Data will be most quickly available from_____.
   A. disk
   B. register
   C. memory
   D. CD

3. The CRT technology that hits the phosphorus coated screen with an electron beam is called_____.
   A. monitoring
   B. raster scanning
   C. imaging
   D. copying

4. A fast computer with a limited instruction set is called_____.
   A. CISC
   B. ALU
   C. ASCII
   D. RISC

5. The technology that lets businesses send business forms electronically is called_____.
   A. Electronic Data Interchange
   B. Voice Mail
   C. File Transfer Software
   D. EFT POS

6. The network topology that connects each device to a central computer (or hub) is_____.

A. tree

B. ring

C. star

D. bus

7. Unauthorized software copying is known as_____.

A. hacking

B. piracy

C. licensing

D. privacy

8. Modulation and demodulation are the processes of a_____.

A. connection device

B. node

C. modulator

D. modem

9. The layout of the computers and devices in a communications network is called a(n)_____.

A. network architecture

B. network topology

C. communications satellite

D. intelligent network

10. The general form of the IF function is_____.

A. =IF(logical_test, value_if_true, value_if_false)

B. =IF(logical_test, value_if_false, value_if_true)

C. =IF(value_if_true, logical_test, value_if_false)

D. =IF(value_if_false, logical_test, value_if_true)

11. A cookie is a small file that is stored on_____.

A. the server

B. the E-Commerce site

C. the user's computer

D. a CDROM

12. The unique address of a web page or file on the Internet is called a(n)_____.

A. browser

B. URL

C. domain

D. applet

13. The tag used to enclose the content of the page is the_____.

A. body tag

B. anchor tag

C. image tag

D. content tag

14. Which piece of code is missing at the start of the following line of HTML code?
    = http://www.google.com>Retrieve Information</A>_____.

    A. <A SRC

    B. <A HREF

    C. <A IMG

    D. <A TBL

15. Electronic Commerce is best described as:_____.

    A. Business transactions across the World Wide Web

    B. Business transactions across the Internet

    C. Business transactions using any electronic media

    D. Buying and selling over the network

16. To create a background for your Web Page the following code should be used:_____.

    A. <BODY bg="motors.jpg">

    B. <BODY bgcolor="motors.jpg">

    C. <BODY background="motors.jpg">

    D. <BODY back ground="motors.jpg">

17. Databases integrity refers to_____.

    A. a relational database

    B. the integrity of the user

    C. an accurate database

    D. a redundant database

18. Before data can be entered to a database, a user must plan its_____.

    A. model

    B. operator

    C. logic

    D. structure

19. The domain name is the name of the_____where the resource is located.

    A. URL

    B. server

    C. client

    D. language

20. _____E-Business model often resembles the electronic of the classifieds ads or an auction.

    A. C2C

    B. B2C

    C. B2B

    D. C2B

SECTION B

1. Briefly explain the following computing terms: ASCII, flash memory and virtual reality.

2. Describe the concept of RAID storage.

3. Write short notes on the following: fibre optics, broadband and timesharing.

4. Explain what is meant by the following terms in the context of web authoring: editor, browser and TABLE tag.

5. TCP/IP is an example of a protocol used on the Internet. Describe how TCP/IP works.

6. What is meant by half-duplex data transmission? Give an example of this type of data transmission.

7. What sort of transmission (simplex, half-duplex or full-duplex) is required for the following situations?____

 (a) CB radio

 (b) Television

 (c) MSN Chat

8. Health Care Inc., which is made up of a team of 10 doctors, has decided to computerise and network all their patient records and consultation details. Assuming that there will be a computer in each consultation room and an additional computer for use at the reception area, list the hardware and software required to set up this network. What topology would you recommend for this network and why?

9. Distinguish between primary storage and secondary storage, with some examples. What data does each type store?

10. This Question Comprises Four Parts.

 (a) How many bits in 4MB?

 (b) Calculate the average access time of a hard drive which has a seek time of 40 milliseconds and rotates at 9600 rpm. Assume negligible data transfer time.

 (c) A company wishes to store records on its 3000 customers and 500 stock items. Each customer requires 20KB, each stock item 30KB and each sales order 10KB. Each customer has on average 30 sale orders per year. Calculate the size of the files over a one-year period (in GB with two decimals).

 (d) Suppose you send a block of 2000 bytes of data using asynchronous data transmission at 9600 bps. The asynchronous format uses 1 start bit, 1 parity bit and 1 stop bit for each byte of data. How long (in seconds with two decimals) will it take to send the data?

SECTION C

1. Windows Operations

You are assigned as a DB Administrator for Multimedia Technology Co., Ltd. In the era of Big Data, where companies are able to quickly make sense of larger quantities of data than ever, information is finally recognized as a critical business asset.

You have been given the responsibility for the collection and distribution of information inside and outside of the company.

(a) Your company provides many kinds of multimedia data for different purposes: word processing, presentation, and web-authoring. Design and implement the suitable directory structure and files as shown below that will be capable of storing the files for each purpose. In a well-organised and efficient way. Print a screenshot of the directory structure.

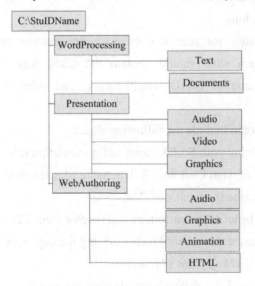

(b) Your lecturer will provide you with the following files:
- .txt, and .docx files for WordProcessing folder.
- .mp3, .mp4, .wav, .jpg, .mov, and .avi files for Presentation folder.
- .swf, .jpg, .wav, .gif, .png, and .htm files for WebAuthoring folder.

Copy each of the files provided into the appropriate folder in the directory structure you created. Print a screen shot of the contents of the Video-Presentation folder.

(c) Try to copy CBS.LOG in drive C into the WebAuthoring folder, and rename it as winlog.dat.

(d) Display the hidden files in file explorer with the large icons. Print a screenshot for the option settings.

(e) Compress all the files in the folders of Audio-Presentation, Animation-WebAuthoring, and Graphics-WebAuthoring into the WebAuthoring folder, named as multimedia.rar with the password 123. Take a screenshot of the password set up screen, a screenshot of the contents of the WebAuthoring folder, and a screenshot of the contents of multimedia.rar.

2. Word Operations

Recall the file saved as "Enterprise Resource Planning.docx" and edit it as follows:

(a) Insert automatic page number at the bottom right of the report.

(b) Insert "IT in Business" as a header.

(c) Insert your StuID, name and today's date (update automatically) as a footer.

(d) Insert the picture "ERP.jpg" into the the body text of Definition of ERP section and position it in the middle right with square text wrapping.

(e) Insert an automatic table of contents (on a separate page).

(f) Insert a front cover and include an appropriate report title, your name and the ERP logo.

(g) Tide up the report as required.

3. PowerPoint Operations

Using suitable websites for reference create a presentation on Customer Relationship Management (CRM). Each slide should contain the CRM logo. On completion save the presentation with the name of StuIDNamePPT.pptx, and print it (6 slides per page) to StuIDNamePPT.pdf.

Your presentation should include the following slides:

Slide 1: Title slide giving your StuID, name and presentation title.

Slide 2: Explanation of what Customer Relationship Management is.

Slide 3: List and describe the types of CRM.

Slide 4: List several benefits the customers can receive from CRM.

Slide 5: List and describe the primary forces driving the explosive growth of CRM.

Slide 6: List several CRM systems in practice.

Slide 7: Copy results of Sales Analysis and display as a graph.

Slide 8: Three hyperlinks to websites giving further information on CRM.

Slide 9: Closing image on CRM.

4. Spreadsheet Operations

The following spreadsheet displays the projected figures for the Cambridge Bricks Company over the next 5 years. The set up costs are $100 000 while the fixed yearly costs are $12 000 per year. Six staff members are required to run the business at $20 000 per year per staff member unless the number of pellets sold per year is greater than 2000 in which case 2 extra staff are required. The rent for the premises is $20 000 per year increasing by 8% per year. Pellets are sold for $180 per pellet increasing by $20 per year while it costs $150 to make pellets and increases by $9 per year.

	A	B	C	D	E	F
1		**Cambridge Bricks**				
2						
3	Set up costs	$100,000.00			Year	No. of pellets sold
4	Fixed yearly costs	$12,000.00			1	1000
5	Staff costs	$20,000.00	per staff menber		2	1200
6	Staff members	6			3	1600
7	if no of pellets>	2,000	then 2 extra staff		4	2000
8	Rent	$20,000.00	per year		5	2600
9	Rent increase	8%	per year			
10	Retail price	$180.00	per year			
11	increase by	$20.00	per year			
12	Cost to make	$150.00	per year			
13	increase by	$9.00	per year			
14						
15	Year	1	2	3	4	5
16	**Income**					
17	pellet sales	1000	1200	1600	2000	2600
18	price/pellet	$180.00	$200.00	$220.00	$240.00	$260.00
19	total sales income	$180,000.00	$240,000.00	$352,000.00	$480,000.00	$676,000.00
20	**Cum income**	$180,000.00	$420,000.00	$772,000.00	$1,252,000.00	$1,928,000.00
21						
22	**Costs**					
23	setup	$100,000.00				
24	yearly	$12,000.00	$12,000.00	$12,000.00	$12,000.00	$12,000.00
25	rent	$20,000.00	$21,600.00	$23,328.00	$25,194.24	$27,209.78
26	staff	$120,000.00	$120,000.00	$120,000.00	$120,000.00	$160,000.00
27	costs to make	$150.00	$159.00	$168.00	$177.00	$186.00
28	**Total costs**	$252,150.00	$153,759.00	$155,496.00	$157,371.24	$199,395.78
29						
30	**Profit/loss**	-$72,150.00	$86,241.00	$196,504.00	$322,628.76	$476,604.22

Write down the formula that would be in each of the following cells.

B17		C24	
B18		C25	
C18		C26	
C19		C28	
C20		C30	

5. HTML Operations

(a) Draw the web page created by the following HTML code.

```
<HTML>
<HEAD>
    <TITLE>Katherine's Page</TITLE>
</HEAD>
<BODY>
    <H1><B><FONT COLOR = "red">Katherine's Web Page</FONT></B></H1>
```

```
      <H2>My Family</H2>
      <TABLE BORDER=5  WIDTH=80%  BGCOLOR="orange">
        <TR>
           <TD WIDTH=20% BGCOLOR="yellow" > Dad </TD>
           <TD COLSPAN=2 BGCOLOR="pink" > Mum </TD>
        </TR>
        <TR>
           <TD>Cindy</TD>
           <TD>Mary</TD>
           <TD>Fido the dog</TD>
        </TR>
     </TABLE>
  <H2>And some other things</H2>
  <UL>   <LI><A HREF="friends.html">My friends</A>
         <LI><A HREF="interests.html">My interests</A>
         <LI><A HREF="http://www.yahoo.com.cn">My favorite sports</A>
  </UL>
</BODY>
</HTML>
```

(b) Write the full HTML code to display the following web page. Assume that the first 2 hyperlinks refer to the following html files stored on the root directory of your account (specials1.html and specials2.html respectively), the third hyperlink refers to the email address: Melissa@hotmail.com.

(c) Write the complete HTML code so that only a picture of a lovely dog("snoopy.gif") will appear on the web page with the bold words "This is my favorite pet" above it.

(d) Write the full HTML code to insert your email address onto your web page. Assume that your email is available from pine_yu@hotmail.com and that the link to your email is My Email.

6. Access Operations

The following two tables represent respectively records from the **JOB** table and **PAYROLL** table of **JOBS** database.

JOB

JOBCODE	STORE	HOURSPERWEEK	POSITION	EXPERIENCE
1001	Big W	10	Sales Assistant	No
1002	Myer	22	Manager	Yes
1003	Safeway	40	Assistant Manager	Yes
1004	Target	20	Supervisor	Yes
1005	Myer	18	Cashier	No
1006	Big W	25	Supervisor	No
1007	Target	35	Manager	Yes
1008	Myer	30	Assistant Manager	No
1009	Big W	19	Cashier	No
1010	Coles	15	Deli Manager	Yes

PAYROLL

JOBCODE	Bonus
1001	$1,234
1002	$3,426
1003	$6,598
1004	$1,290
1005	$5,920
1006	$3,741
1007	$2,345
1008	$4,521
1009	$4,193
1010	$3,218

(a) The following represents a query that was entered into ACCESS. Write down the output of the query.

Field:	JOBCODE	STORE	POSITION	HOURSPERWEEK
Table:	JOB	JOB	JOB	JOB
Sort:		Ascending	Ascending	
Show:	✓	✓	✓	☐
Criteria:				>=20 And <40

(b) Construct a query that lists the information about the jobs which do not require experience or HOURSPERWEEK is between 20 and 30. The query should list the JOBCODE, POSITION, and Salary (assuming the pay_rate is $30 per hour), in descending HOURSPERWEEK order within each POSITION (POSITION sorted in ascending order). Save this query with the name of Query3-b.

(c) Write down the output of the query in 6(b).

(d) Construct a query that lists the information about the required number of employee, average working hours per work, and average bonus for each position. The query should list the POSITION, number of employee (shown as HeadCount), average working hours per work (shown as AvgHour), and average bonus (shown as AvgBonus), in order from the lowest average bonus to the highest. Save this query with the name of Query3-d.

(e) Write down the output of the query in 6(d).

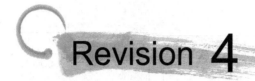

Revision 4

SECTION A

1. What is the average access time of a hard drive which has a seek time of 10 milliseconds and rotates at 7200 rpm? Assume that the data transfer time is negligible._____
    A. 11.17 msec
    B. 14.17 msec
    C. 18.33 msec
    D. 22.33 msec

2. The time while desired data on a disk sector moves under the read/write head is known as ____.
    A. rotational delay
    B. longitudinal delay
    C. seek time
    D. transfer time

3. Conventional computers with complete instruction sets are referred to as_____.
    A. SRAM
    B. CISC
    C. ASCII
    D. RISC

4. A company wishes to store records on its 2000 customers and 100 stock items. Each customer requires 5KB, each stock item 20KB and each sales order 20KB. Each customer has on average 30 sale orders per year. Estimate the size of the files over a one year period.
    A. 0.6 GB
    B. 0.8 GB
    C. 1.0 GB
    D. 1.2 GB

5. Another name for asynchronous data transmission is_____.
    A. go/no go

B. upload/download

C. yes/no

D. start/stop

6. The number of frequencies that can fit on a communications link at the same time is its_____.

   A. modulation

   B. topology

   C. demodulation

   D. bandwidth

7. The card inserted in a slot in a personal computer to allow it to send and receive messages from the LAN is the_____.

   A. CSMA/CD

   B. NIC

   C. TCP/IP

   D. EFT

8. In which kind of network do all computers have equal status and share files and peripheral devices?_____

   A. bus

   B. peer-to-peer

   C. ring

   D. star

9. =SUM(D4:F15) would find the sum of the contents of how many cells?

   A. 15

   B. 28

   C. 36

   D. 39

10.
```
<TABLE>
<TR>
    <TH>  data content  </TH>
    <TH>  data content  </TH>
    <TH>  data content  </TH>
</TR>
<TR>
    <TD>  data content  </TD>
    <TD>  data content  </TD>
    <TD>  data content  </TD>
</TR>
</TABLE>
```

The above HTML code would display a table of_____.

   A. 2 rows and 3 columns

B. 1 row and 2 columns

C. 1 row and 3 columns

D. 2 rows and 2 columns

11. _____has to regularly make changes to a particular field.

   A. Data redundancy

   B. Data sharing

   C. Data integrity

   D. Data maintenance

12. Data sorted in_____order is arranged alphabetically A to Z or numerically 0 to 9.

   A. descending

   B. decreasing

   C. ascending

   D. increasing

13. Documents on the Web are called_____.

   A. Web pages

   B. Web sites

   C. Web communities

   D. Web tags

14. To create a background color for your Web Page the following code should be used:_____.

   A. <BODY bg="blue">

   B. <BODY bgcolor="blue">

   C. <BODY background="blue">

   D. <BODY back ground="blue">

15. The number system that has just two unique digits, 0 and 1, is called the_____.

   A. digital system

   B. bit system

   C. analog system

   D. binary system

16. The large, sealed glass tube inside a CRT monitor is called a_____.

   A. liquid crystal display

   B. graphics card

   C. cathode-ray tube

   D. graphics processing unit (GPU)

17. A_____port connects a device to the system unit by transferring data more than one bit at a time.

   A. LPT

   B. RS-422

C. USB

D. RS-232C

18. Fast memory that stores frequently used instructions and data is called_____.

A. RAM

B. SRAM

C. cache

D. DDR RAM

19. You are performing a_____when turning on a computer that has been powered off completely.

A. cold boot

B. backup

C. warm boot

D. virus scan

20. In a MS Access query, what would be the criteria for displaying all employees whose pay rate is more than $9.00 OR who are in the Union (whose data type is Boolean)? ____

A.
Field:	Pay rate	Union
Criteria:	<9.00	Yes
Or:		

B.
Field:	Pay rate	Union
Criteria:	>9.00	
Or:		"Yes"

C.
Field:	Pay rate	Union
Criteria:	>9.00	Yes
Or:		

D.
Field:	Pay rate	Union
Criteria:	>9.00	
Or:		Yes

SECTION B

1. Write short notes on the following terms: client-server network, star topology.

2. Briefly describe the following WWW terms: HTML, server, and cookies.

3. Briefly describe the following computing terms: cache memory, RAM, and ASCII character set.

4. Compare the capacity of floppy disk, CD ROM, and DVD ROM. Discuss the technology each uses to explain the difference.

5. Describe the difference between fragmentation and defragmentation of a disk. Why does a disk become fragmented and how do you fix it up?

6. The Jones family has just purchased 5 PCs, one for each member of the family. They have one printer which is kept in the study. Being a family friend studying at ECNU University, you have been asked to network their computers together so that they can print from any machine and can share files and resources with each other. Make a list of the hardware and software they will need to purchase to do this and design a plan of how you would complete this project.

7. One of the requirements needed to construct and display a web page is a writing vehicle. In this course we have used Notepad. What are the other 2 requirements needed to construct and display a web page?

8. What is Local Area Network? How would you set up a LAN for a small business with 8 staff members? What are the benefits to the small business of having a LAN? What is a Wide Area Network? What additional benefits can be gained by this small business if connected to a WAN?

9. Describe the concept of virtual reality. Give three examples of where it is used.

10. This question comprises four parts.

    (a) Convert 2 GB (gigabytes) to MB (megabytes)

    (b) Convert 20 480 bits to KB (kilobytes).

    (c) How much memory (in KB) is required for a display containing 800 × 600 pixels having 1024 colors?

    (d) Suppose you send a block of 5000 bytes of data using synchronous data transmission at 9600 bps. The synchronous data block begins with 2 SYN characters, 14 bytes of control information, and then the 5000 bytes of data and finishes with 8 bytes of error checking. How long (in seconds with two decimals) will it take to send the data?

SECTION C

1. Windows Operations

You are employed as a Marketing Assistant with Global Fitness Club. The company has decided to expand the business by issuing franchise licenses. You have been given the responsibility for the collection and distribution of information to and from franchisees.

(a) You are involved with the setting up of three of the new Fitness franchises. You will need to be able access relevant files quickly. In order to do this, design and implement a suitable directory structure as shown below that will be capable of storing your files for each of the franchises in a well-organised and efficient way. Print a screenshot of the directory structure.

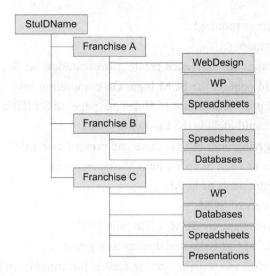

(b) Your lecturer will provide you with the following files:
- Franchise A Analysis Report
- Franchise A Cash Budget
- Franchise A Web Page
- Franchise B Staff
- Franchise B Cash Budget
- Franchise C Cash Budget
- Franchise C Analysis Report
- Franchise C Presentation

Copy each of the files provided into the appropriate folder in the directory structure you created. Print a screen shot of the contents of the Franchise B - Database folder.

(c) Let the author of ITF_Franchise_C_Analysis_Report.docx be your name. Print a screenshot of the author editing dialog.

(d) Try to copy setupact.log in drive C into the Franchise C folder, and rename it as cfglog.dat. Print a screenshot of the contents of the Franchise C folder.

(e) Compress all the files in the folders of WebDesign-Franchise A, Spreadsheets-Franchise, B and Presentation-Franchise C into the Franchise A folder, named MyDocs.rar with the password 123456. Take a screenshot of the password set up screen, a screenshot of the contents of MyDocs.rar, and a screenshot of the contents of the Franchise A folder.

2. Word Operations

Recall the file saved as "Information Security.docx" and edit it as follows:

(a) Insert automatic page number with the format of Page XXX at the bottom of the report.

(b) Insert "Fundamentals of Information Technology" as a header.

(c) Insert your StuID, name and today's date (update automatically) as a footer.

(d) Insert the picture "Security.jpg" into the the body text of What is Information Security section and position it in the middle right with square text wrapping.

(e) Insert an automatic table of contents (on a separate page).

(f) Insert a front cover and include an appropriate report title, your name and the IT logo.

(g) Tide up the report as required.

3. PowerPoint Operations

Using suitable websites for reference create a presentation on Supply Chain Management (SCM). Each slide should contain the SCM logo. On completion save the presentation with the name of StuIDNamePPT.pptx, and print it (6 slides per page) to StuIDNamePPT.pdf.

Your presentation should include the following slides:

Slide 1: Title slide giving your StuID, name and presentation title.

Slide 2: Explanation of what Supply Chain Management is.

Slide 3: List the components of SCM.

Slide 4: List the drivers of SCM.

Slide 5: The differences among SCM, CRM, and ERP.

Slide 6: Copy results of Car Sales and display as a graph.

Slide 7: Three hyperlinks to websites giving further information on SCM.

Slide 8: Closing image on SCM.

4. Spreadsheet Operations

Tammy works as a podiatrist in her own business. The number of consultations varies per month as indicated in the table. The cost of each consultation is $35. On average 40 surgical procedures are performed per month at a cost of $120 each. Expenses are the rent at $1500 per month, office staff at $2500 per month, stationery at $280 per month increasing by 8% per month, miscellaneous expenses at $600 increasing by $20 per month. An extra podiatrist is brought in to help out if the number of patients per month is greater than 200 patients. The cost of the extra podiatrist is $7000 per month.

	A	B	C	D	E	F	G
1			**Tammy's Podiatry Clinic**				
2							
3	Income Details					Month	Patients
4	Average consultation fees	$ 35.00	per patient			Jan	80
5	No of surgical procedures	40	per month			Feb	120
6	Surgical procedure costs	$ 120.00	per procedure			March	220
7						April	140
8	Expenses					May	170
9	Rent	$ 1,500.00	per month			June	170
10	Office staff	$ 2,500.00	per month			July	180
11	Stationery	$ 280.00	per month			Aug	210
12	increase by	8%	per month			Sept	210
13	Miscellaneous	$ 600.00	per month			Oct	150
14	increase by	$ 20.00	per month			Nov	130
15	Extra Podiatrist wage	$ 7,000.00	per month			Dec	100
16	if patients per month>	200					
17							
18		Jan	Feb	March	April	May	June
19	Income						
20	Consultations	$ 2,800.00	$ 4,200.00	$ 7,700.00	$ 4,900.00	$ 5,950.00	$ 5,950.00
21	Surgical Procedures	$ 4,800.00	$ 4,800.00	$ 4,800.00	$ 4,800.00	$ 4,800.00	$ 4,800.00
22	Total Income	$ 7,600.00	$ 9,000.00	$ 12,500.00	$ 9,700.00	$ 10,750.00	$ 10,750.00
23							
24	Expenses						
25	Rent	$ 1,500.00	$ 1,500.00	$ 1,500.00	$ 1,500.00	$ 1,500.00	$ 1,500.00
26	Office Staff	$ 2,500.00	$ 2,500.00	$ 2,500.00	$ 2,500.00	$ 2,500.00	$ 2,500.00
27	Stationery	$ 280.00	$ 302.40	$ 326.59	$ 352.72	$ 380.94	$ 411.41
28	Miscellaneous	$ 600.00	$ 620.00	$ 640.00	$ 660.00	$ 680.00	$ 700.00
29	Extra Podiatrist	$ -	$ -	$ 7,000.00	$ -	$ -	$ -
30	Total Expenses	$ 4,880.00	$ 4,922.40	$ 11,966.59	$ 5,012.72	$ 5,060.94	$ 5,111.41
31							
32	Profit/Loss	$ 2,720.00	$ 4,077.60	$ 533.41	$ 4,687.28	$ 5,689.06	$ 5,638.59

Write down the formula for the following cells.

C20		C27	
C21		C28	
C22		C29	
C25		C30	
C26		C32	

5. HTML Operations

   (a) Draw the resulting page(s) that will be shown on the browser with the following HTML files.

Main.html

```
<HTML>
<HEAD>
    <TITLE>Tournament of Minds</TITLE>
</HEAD>
<BODY>
    <H1 ALIGN=CENTER><B>Greatwall Tournament of Minds Club</B></H1>
    <H2>President: William Wang</H2>
    <UL> <LI><A HREF="other.html">Interesting links</A>
            <LI>members
            <LI>fees
</UL>
</BODY>
</HTML>
```

other.html

```
<HTML>
<HEAD>
    <TITLE>Main's Link</TITLE>
</HEAD>
<BODY>
    <H1> Interesting Links</H1>
    <TABLE BORDER= 3   WIDTH= 90% >
       <TR>
          <TD WIDTH=65% BGCOLOR="pink"> Past champions </TD>
          <TD WIDTH = 35%  BGCOLOR = "cyan"> Best Judges </TD>
       </TR>
       <TR>
          <TD WIDTH=65% BGCOLOR="yellow"> Scoring </TD>
          <TD WIDTH = 35%  BGCOLOR = "orange"> Schools </TD>
       </TR>
    </TABLE>
</BODY>
</HTML>
```

(b) Write the full HTML code to produce the following web page. The words **Enjoy My Picture** (with H1 format) will appear on your web page and below it will be a picture of a lovely angel. (The picture will be saved in a file called *pic.gif*). Below this will be two links with the words **Go to My Links (**linking to a page called *links.html* saved in your directory**)** and **Email to Me (**this hyperlink refers to the email address: admin@cc.ecnu.edu.cn**)** respectively.

6. Access Operations

The following two tables represent respectively records from the **Wooden Crafts** table and **Selling** table of **CraftShop** database.

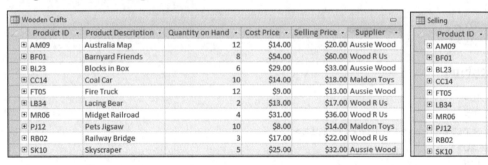

(a) Write down the data displayed for the following query.

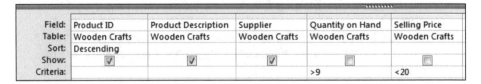

(b) Construct a query that lists all products which have a Quantity on Hand of not more than 5 or the difference between Selling Price and Cost Price of not less than $7. The query should list the Supplier, Product ID, Product Description and Difference, in descending Supplier followed by ascending Difference order. Save this query with the name of Query4-b.

(c) Write down the output of the query in 3(b).

(d) Construct a query that lists the sales volume and sales amount for each supplier. The query should list the Supplier, sales volume (shown as SalesVolume), and sales amount (shown as SalesAmount), in order from the highest sales volume to the lowest. Save this query with the name of Query4-d.

(e) Write down the output of the query in 6(d).

参 考 文 献

1. 江红，余青松．计算机专业英语教程．北京：清华大学出版社，2012.
2. Morley, D.，Parker, C.S.．Understanding Computers: Today and Tomorrow, Comprehensive 15th Edition，Thomson Course Technology. Boston: MA，2014.
3. Timothy J. O'Leary, Linda I. O'Leary．Computing Essentials 2016: Comp (26th Edition)．McGraw Hill Higher Education，2016.
4. June Jamrich Parsons．New Perspectives on Computer Concepts 2016, Comprehensive．Course Technology，2016.
5. 金志权等．计算机专业英语教程.6 版．北京：电子工业出版社，2015.
6. 张强华，司爱侠等．计算机英语使用教程.4 版．北京：清华大学出版社，2014.
7. Faithe Wempen, Office 2016 for Seniors for Dummies, John Wiley & Sons Inc., 2015.

图书资源支持

感谢您一直以来对清华版图书的支持和爱护。为了配合本书的使用,本书提供配套的素材,有需求的用户请到清华大学出版社主页(http://www.tup.com.cn)上查询和下载,也可以拨打电话或发送电子邮件咨询。

如果您在使用本书的过程中遇到了什么问题,或者有相关图书出版计划,也请您发邮件告诉我们,以便我们更好地为您服务。

我们的联系方式:

地　　址:北京海淀区双清路学研大厦 A 座 707

邮　　编:100084

电　　话:010-62770175-4604

资源下载:http://www.tup.com.cn

电子邮件:weijj@tup.tsinghua.edu.cn

QQ:883604(请写明您的单位和姓名)

用微信扫一扫右边的二维码,即可关注清华大学出版社公众号"书圈"。

扫一扫
资源下载、样书申请
新书推荐、技术交流